Springer Remote Sensing/Photogrammetry

The Springer Remote Sensing/Photogrammetry series seeks to publish a broad portfolio of scientific books, aiming at researchers, students, and everyone interested in the broad field of geospatial science and technologies. The series includes peer-reviewed monographs, edited volumes, textbooks, and conference proceedings. It covers the entire area of Remote Sensing, including, but not limited to, land, ocean, atmospheric science and meteorology, geophysics and tectonics, hydrology and water resources management, earth resources, geography and land information, image processing and analysis, satellite imagery, global positioning systems, archaeological investigations, and geomorphological surveying.

More information about this series at http://www.springer.com/series/10182

Andreas Laake

Remote Sensing for Hydrocarbon Exploration

Springer

Andreas Laake
Schlumberger
Aachen, Germany

ISSN 2198-0721 ISSN 2198-073X (electronic)
Springer Remote Sensing/Photogrammetry
ISBN 978-3-030-73318-6 ISBN 978-3-030-73319-3 (eBook)
https://doi.org/10.1007/978-3-030-73319-3

This Springer imprint is published by the registered company Springer Nature Switzerland AG
The registered company address is: Gewerbestrasse 11, 6330 Cham, Switzerland

For Elisabeth, Daphne and Viola

Foreword

When observing a landscape, we have different feelings: on one side we see ourselves as part of the world and we appreciate its perfection; at the same time, we are awed of its greatness. Artists have tried to pass to us these feelings through innumerous paintings: in some, the observer is part of the landscape, in others he is submerged. These are the sensations that come to my mind when studying the stupendous book by Andreas Laake on the advances of remote sensing when directed to the goal of hydrocarbon exploration.

The ancients had no knowledge of geology and the fossils were, Aristoteles ipse dixit, tentative of the gods to forge animals or plants. But in the painting (Fig. 1a) of Joachim Patinir we see the meandering of the rivers in a limestone (1521–1524) landscape as if the local geology was familiar to him. Similarly, Caspar David Friedrich imaged landscape in a painting (1818, Fig. 1b) where the awe of the observer is deeply felt in a perfectly described environment. And then we see how Max Ernst had already figured out, in his artistic and surrealistic perspective, what the buried layers should look alike, exactly as he was watching one of the seismic sections that would be created several decades later (1935, Fig. 1c).

In this book we see our world and we also watch our terrestrial landscapes, be they deserts or iced surfaces. At the same time, beyond the intuition of the ancient masters, we have now the author who accompanies us in the understanding of what we see. He leads us in the explanation of the rationale of the landscape, why the landscape lines we either see or hypothesize are there, which are their causes, and finally their impact on oil exploration or on the logistics. It is this interaction between the beauty and the perfection of our world and then the science to explain it that moved humans first to understand the world, and then to apply this understanding to improve it. In this case, the goal is the improvement of the energy availability to gain better life quality. Nowadays, modern technologies that extract energy from renewables, be they sun, wind, or sea waves, allow the aesthetical appreciation to prevail on the economical one. But both are present in the book.

In the last century, oil has been one of the drivers of the world economy: its prospection has always been the domain of a group of extremely advanced engineers and scientists. The technologies used are and have been at the edge, and also

Fig. 1 Representation of geology in art through the centuries. (**a**) Joachim Patinir 1521–1524 "Taufe Christi," (**b**) Caspar David Friedrich 1818 "Kreidefelsen auf Rügen," (**c**) Max Ernst 1935 "Paesaggio"

the community of the participants is tight and cooperative. The usual technique for prospection is the seismic method, as acoustical sources at the surface "illuminate" the subsurface, their echoes are recorded, and then processed and interpreted. This prospection technique is very expensive: on sea, gigantic vessels trail cable formations kilometers long and so wide to occupy an area of the dimension of the Manhattan Island. On land, trucks carrying vibrating masses weighing tons or strong explosive devices are used for the illumination. Thousands of receivers are then arranged to record the echoes. The selection of the areas where to prospect is dependent indeed on the general geological setting, but the choice of instrumentation is also dependent on the characteristics of the shallow subsurface that acts as a barrier between the sources and the buried layers to be investigated.

Here remote sensing, clearly less costly than dispatching explorers, and encompassing entire regions at will, comes to the help of the prospectors. The state of the surface is examined, its geology understood, and then the prospection survey is properly designed and dimensioned. Things may be easier somewhere, but many difficulties may arise and here the expertise of scientists as Andreas Laake comes to the help of the surveyor.

This book is thus born from the necessity of providing expertise on the state of the surface in order to optimize its prospection for oil, but it accompanies this substance with a splendid form of many images, which recall the creations of the ancient masters. Further, the deep geological analysis always provided by the author helps the newcomer to understand the content of the images that are shown even to be able to predict whether a heavy vibrator will sink into the rocky crust of an otherwise impassable Arabian desert. The scientific tools needed to achieve such a feat are discussed at the beginning.

The first section of the book is dedicated to the basics of remote sensing. It can be passive, say using the reflectance of the solar radiation in the optical band: color vision implies the use of three channels usually synthesized as red, green, and blue. Optical satellites exist nowadays providing world images for free and with sub-daily revisit time, with sub-meter resolution. Further, multispectral and now hyper-spectral vision from satellites, airplanes, and drones provide up to 256 channels that sample the electromagnetic spectrum from optical down to infrared and thermal frequency bands. Remote sensing can also be active, with the radar satellites that illuminate the targets and work night and day, penetrating the cloud cover. The frequency of the electromagnetic waves now is lower and we achieve penetration in the dry rocks and we see targets under the forest canopy. It is constant care of the author to provide practical examples (even in the title of the book) for each methodology mentioned, so that the reader immediately grasps the impact on the final objective, namely oil exploration, never forgotten. Satellite altimetry is also mentioned as it is useful for marine gravity, as the sea surface maps an equipotential gravity surface and thus the details of the subsurface density spatial variations are fully evidenced.

Using these tools, the author describes the available products, starting with digital terrain models that can be obtained photogrammetrically when optical, or by exploiting radar or LIDAR altimetry and thus the timing of the returns. Terrain model and their rendering are shown, with different spatial resolutions.

In the second section we meet geology and geomorphology: we begin our navigation in the landscape, with the support of the new tools and the explanation of what we see. The reader should be careful here: geology is compressed in single lines informing the newcomers that hard rocks weather less than shales, as this shapes the landscape. Tectonic deformation becomes visible and the Patinirs and Ernsts should by now know the whys of what they are imaging. Every image is a story by itself, and of these we have scores and scores: the river valleys incised by faults in the Appalachians, the fractal geometry of the Siberian rivers, the Grand Canyon.

The following section is dedicated to hydrocarbon exploration: again, readers beware! Just one figure is dedicated to Alfred Wegener continental drift theory, but it is telling because the spreading Atlantic is mapped with the magnetic traces and a theory so ridiculed at the beginnings is demonstrated in just that slide. Then, coming to the objective, few lines describe how the opening of the Atlantic Ocean created two parallel but independent basins where oil could be found, and tells us how gravity from satellite altimetry led to the discovery of new oil in the offshore Brazil and Ghana. Thus, we have the 4D images of the temporal evolution of the tectonic margins. If the surface is dry, then radar satellites come to help beyond the digital terrain map, the roughness of the surface and its gradients to identify faults and the basin outlines. The kerogen formation is finally achieved from the regional heat flows. We move from Libya to the Red Sea and the Gulf of Suez, all in a few deep and "penetrating" pages. New science is also found in the correlation between synthetic aperture radar backscatter characters and seismic waves velocity. A demonstrative tour of a volcanic system in Libya tells us about the impact on oil formation.

The next section is dedicated to the logistics of survey planning: once identified the area for exploration, seismic crews have to be dispatched to identify if and where to drill the expensive wells. The area is volcanic? Beware of steep slopes and big boulders. And then you have other precautions to take in sabkha (the salty flats in the Arabian desert) be it wet or dry (your vibrator might sink), or for the sand dunes (your vibrators might rollover or the generated elastic waves could be distorted). The prediction of the seismic quality may be a delicacy for those professionals at the edge, but the amount of information the author can squeeze out of remote sensing is impressive.

The following section discusses the precautions to be taken for development of oilfields infrastructure. What if the permafrost thaws in Alaska? How likely the humid floodplains in Arkansas will be flooded? How the climate change will affect the costal settings in Louisiana or the drying up of Lake Chad? And the sand migration in the Erg in Tunisia? It is your choice to follow the path of the oil seeking wanderer much beyond the dreams of any painter.

The final section is a very useful geology primer, proposing the study of geological analogs to be able to understand the mechanics of deposition affected by tectonics and erosion. Relay ramps to delineate and evaluate fault sealing, karsting in Algeria, tectonic of the Mount Zagros in Iran, diapirs and salt domes in Louisiana, and carbonate reefs and more and more. I counted 17 examples equally distributed among structural, depositional, erosional examples, all with their dote of wonderful images.

After reading the book, my conclusion is that while the voyage we had under the guide of Andreas Laake is indeed for the quest of oil, we were lucky enough to be able to watch and understand a big chunk of the world we live in. Just for that, even if we are not oil prospectors and explorers, we have to thank him very deeply. As an old hand in remote sensing, I have to say that this is the first time I see such a long and fruitful experience gifted to us, and such a systemic and profoundly competent use of remote sensing. This time not for technology push, but for user's pull. And what a skilled user!

The professionals will be very happy too, but then they know each other and so they know what to expect from the author.

Thanks Andreas, it was a wonderful trip!

Fabio Rocca
Telecommunication, Politecnico di Milano
Milan, Italy

References

Friedrich, C. D. (1818). *Kreidefelsen auf Rügen*. Kunst Museum Winterthur – Reinhardt am Stadtgarten. http://kmw.zetcom.net/de/collection/item/15182/

Patinir, J. (ca. 1521–1524). *Taufe Christi*. Kunsthistorisches Museum. Gemaeldegalerie. https://www.khm.at/de/object/56352739f5/

Preface

Remote sensing, and particularly satellite remote sensing, gives humans the data to understand the world in which they are living. The first time the author came in contact with satellite imagery was a paper printout of a satellite image of the sand dunes near the Umm As Samim sabkha in western Oman in the mid-1990s (Fig. 2).

Fig. 2 Sabkha Umm As Samim and sand dunes in the Bizat area, West Oman. Landsat 5 432 RGB. (Satellite imagery courtesy U.S. Geological Survey)

Landsat 5 provided enough spectral bands to create a multiband composite showing the exceptional beauty of the hostile and dangerous wet sabkha of Umm As Samim in dark blue tones contrasting sharply against the delicate bands of up to 300 m high sand dunes of the area called Bizat in front of a gray-brown backdrop of gravel that was decorated locally with bright white evaporites like gypsum. The physical properties used in the acquisition of spectral images, their processing, and composition into a multiband RGB image yield a colored satellite image that can easily be interpreted for surface geologic features.

This image ignited the author's interest in remote sensing techniques, the application of which was subsequently fostered by a fruitful multi-year collaboration with the Earth Observation Market Development team of the ESA in Frascati near Rome. The author's travels to many desert areas yielded the understanding of the surface geology which is essential for the ground truthing of remote sensing data.

The essential breakthrough came with the opportunity to integrate geophysics subsurface and remote sensing surface data which culminated in using the concept of geomorphology to connect the two disciplines.

Parallel to the dramatically increasing amount of satellite imagery, the development of the internet accelerated the accessibility of remote sensing data which in turn allowed fast application and integration of remote sensing data with other data from a large number of disciplines.

In recent years, multi-temporal remote sensing data of sufficient quality became available for time-lapse studies that provided insights into the dynamic processes of the world around us. Remote sensing data evolved to become one of the most important tools for environmental monitoring.

This book introduces easily accessible remote sensing data, shows which data to use for what task, how to process them, and what rendering technique to use to provide easily interpretable results. Apart from the geoscience aspect of remote sensing, the author intends to inspire in the reader the art of unlocking the hidden information from the remote sensing data.

Hannover, Germany Andreas Laake
March 2021

Acknowledgments

The rise of remote sensing within the last 20 years has led to the creation of publicly available remote sensing archives of up to 50 years of continuous coverage at ever-increasing quality and resolution. Assisted by the rapid expansion and increasing speed of the worldwide web, USGS, NASA, and ESA organized a significant amount of satellite data in web-based satellite data archives that allow downloading for a large variety of tasks.

My deep gratitude goes to the USGS that provides critical digital relief models and satellite data from the Landsat program, the longest running consistent earth observation satellite. The USGS also coordinates access to satellite and airborne as well as ground and ship-based geophysical data from further space organization such as ESA (Europe) and METI (Japan). The author thanks CSA (Canada) for making available Radarsat-1 microwave radar data that allows looking into the near surface. Special thanks go to the numerous academic research institutions that collect, process, and make available through websites their interpreted results and models. The author appreciates particularly the global gravity and bathymetry program by the University of California in San Diego and the collection of arctic digital relief data that is coordinated by the University of Minnesota in St. Paul. The author's appreciation also goes to public institutions such as NOAA (USA) and BGS (UK) for making available geoscience data and the state governments of Arkansas and Alaska for publicly sharing their LiDAR high-resolution digital elevation models.

The author also thanks the EAGE and the SEG for the permission to use figures from previous publications to support the integration of surface geoscience interpretation from remote sensing data with subsurface seismic data from surface geophysical mapping.

The integration of all data has been assisted significantly by the modeling capabilities of the Petrel® geomodelling software, access to which was thankfully granted by Schlumberger.

The author would like to close the acknowledgments with thanks to Prof. Fabio Rocca of Politechnico di Milano, who has been a long-term mentor and inspired the creation of this book.

About the Book

Remote sensing data provide a unique insight into the earth composition, structure, and texture, however, at first sight only at its surface. This aspect is reflected in most remote sensing literature that focuses on the earth surface and its interaction with the hydrosphere and atmosphere. Introducing geologic processes and their imprint on the earth surface enables looking below the earth surface. This deeper look is what explorationists are using when they explore for hydrocarbon, mineral, and water resources.

This book provides insights into the benefits of using remote sensing data from a geoscientist's point of view: the integration of data and understanding of surface and subsurface.

It presents the usage of remote sensing data in exploration in a structured way using the detection of individual geologic features as building blocks for complex geologic systems that are studied in exploration. This concept aims at enabling readers to build their own workflows for the assessment of complex geologic systems using combinations of remote sensing data.

The book is composed of three sections that are built on the following structure:

1. **Basics of remote sensing**
   After a brief history of remote earth observation, a short overview of the electromagnetic spectrum explains the physical foundations, limitations, and applications of the bands that are used for exploration. For the complete physics foundation, the book refers to available textbooks. The second part of the introduction covers various methods of image processing with focus on those methods that provide high resolution and high discrimination of geological features. Special attention is paid to various digital relief models that are essential for the understanding and rendering of remote sensing data.
2. **Primary application for geosciences**
   The foundation for the rendering and interpretation of remote sensing data is the concept of geomorphology, i.e., the joint analysis of the earth topography and physical surface properties. This section describes in detail the interpretation

methods for individual categories of geologic features that are introduced as building blocks of more complex geologic systems which are the focus of exploration.

The section explains the different techniques for rendering remote sensing data in 2D and 3D. The second part is dedicated to the extraction of geological features from remote sensing data, discussing structural elements as well as depositional features in arctic and arid desert settings. Particular focus is laid on the interaction between subsurface geology and surface landforms.

3. **Remote sensing for hydrocarbon exploration**
This section presents dedicated workflows extracting geological features from remote sensing data to use them for surface and subsurface model building. The structure of this section follows the traditional exploration workflow starting with frontier exploration at regional scale and reducing the scope via basin scale to field scale. The second thread weaving through this section follows the life cycle of a field from exploration using seismic data acquisition as the main 3D subsurface imaging technique via near-surface and subsurface modeling to aspects of field development and infrastructure planning.

The section closes with a series of geologic analogs using surface features and their representation in remote sensing data in subsurface exploration. The case studies are grouped into structural, depositional, and erosional analogs.

Contents

List of Figures

List of Tables

Abbreviations

2D	Two-dimensional representation of data in map format
3D	Three-dimensional representation of data
4D Image	Four-dimensional image of three space and one time components
ArcticDEM	Arctic digital elevation model based on optical stereo imagery, project located at the University of Minnesota, USA
ASTER	Advanced Spaceborne Thermal Emission and Reflection Radiometer, sensor aboard the TERRA satellite operated by METI (JAPAN)
ASTER GDEM	Global digital elevation model based on stereo imagery from ASTER sensor
B.P.	(Time) before present
BGS	British Geological Survey (UK)
BM	Basin model
C-band	Electromagnetic spectrum in microwave range 4.0 to 8.0 GHz
CoNED	Coastal National Elevation Database (USGS USA)
CSA	Canadian Space Agency
DBM	Digital bathymetry model
DEM	Digital elevation model
DLR	Deutsches Zentrum fuer Luft- und Raumfahrt (Germany)
DSDP	Deep Sea Drilling Project
DSM	Digital surface model
EAGE	European Association of Geoscientists and Engineers
ESA	European Space Agency
ETM+	Enhanced Thematic Mapper, sensor of Landsat 7
ETOPO	Global relief model produced by the NGDC
GIS	Geographic Information System
GLOBE	Global Land One-Kilometer Base Elevation (model)
GMTED2010	Global Multi-resolution Terrain Elevation Data 2010 (USGS, USA)

GPS	Global Positioning System
GR	Ground roll, seismic surface wave mode
GTOPO	Global digital elevation model (USGS, USA)
IFSAR	Deprecated for Interferometric Synthetic Aperture Radar (InSAR)
IODP	International Ocean Discovery Program
ka	Thousand years
LGM	Last glacial maximum extent of the Pleistocene ice sheet
LiDAR	Light Detection and Ranging
LS	Landsat
LS8	Landsat 8
MAR	Mid-Atlantic Ridge
METI	Ministry of Economy, Trade and Industry (Japan)
MODIS	Moderate-resolution Imaging Spectroradiometer, climate sensor aboard the TERRA satellite operated by METI (Japan)
MSL	Mean sea level
MSS	Multispectral Scanner, sensor of Landsat 1 to 5
MW	Microwave spectrum
NAIP	National Agriculture Imagery Program (USGS USA)
NASA	National Aeronautics and Space Administration (USA)
NDVI	Normalized difference vegetation index
NED	National Elevation Dataset (USGS USA)
NGA	National Geospatial-Intelligence Agency (USA)
NGDC	National Geophysical Data Center (CO, USA)
NGOM DEM	Northern Gulf of Mexico Digital Elevation Model, part of CoNED (USGS USA)
NIR	Near infrared spectrum
NOAA	National Oceanic and Atmospheric Administration (USA)
NSF	National Science Foundation (USA)
ODP	Ocean Drilling Program
OLI	Operational Land Imager, sensor of Landsat 8
PSM	Petroleum system model
QC	Quality control
RADAR	Radio Detection and Ranging
RGB	Red-green-blue color image
SEG	Society of Exploration Geophysicists (USA)
SRTM DEM	Shuttle Radar Tomographic Mission digital elevation model
SWIR	Short wave infrared spectrum
TCP/IP	Transmission Control Protocol/Internet Protocol
THD	Total harmonic distortion of signals (vibroseis source)
TIRS	Thermal Infrared Sensor, sensor of Landsat 8

TIR	Thermal infrared spectrum
TM	Thematic Mapper, sensor of Landsat 4 and 5
UNESCO	United Nations Educational, Scientific and Cultural Organization
USACE	US Army Corps of Engineers
USGS	United States Geological Survey
VIS	Visible spectrum
VNIR	Very near infrared spectrum
WDMAM	World Digital Magnetic Anomaly Map by the Commission for the Geological Map of the World

About the Author

Andreas Laake is a physicist and geoscientist with over 30 years of industry experience in processing, interpretation, and integration of geological, geophysical, and remote sensing data from acquisition planning and data acquisition, to geological modeling from prospect to global scale. He is a Geophysical Advisor for Schlumberger Digital and Integration in Aachen, Germany, where he develops and executes digital exploration projects using analysis and interpretation of seismic data for geological modeling and reservoir characterization, integration of seismic, non-seismic, satellite and well data, and geological model building from basin to global scale.The author envisages processing, interpretation, and integration of geoscience data as an art to unlock the information that is hidden in the raw data. The author's approach to the visualization of the results aims at the co-rendering of multiple data types in a form that inspires the intuitive interpretation of the viewer to draw comprehensive conclusions that are supported by the context of multiple data types, each of which provide a different facet of the studied features.

Part I
Basics of Remote Sensing

Chapter 1
Basics of Remote Sensing

Abstract This section gives a brief history of the development, implementation and utilization of remote sensing in the context hydrocarbon exploration. An important historic aspect is the long-term continuation of remote sensing programs to ensure evolutionary studies can be executed.

We summarize the physical foundations of remote sensing comprising the propagation of electromagnetic waves through the atmosphere and into water and soil and their impact on the spectra useable for remote sensing. Typical use cases comprise textural studies of the earth surface, landuse and lithological applications. Special cases cover subsurface mapping in shallow water as well as in hot arid deserts.

Keywords Remote earth observation · Electromagnetic bands · Absorption · Reflection · Optical · Near infrared · Short-wave infrared · Thermal infrared · Microwave radar

1.1 Introduction

In this section we will introduce the basics of remote sensing. To illustrate how the earth surface and its characteristics are expressed in the different spectral bands that are used in remote sensing we use examples from different geological and climatic settings. Arid deserts and volcanic terrain expose lithology and erosional landforms, in tropical climate vegetation dominates. Floodplains and karst features are expressed as subtle features in the digital elevation model. Figure 1.1 shows on a map the examples used in this section and Table 1.1 gives an overview of their locations.

A. Laake, *Remote Sensing for Hydrocarbon Exploration*, Springer Remote Sensing/Photogrammetry, https://doi.org/10.1007/978-3-030-73319-3_1

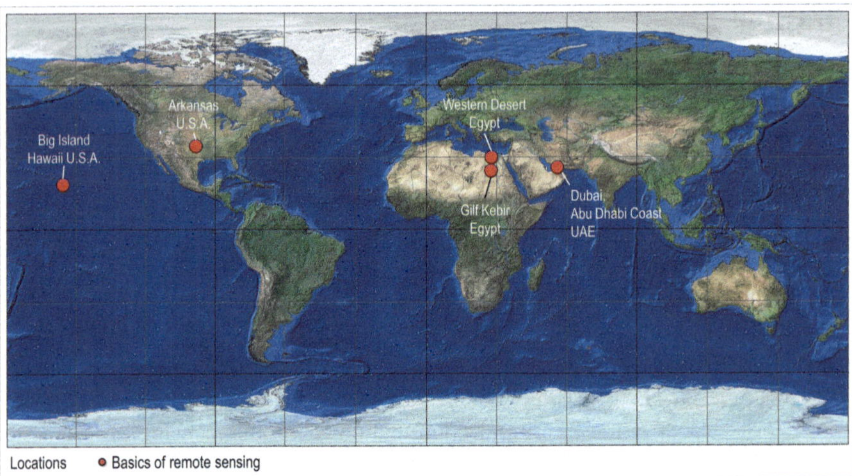

Fig. 1.1 Location of remote sensing examples used to illustrate the basics of remote sensing. (Satellite imagery courtesy of U.S. Geological Survey)

Table 1.1 Remote sensing examples used to illustrate the basics of remote sensing

Section/ Chapter	Geologic setting	Area of interest	Remote sensing input for
1.1.2	Arid desert coast	Dubai, UAE	Mapping of city development
1.1.4	Arid desert inland	Gilf Kebir, Egypt	Expression of topography and lithology in remote sensing data
2.1	Volcano in tropical climate	Big Island, Hawaii U.S.A.	Expression in remote sensing data
2.1	Arid desert inland	Gilf Kebir, Egypt	Expression in remote sensing data
2.1	Arid desert coast	Abu Dhabi coast, UAE	Expression in remote sensing data
3.4	Surface limestone karst	Western Desert, Egypt	Expression in band ratio and band difference
3.5	Arid desert and basement rock	Western coastal plains, Gulf of Suez, Egypt	Expression in RGB composites and inverted RGB composites
Chapter 4	Meandering river channels	Eastern Ozark Mountains, White River floodplain, AR U.S.A.	Expression in high-resolution digital elevation models

1.2 Scope of Remote Sensing

Remote sensing of the earth is a technique for gathering information about the surface and subsurface of the earth without physically placing measurement devices on the earth. The technique is particularly useful in terrain that is difficult to access by humans or machines that operate on the ground. Examples for such terrain comprise rugged terrain like mountains and arctic ice fields and areas without infrastructure such as deserts. Particularly challenging is densely forested terrain since the earth surface is covered by vegetation. Remote sensing is also employed for global continuous earth observation, especially when temporal changes need to be monitored.

Remotely operated sensors use two physical concepts for remote detection: reflection as well as back-scattering of electromagnetic waves and gravity effect on satellite orbits. The advantage of electromagnetic waves is the broad spectrum that is available for the investigation of the earth. Therefore, most remote sensing methods employ satellites that either measure passively the spectral reflectance of the earth surface or actively emit electromagnetic waves and measure the backscattering and absorption by the earth. Gravity-based methods take advantage of the changes in the density of the rocks in the earth's upper layers and their gravity effect on the orbits of satellites.

1.3 History of Remote Earth Observation

The invention of flight enabled remote observation of the earth's surface for the first time (Fig. 1.2). The first remote photograph was taken by Gaspard-Félix Tournachon from a hot air balloon, an aerial photo of Paris. During the two world wars, remote earth observation was used routinely for military reconnaissance through the acquisition of aerial photography (Baumann, 2009).

Soon after the invention of the laser, the mapping method of Light Detection and Ranging (LiDAR) was implement on airplanes allowing to map large areas. The method was extended to bathymetric LiDAR mapping in 1968 (Neff, 2018).

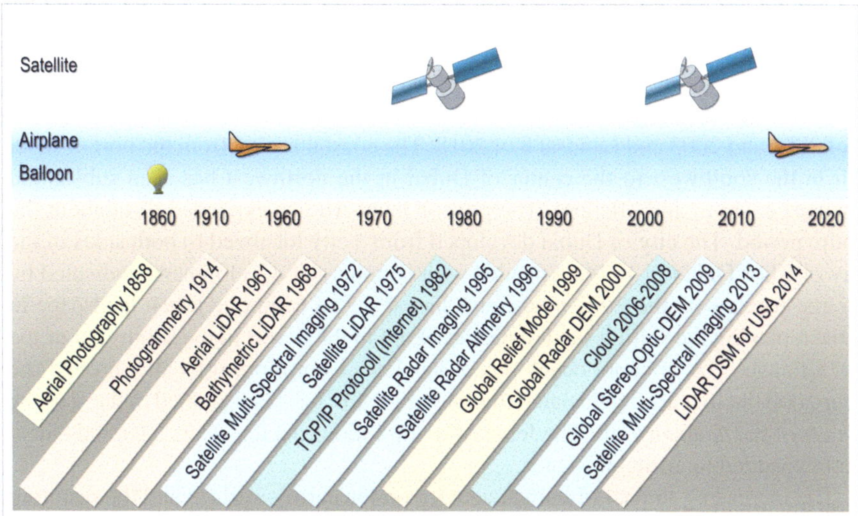

Fig. 1.2 Milestones in the history of remote sensing

Human space flight and the advent of weather satellites in the 1960 laid the basis for earth observation from civilian satellites which started with the Landsat program in 1972, the launch of Landsat 1 (Rocchio, 2020). Since then the Landsat family of satellites has provided continuous mapping of the earth surface with ever improving specifications of sensors and spatial coverage. The first systematic collection of remote sensing images and their interpretation has been published by the NASA in 1976 (Short et al., 1976). The authors outline applications of remote sensing using Landsat 1 for vegetation analysis, landuse mapping, geology, water resources, oceanography and marine resources as well as for the environment. LiDAR was implemented on satellites in 1975 (Neff, 2018).

Information technology had developed from the 1960s reaching commerciality with the definition of the TCP/IP protocol in 1982, which made the global implementation the internet possible thus preparing the basis for the global exchange of remote sensing data. Satellite radar imaging became available in 1995 (Raney et al. 1991a). One year later satellite radar altimetry was implemented. This technology enabled to enhance the global relief model GLOBE (Hastings & Dunbar, 1999) that had been composed at 5 km spatial resolution from all available resources. A dual radar device on board of Shuttle Endeavour (Farr et al. 2007a) provided the first global digital elevation model at 90 m spatial resolution. Information technology developed the cloud starting in 2006. Big data sets could now be processed and made available to users worldwide.

The quality of the shuttle radar digital terrain model was complemented with a new model derived from stereo-optic image processing in 2009 (Abrams & Crippen, 2019). Satellite multi-spectral imaging was further enhanced by integrating the mapping possibility in 2013 (USGS, 2019a). In 2014 the first countrywide LiDAR DEM was initiated in the U.S.A.

Figure 1.3 illustrates the advancement of remotely sensed imagery from Landsat satellites from 1978 to 2018 at the example of Dubai on the United Arab Emirates coast. In addition to the enhancement of the resolution and image quality, the development of the infrastructure is highlighted. The sequence is a series of false color images using the red, green and near-infrared bands from Landsat 4 of 1978, Landsat 7 of 2001 and 2003 and Landsat 8 of 2018. The coastal region from the port of Jebel Ali in the southwest to the center of Dubai in the northwest has seen substantial changes onshore and offshore. The port of Jebel Ali has expanded throughout the entire period. The city of Dubai developed from a city localized to both sides of the bay of Khor Dubai in 1978, expanded with commercial developments indicated by the red patches on the 2001 image and spread out to form one urban landscape in 2018 replacing the rolling sand dunes observed in the desert in the southwest of the 1978 image. The offshore development has been even more dramatic: around the year 2000 the first artificial island, the Jumeira Palm, has been created in the shallow waters of the Arabian Gulf. The Jebel Ali Palm followed a few years later and Dubai Port expanded to artificial islands.

The increasing impact of human activity on the earth's surface and the resulting climate change have increased the interest of the general public in thematic mapping using multi-spectral satellite imagery. Figure 1.3 shows also another important

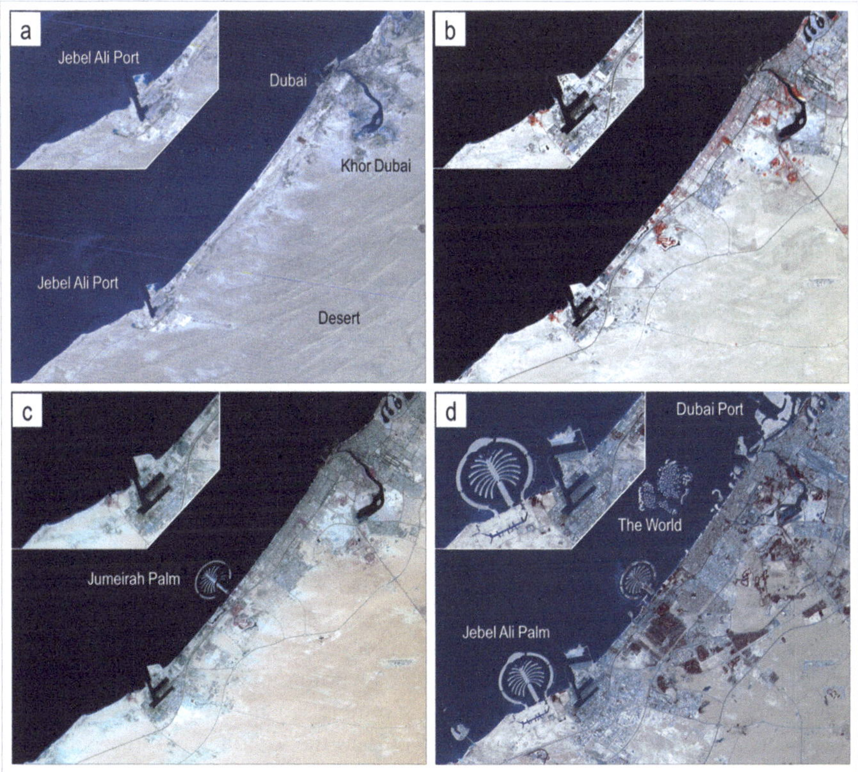

Fig. 1.3 Evolution of false color satellite images covering the city of Dubai in the Arab Emirates. (**a**) Landsat 4 321 RGB acquired on November 22, 1978, (**b**) Landsat 7 432 RGB acquired on May 22, 2001, (**c**) Landsat 7 532 RGB acquired May 28, 2003, (**d**) Landsat 8 543 RGB acquired on March 26, 2018. (Source of Landsat imagery: U.S. Geological Survey)

aspect of multi-spectral sensing: the comparability of data from different satellites over long time periods. To achieve this, the Landsat program has designed the latest satellites along the concept of a data continuity mission (Rocchio, 2020) that ensures a fixed set of spectral bands is included in the design of all sensors for satellites since Landsat 4.

1.4 Remote Sensing Methods

In the context of hydrocarbon exploration, remote sensing plays an important role from global basin mapping and assessment through opportunity identification and maturation to development of fields and infrastructure as well as the monitoring of production and its environmental impact. Basin modeling uses basement maps onshore and offshore, that are based on satellite radar data. The acquisition of geophysical data for subsurface geological mapping is supported by remotely sensed information for logistic planning and estimates for the geological and geophysical properties of the near surface. The relevance of geological models of the subsurface can be validated through surface deformation measurements from satellite radar data. The planning of infrastructure during field development and production as well as the transportation logistics of hydrocarbons benefits from remote sensing of topography and surface classification for lithology, water contents and vegetation. The environmental impact of production can be monitored using remote sensing detecting gases in the atmosphere around infrastructure and mapping changes in the surface reflectance as a result of the operation.

For satellite remote sensing of the earth's surface, solar radiation is used as the source for illumination. The solar radiation passes through the atmosphere, is reflected or scattered at the earth's surface. Part of the incident solar radiation is returned into space after passing through the atmosphere. An imager aboard an earth-orbiting satellite is recording the returned radiation, converts its signals to electromagnetic signals that are transmitted to a ground receiving station at the surface of the earth where the signals are decoded and processed to imagery.

The imager typically uses a scanner to sweep across an area of the earth's surface whilst flying on its orbit leaving a path of scanned surface the imagery of which becomes available as a satellite image after processing at the ground receiving station. The method described so far uses passive recording of the earth's illumination by the sun and is employed in passive optical remote sensing (Fig. 1.4).

For active microwave remote sensing (Fig. 1.5) the satellite comprises also the source for the radiation. The radiation emitted from the satellite passes through the

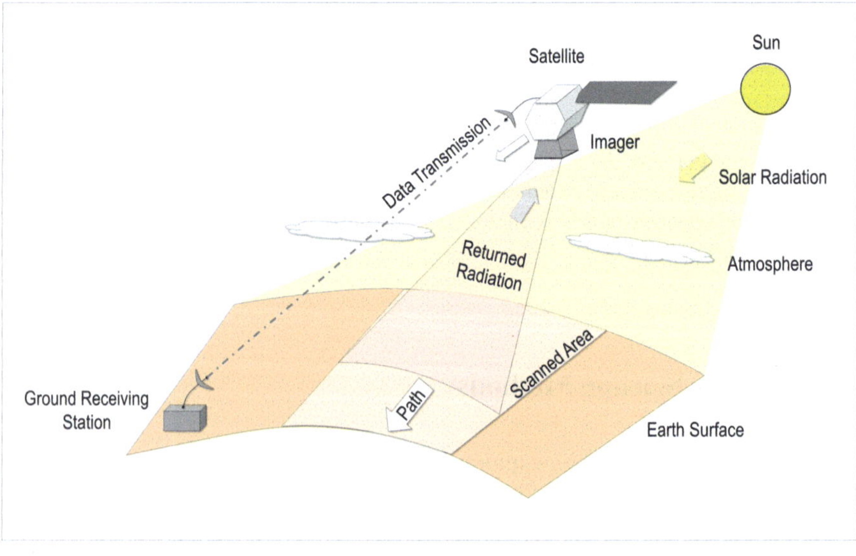

Fig. 1.4 Principle of passive optical satellite remote sensing

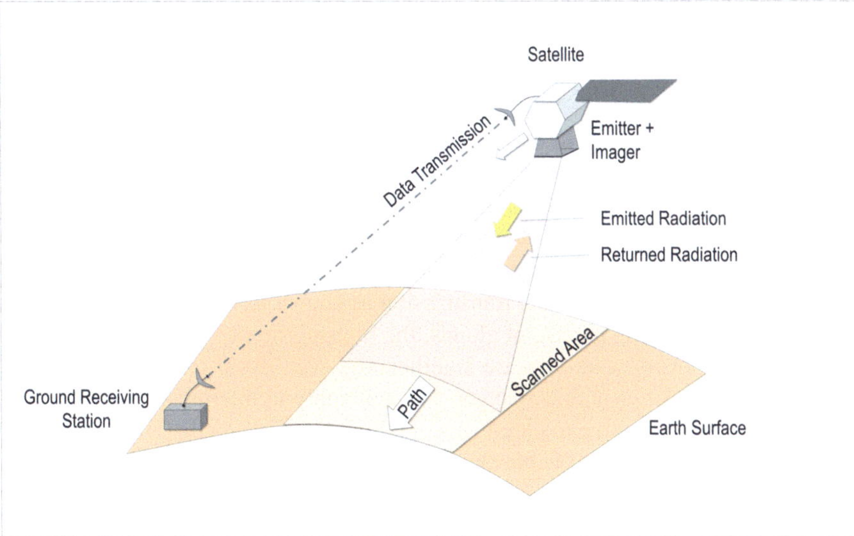

Fig. 1.5 Principle of active microwave satellite remote sensing

atmosphere, which does not attenuate microwave electromagnetic radiation and has therefore been omitted from the figure. The returning radiation from the earth is recorded by the imager aboard the earth-orbiting spacecraft. The transmission and decoding part of microwave remote sensing is identical to the passive case.

1.5 Atmospheric Absorption and Wavelength Bands for Remote Sensing

Electromagnetic radiation emitted by the sun or from a satellite orbiting outside the atmosphere needs to travel through the atmosphere to deliver measurements of the earth surface and subsurface. The atmosphere attenuates the transmitted radiation through reflection, absorption and scattering with a spectral characteristics that is determined by the gases that compose the atmosphere. Water vapor (H_2O) contributes most to this attenuation followed by carbon dioxide (CO_2) and ozone (O_3). Since the absorption by these gases occurs preferably at certain wavelengths, the combination of the absorption spectra defines certain windows in the electromagnetic wavelength spectrum that provide sufficient transmission for the remote sensing of the earth surface using satellites. Figure 1.6 shows the transmission spectrum of the earth's atmosphere with indications which gases are mainly responsible for atmospheric absorption.

The atmospheric absorption and physical limitations define the wavelength regions and bands that can be used for remote sensing (Sabins, 2000). The total absorption of wavelengths less than 0.4 μm renders ultra-violet and shorter wavelengths unusable for remote sensing and thereby sets the low wavelength boundary. The long wavelength boundary is defined by wavelengths that are too long to allow mapping of surface features, which applies to wavelengths beyond 1 m. Preferred wavelength windows occur in the visible, infrared and microwave regions. Visible wavelengths show the earth in natural colors as the human eye sees them. Reflected infrared wavelengths are used for the mapping of vegetation and surface rocks. Thermal infrared wavelengths provide insight into the earth's surface properties using the conversion of solar radiation heat.

It is important to note that the transmission as shown in Fig. 1.6 assumes clear sky and no cloud cover. In case of clouds, the transmission of wavelength used in optical remote sensing can decrease significantly, even down to total absorption. This sensitivity to atmospheric changes does not apply to microwave radar waves. They are largely insensitive to cloud cover and other processes that limit the penetration of wavelength in the visible and infrared.

Table 1.2 summarizes the wavelength regions and bands that are used in remote sensing.

The sensors aboard the remote sensing satellites are designed such that their sensitive wavelength in these bands provides a representative spectral response of the earth's surface. Historically, the Multispectral Scanner (MSS) used in Landsat 1–3 focused on the optical and near infrared bands to map the earth's surface cover including plants (Short et al., 1976; USGS, 2020a). The Thematic Mapper (TM)

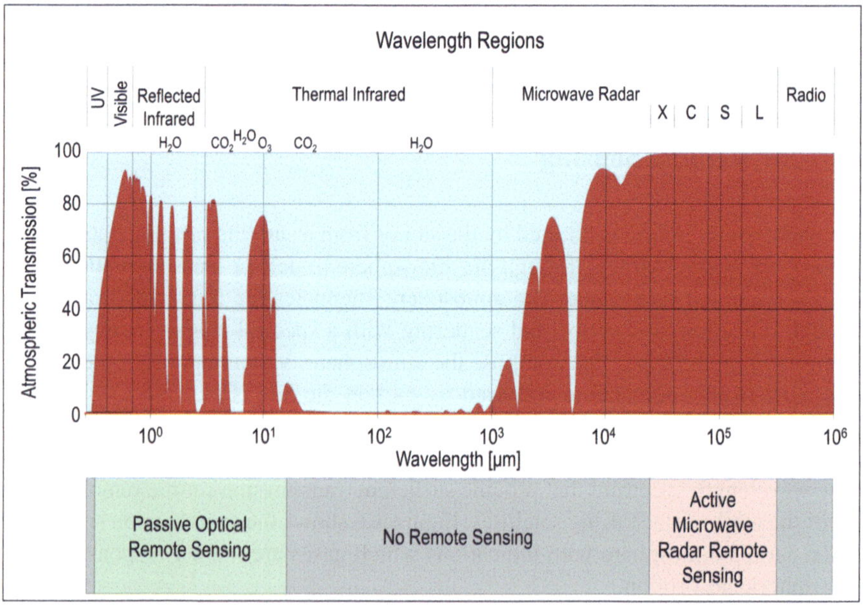

Fig. 1.6 Atmospheric transmission spectrum for electromagnetic waves and wavelength regions for remote sensing applications. (Spectral data courtesy of U.S. Geological Survey)

Table 1.2 Wavelength regions and bands used for remote sensing

Wavelength region	Wavelength (μm)	Remark
Ultra-violet region	0.03–0.4	Completely absorbed by atmosphere
Visible region	0.4–0.7	Visible images, no surface penetration other than into water
Infrared region	$0.7–10^3$	
Reflected infrared band	0.7–3.0	Surface reflected infrared, no penetration
Near infrared	0.7–1.3	Focus on vegetation
Shortwave infrared	1.4–3.0	Focus on lithology
Thermal infrared band	$7–10^3$	Emission of radiation converted to thermal waves
Microwave region	10^3 to 10^8	Penetrates clouds, fog, rain
Radio wave region	$>10^8$	Wavelength too long for imaging

starting with Landsat 4 expanded the spectrum into the infrared bands by offering two bands in the short wave infrared and one in the thermal long wave infrared. The Enhanced Thematic Mapper (ETM+) introduced a panchromatic band covering the wavelength range from 0.52 to 0.9 μm which provided a grayscale image at higher spatial resolution intended for sharpening the other bands. The present Landsat 8 continued to use the same band composition for continuity reasons with some amendments: a coastal and aerosol band at 0.43–0.45 μm, i.e. at a wavelength below the blue band of the previous sensors was added to improve the mapping in shallow water. This band is also sensitive to some aerosols. The extent of the panchromatic band was limited to just green and red to retain the spectral response during the panchromatic sharpening process.

Two thermal infrared bands give better insight into the surface temperature than the single band used in the previous sensors. And, finally a very narrow short-wave infrared band at 1.36–1.38 μm was added to specifically detect cirrus clouds (USGS, 2019b).

Figure 1.7 shows the spectral bands that have been employed by Landsat satellites in their context with the atmospheric transmission spectrum and the atmospheric gases dominating the atmospheric absorption.

The bands for the visible and infrared sensors of Landsat 8, their wavelength bands and potential applications are listed in Table 1.3, examples for a hot arid desert in southwest Egypt is shown in Fig. 1.8 along with an example of the C-band of microwave radar.

The shorter wavelength of the coastal band 1 is subject to less attenuation compared to the traditional blue band of previous Landsat sensors and therefore provides deeper penetration into shallow water. The seafloor can be mapped at greater water depth and with better resolution of its texture. This is important for the operations planning in shallow water. The visible bands 2–4 image the earth surface in the visible range and can be used in combination with visible aerial photography. The near infrared band 5 is sensitive to chlorophyll and, especially when combined with

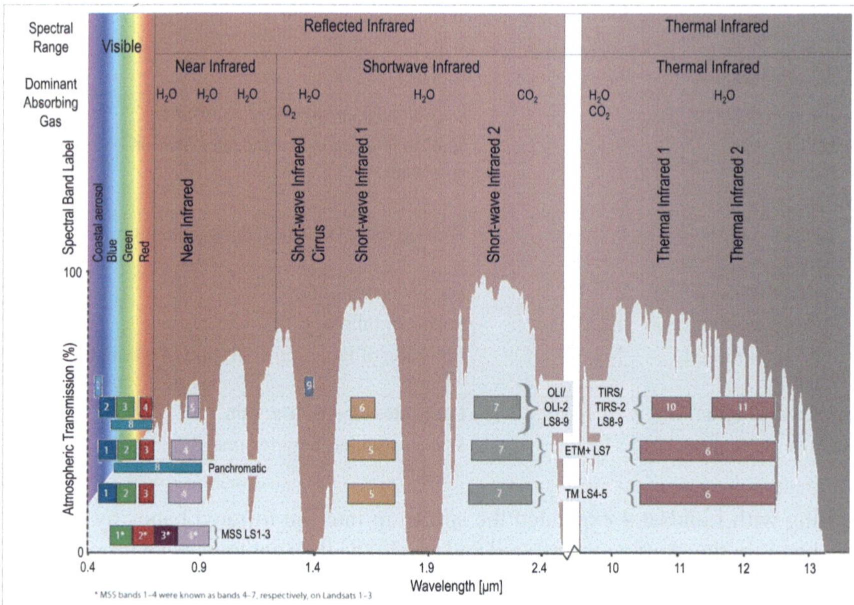

Fig. 1.7 Visible and infrared atmospheric transmission spectrum and spectral bands for remote sensing by different generations of Landsat satellites (after USGS, 2018a), *LS* Landsat, *OLI* operational land imager (LS8), *TRS* thermal infrared sensor (LS8), *ETM+* enhanced thematic mapper (LS7), *TM* thematic mapper (LS4–5), *MSS* multispectral scanner (LS1–3). (Spectral data courtesy U.S. Geological Survey)

the visible green band 3, allows the study of vegetation, for example seasonal changes of growth cycles and impact of forest fires. The two short wave infrared bands 6 and 7 are located in spectral bands where common minerals in surface soils and rocks show the largest differences in spectral response. The panchromatic band 8 is sampled at twice the spatial resolution across the visible region of the spectrum to sharpen images from visible bands. When applying the sharpening it is important to note that the uplift in sharpness of the image only applies to visible bands and not infrared bands (see Sect. 3.2). The application of the cirrus band 9 is mainly in detecting the presence of sometimes subtle distortion of the spectral response due to cirrus clouds. The signature of band 9 allows to delineate affected areas and to correct for the distortion.

Bands 1–9 do not deliver information about the subsurface with the exception that bands 1 and 2 show significant penetration into water. This means that the information obtainable from these bands is restricted to the immediate surface. No penetration into the subsurface is possible.

The two thermal infrared bands 10 and 11 employ the conversion of the sunlight into heat radiation. Thereby, some information about near surface layers can be obtained. Typically, the first 1 m of soil below the earth surface reacts on the diurnal changes in sun light illumination. In this depth interval the solar radiation can

Table 1.3 Landsat 8 sensor characteristics using information from USGS (2018b)

Sensor	Band	Description	Wavelength [μm]	Useful for mapping	
OLI	1	Coastal aerosol	0.43–0.45	Coastal and bathymetric studies	Penetration into water only
	2	Blue	0.45–0.51	Coastal studies, distinguishing soil from vegetation, lithology studies	
	3	Green	0.53–0.59	Emphasizes peak vegetation, discriminating vegetation from rocks	No penetration into subsurface
	4	Red	0.64–0.67	Discriminates vegetation slopes	
	5	Near Infrared (NIR)	0.85–0.88	Emphasizes biomass content and shorelines	
	6	Short-wave Infrared (SWIR) 1	1.57–1.65	Discriminates moisture content of soil and vegetation; discriminates lithology	
	7	Short-wave Infrared (SWIR) 2	2.11–2.29	Improved moisture content of soil and vegetation; discriminate lithology	
	8	Panchromatic	0.50–0.68	15 m resolution, sharper definition for visible band images	
	9	Cirrus	1.36–1.38	Improved detection of cirrus cloud contamination to improve image interpretation	
TIRS	10	TIRS 1	10.60–11.19	100 m resolution, thermal mapping of lithology and estimated soil moisture	Subsurface information
	11	TIRS 2	11.50–12.51		

be converted to heat, which is emitted back to space and can be recorded by a satellite-based thermal infrared sensor. Using this approach, buried changes in the very shallow near surface can be detected using thermal infrared waves provided they have an extent of several 100 m. Examples for such buried changes are tar lakes that are covered by a thin sand layer, thin local sand sheets covering intermittent rock outcrops and extended fault systems where open fractures are cooler than the host rock.

Microwave radar penetrates clouds and is, therefore, not affected by the distortion induced by cirrus clouds as evident from the comparison of Landsat band 9 and the microwave C-band in Fig. 1.8. Furthermore, microwave radar highlights the contrast of outcropping hard rock areas (dark areas) and valleys filled with soft sand (light areas).

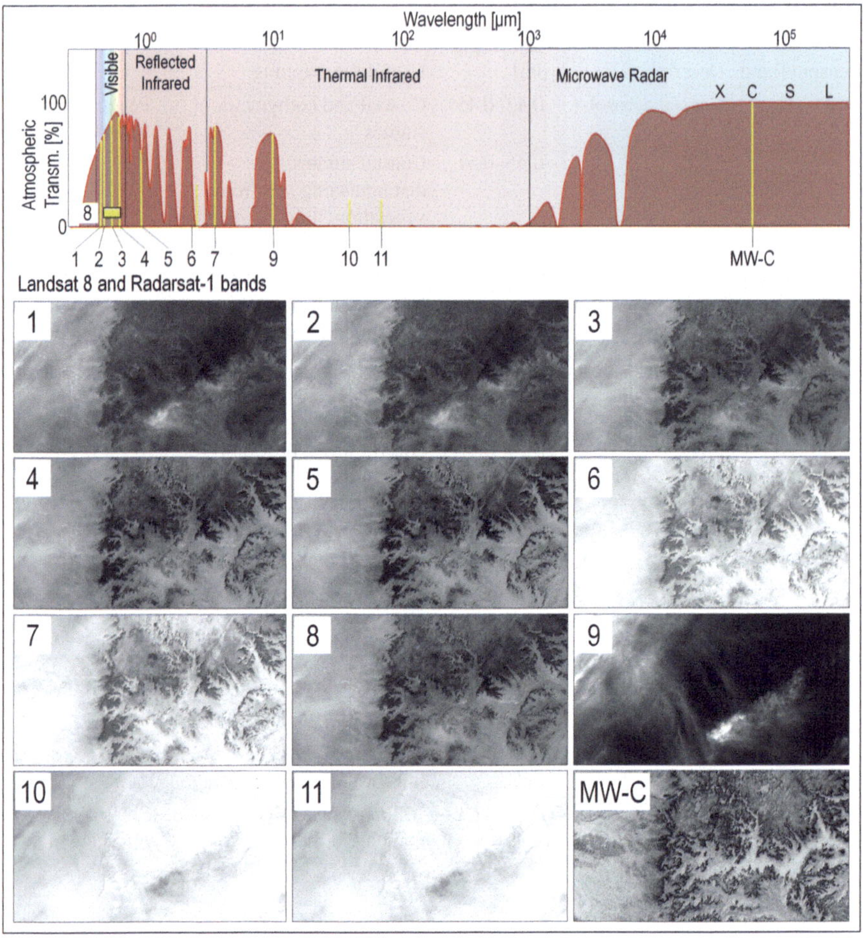

Fig. 1.8 Examples of Landsat and RADARSAT-1 images for a desert terrain in southwest Egypt. (Spectral data courtesy U.S. Geological Survey, satellite imagery courtesy U.S. Geological Survey and Canadian Space Agency)

1.6 Interaction of Electromagnetic Radiation with Earth and Water

After transmission through the atmosphere, the electromagnetic radiation interacts with the earth's surface through reflection, back-scattering or transmission into the earth. Different materials of the earth's surface yield different contributions of these three processes in a way that is characteristic for these materials. In Table 1.4 the most common types of interaction of electromagnetic radiation with the earth surface are listed, examples for the geological settings are shown in Fig. 1.9.

Table 1.4 Types of interaction of electromagnetic waves with the earth surface

Type of interaction	Surface	Wave type	Example
Surface reflection			
Specular	Smooth	Optical, radar	Water
Diffuse	Rough	Optical, radar	Rock, vegetation, infrastructure
Scattering			
Surface backscatter	Hard	Optical, radar	Rock
Partial backscatter and partial transmission	Soft	Radar	Sand
Subsurface backscatter	Soft above hard layer	Radar	Buried channel
Absorption (transmission without backscatter)	Thick soft layer	Optical, radar	Thick sand

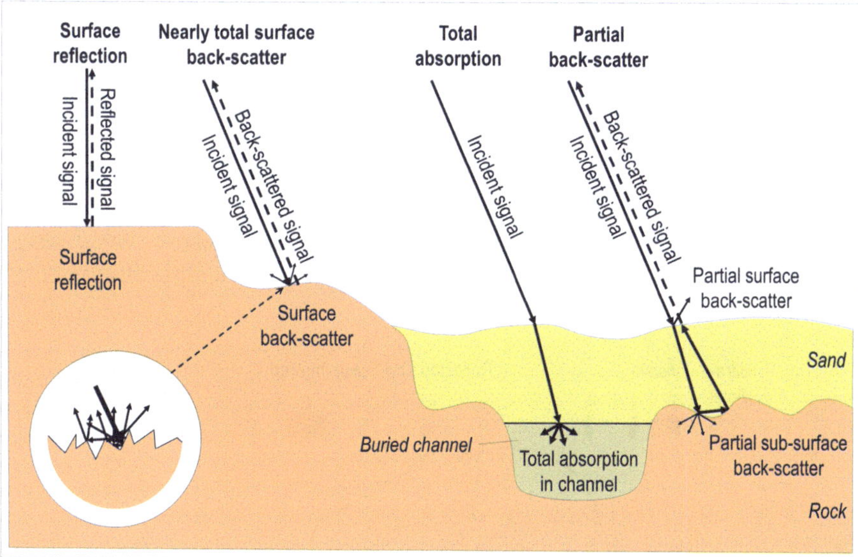

Fig. 1.9 Interaction of electromagnetic radiation with the earth's surface

Surface reflection and absorption are the most common types of interaction of optical radiation with the earth surface. Since the earth surface is mostly rough, diffuse surface reflection prevails. Surface backscatter of optical radiation also occurs but is much weaker that reflection. Therefore, for optical remote sensing reflection spectra are the most frequently used mapping tool.

The interaction of microwave radar with the earth surface is more complex because a considerable portion of the incident radiation can be transmitted into the ground. Specular reflection occurs typically at water surfaces. When a thin oil film smoothes out even minor ripples, the difference in specular reflection at the oil film from the slightly diffuse reflection of the non-contaminated water can be used to detect subsea oil seeps (Zhang et al., 2017). The amount of diffuse reflection at rocky surfaces may be used for the discrimination of hard rough rocky surfaces from soft sandy surfaces (Laake & Strobbia, 2013a). When partial back-scattering occurs at sandy surfaces, a portion of the incident waves is transmitted into the subsurface. Sand covered buried channels near the surface absorb the transmitted wave energy, whereas surrounding hard rock scatters back the transmitted wave energy in the subsurface. This partially back scattered energy propagates through the soft overlaying sand and back into the atmosphere where it can be recorded by the satellite-based radar receiver. Using this method buried river channel systems have been detected (Laake & Strobbia, 2013b).

The spectral distribution of reflected optical wave energy has been studied from spectroscopic laboratory measurements at the U.S. Geological Survey (Kokaly et al., 2017a). Figure 1.10 shows a selection of optical reflectance spectra of materials commonly encountered in remote mapping of the earth surface in the context of the spectral bands of Landsat 8. Rocks are represented by dolomite, urban infrastructure is road asphalt, oak represents vegetation, and soil is a mix of the clay minerals montmorillonite and illite. The vegetation spectrum shows a strong peak at the wavelength measured by the very near infrared Landsat 8 band 5. Rock, soil and urban spectra show different spectral reflectance for the two short-wave infrared bands 6 and 7, which makes them distinguishable through these two bands. The low spectral reflectance of seawater indicates that most of the Landsat 8 spectral bands will show seawater as black.

The absorption of electromagnetic waves in water strongly depends on the wavelength. Hale and Querry (1973) found that the spectral minimum for absorption of

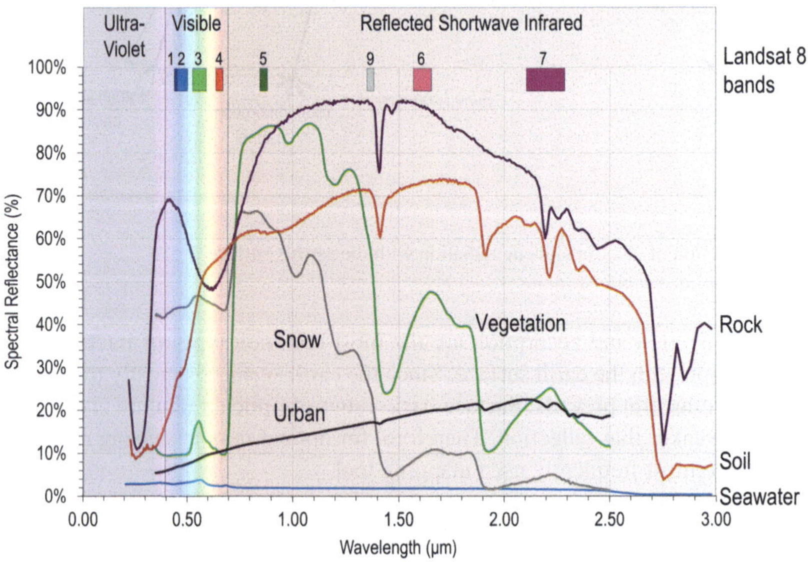

Fig. 1.10 Optical reflectance spectra of common materials at the earth surface. (Data source Kokaly et al., 2017b, U.S. Geological Service Spectral Library Version 7)

electromagnetic waves in seawater is located at about 0.43 μm, i.e. in the spectral band covered by the coastal and blue bands (Landsat 8 bands 1 and 2). For wavelengths below and above this value the spectral absorption increases significantly rendering seafloor mapping applications with Landsat bands 3 and above as well as with microwave radar impossible. Figure 1.11 shows the spectral absorption for electromagnetic waves in seawater in the context of the spectral bands of Landsat 8.

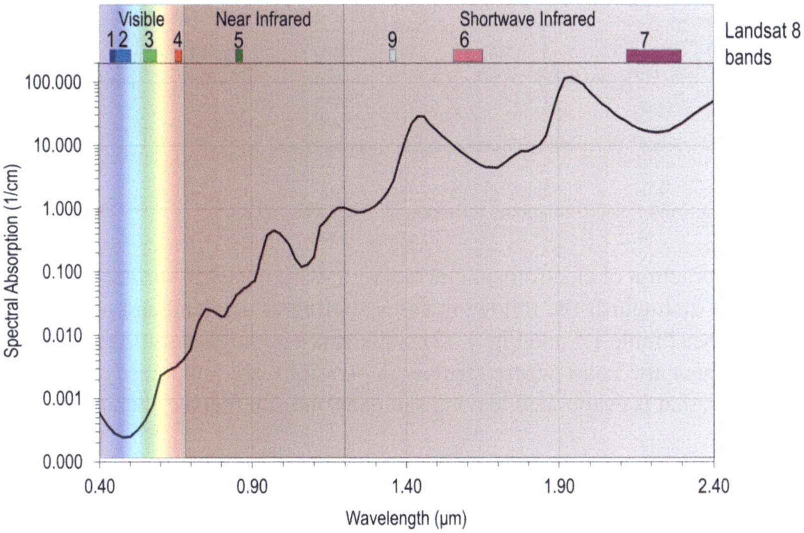

Fig. 1.11 Optical absorption spectrum of seawater. (Source of data Hale & Querry, 1973)

Table 1.5 Interaction of electromagnetic waves and earth surface in coastal setting

Geological setting	Surface interaction	Focus for detection	Optical bands	Microwave bands
Dry land	Diffuse reflection from surface rocks	Lithology discrimination	2–7	Surface roughness C-band
Sabkha	Diffuse reflection from surface salts	Lithology discrimination	2–7	Surface roughness C-band
Mangrove forest	Diffuse reflection from vegetation	Vegetation detection	5–7	Surface roughness C-band
Shallow water seafloor	Diffuse reflection from seafloor	Seafloor texture detection	1–3	
Coral reefs	Diffuse reflection top of reef	Texture of coral reef	1–3	
Oil seeps	Specular vs. diffuse reflection from the oil film at the surface	Detection and delineation of oil film		C-band

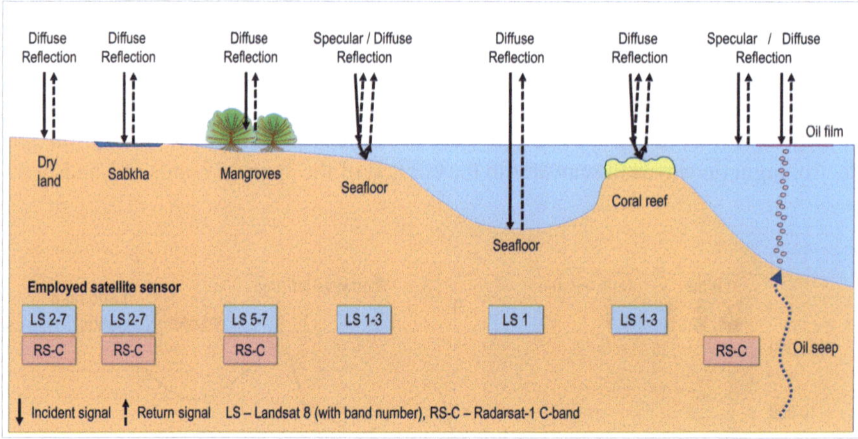

Fig. 1.12 Interaction of electromagnetic radiation with seawater

For the interaction of electromagnetic radiation with the earth surface in shallow water we can distinguish the following cases starting at dry land and propagating into the deep sea (Table 1.5 and Fig. 1.12). Sabkha is a geological setting that occurs for example near the coast where shallow depressions are temporarily inundated with seawater, that is evaporated leaving salt pans and salt marshes behind.

References

General Remote Sensing

Sabins, F. F. (2000). *Remote sensing, principles and interpretation* (3rd ed.). Freeman and Company.

Zhang, Y., Li, Y., & Lin, H. (2017). Oil-spill pollution remote sensing by synthetic aperture radar, Chapter 2. In *Advanced geoscience remote sensing* (pp. 27–50). INTECH Open. https://doi.org/10.5772/57477

History of Remote Sensing

Baumann, P. R. (2009). *History of remote sensing, satellite imagery, Part II*. http://employees.oneonta.edu/baumanpr/geosat2/RS%20History%20II/RS-History-Part-2.html

Rocchio, L. (curator). (2020). *Landsat science*. NASA. https://landsat.gsfc.nasa.gov/

Landsat

Short, N. M., Lowman, D. L., & Freden, S. C. (1976). *Mission to earth: Landsat views the world, NASA SP 360*. National Aeronautics and Space Administration, Scientific and Technical Information Office.

USGS. (2018a). *Landsat 7 (LS7) Data users handbook, USGS LSDS-1927 Version 1.0*. USGS EROS.

USGS. (2018b). *Landsat, continuing to improve everyday life*. USGS, EROS.

USGS. (2019a). *Landsat 8 (LS8) Data users handbook, USGS LSDS-1573 Version 4.0*. USGS EROS.

USGS. (2020a). *Landsat missions*. https://www.usgs.gov/land-resources/nli/landsat/landsat-satellite-missions

Radarsat

Raney, R. K., Luscombe, A. P., Langham, E. J., & Ahmed, S. (1991a). RADARSAT. *Proceedings of the IEEE, 79*(6) (Reprint).

Transmission Through Atmosphere

Kokaly, R. F., Clark, R. N., Swayze, G. A., Livo, K. E., Hoefen, T. M., Pearson, N. C., Wise, R. A., Benzel, W. M., Lowers, H. A., Driscoll, R. L., & Klein, A. J. (2017a). *USGS Spectral library version 7: U.S. Geological Survey data series* (Vol. 1035). USGS. https://doi.org/10.3133/ds1035. 61 p.

Kokaly, R. F., Clark, R. N., Swayze, G. A., Livo, K. E., Hoefen, T. M., Pearson, N. C., Wise, R. A., Benzel, W. M., Lowers, H. A., Driscoll, R. L., & Klein, A. J. (2017b). *USGS Spectral library version 7 data: U.S. Geological Survey data release*. USGS. https://doi.org/10.5066/F7RR1WDJ. https://www.sciencebase.gov/catalog/item/5807a2a2e4b0841e59e3a18d

Interaction with Earth Surface

Laake, A., & Strobbia, C. (2013a). Mapping paleorivers from satellite radar and surface-wave velocities. In *Proceedings 75th EAGE Conference & Exhibition Incorporating SPE EUROPEC 2013, London, UK*.

Interaction with Seawater

Hale, G. M., & Querry, M. R. (1973). Optical constants of water in the 200-nm to 200-microm wavelength region. *Applied Optics, 12*(3), 555–563.

Microwave Radar for Geologic Interpretation and Elevation Models

Farr, T. G., Rosen, P. A., Caro, E., Crippen, R., Duren, R., Hensley, S., Kobrick, M., Paller, M., Rodriguez, E., Roth, L., Shaffer, S., Shimada, J., Umland, J., Werner, M., Oskin, M., Burbank, D., & Alsdorf, D. (2007a). The shuttle radar topography mission. *Reviews of Geophysics, 45*, RG2004. https://doi.org/10.1029/2005RG000183

Electromagnetic Spectral Bands Used for Remote Sensing

USGS. (2019b). *Landsat 8 (L8) data user handbook, version 4.0, LSDS-1574*. EROS.

ASTER GDEM V3

Abrams, M., & Crippen, R. (2019). *ASTER GDEM V3 (ASTER global DEM), User Guide Version 1, Japan's Ministry of Economy, Trade and Industry (METI)*. National Aeronautics and Space Administration (NASA), Jet Propulsion Laboratory/California Institute of Technology.

GLOBE

Hastings, D. A., & Dunbar, P. K. (1999). *Global land one-kilometer base elevation (GLOBE) digital elevation model, documentation, Volume 1.0* (Key to geophysical records documentation (KGRD) 34). National Oceanic and Atmospheric Administration, National Geophysical Data Center.

LiDAR DEM

Neff, T. (2018). *The laser that's changing the world, the amazing stories behind Lidar from 3D mapping to self-driving cars*. Prometheus Books.

Radarsat Integration References Missing So Far

Laake, A., & Strobbia, C. (2013b). Mapping Paleorivers from satellite radar and surface-wave velocities. In *75th EAGE Conference and Exhibiiton 2013, Workshop Paper G06*.

Chapter 2
Electromagnetic Spectral Bands Used for Remote Sensing

Abstract Electromagnetic bands in remote sensing cover the spectrum from visible blue across the optical and infrared bands to microwave radar. With applications from different geological settings and climatic zones we will determine which spectral bands or band combinations are useful for the mapping task at hand.

Optical bands are useful for photographic assessments. The coastal blue band is used for shallow offshore mapping. Very near infrared is sensitive for vegetation, whereas shortwave infrared has its main application for lithology and landuse mapping. Radar images are used for two different tasks: first for surface textural characterization using radar surface backscatter. And second, when microwave radar can penetrate the ground, then volume backscatter and absorption is used to map shallow subsurface features.

Keywords Visible band composite · Natural color composite · Landuse image · Landsat satellites · Water penetration · Surface penetration

2.1 Introduction

This section provides examples for optical and microwave radar imagery for three different environmental settings: tropical volcanic environment of Hawaii Big Island, hot arid rocky setting in southwest Egypt and shallow water desert of the Abu Dhabi coast.

The preferred bands for each setting are schematically shown in Fig. 2.1. Fossil channels in valleys buried under sand can be detected with microwave radar. Terrain edges and escarpment leave a strong signature on microwave radar bands. Desert features related to lithology such as dunes, mesas and volcanic rocks are discriminated best by shortwave infrared (SWIR) bands. Volcanics are best mapped from short wave infrared and thermal infrared bands. Near infrared (NIR) and visible bands show vegetation, whereas SWIR and visible bands map infrastructure. The texture of shallow water seafloor shows up best in the coastal and blue bands of the visible region. For the examples, Landsat 8 (USGS, 2019) has been used for the

A. Laake, *Remote Sensing for Hydrocarbon Exploration*, Springer Remote Sensing/Photogrammetry, https://doi.org/10.1007/978-3-030-73319-3_2

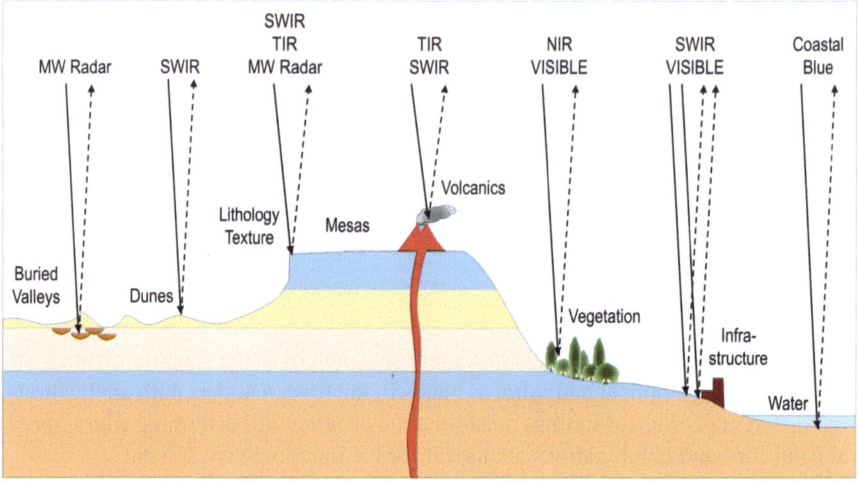

Fig. 2.1 Schematics of the interaction of electromagnetic spectral bands with the earth surface

optical spectral bands and Radarsat 1 (Raney et al., 1991) for the radar bands. Each case is presented by a characteristic multi-band combination of three bands. Field photos illustrate the different environmental settings shown in select areas of the case study.

The first case study covers the northern cost of Big Island, Hawaii (Fig. 2.2). The dominating feature is the Mauna Loa volcano with its lava fields (L) and craters. Due to the tropical climate vegetation (V) covers older, weathered lava fields. The multiband combination providing the best representation of these different environments is the combination of visible bands in a true color 432 RGB combination. The intensity and shade of the green color allows the distinction between different plant species and the status of their health. Brown to black shades are typical for the Mauna Loa volcano and volcanic rocks like lava and ash.

The Pacific Ocean which drops quickly to great depths off the coast appears deep blue. Infrastructure such as roads and houses as well as weathered rock near the coast that is not covered in vegetation appears in a blueish purple tone.

The next case study covers the Gilf Kebir highlands in the far southwest of Egypt at the border with Libya and Sudan (Fig. 2.3). The hot and extremely arid climate prevents any growth of vegetation. Erosion is mainly due to wind and the extreme temperature differences between day and night leaving most of the area covered in heat-weathered rock or gravel. Occasional flashfloods wash out fine debris valleys into the flat foreland. Due the horizontal layering of the formations on the plateau, heat weathering exposes hard sandstone as table mountains or mesas with sharp escarpments where valleys have been eroding the underlying soft sandstones and marls. The shortwave infrared and visible bands provide the best differentiation of rock types through the false color multi-band combination 753 RGB in the lower part of the combined satellite image. The texture of the rock formations, escarpments and dunes show best on the microwave radar image in the upper part of the combined satellite image. Rough sandstone ledges and boulders (S) are identified as

Fig. 2.2 Hawaii case study: Landsat 8 true color 432 RGB image with photographic examples of the different surface cover classes lava, vegetation and infrastructure. (Source of Landsat imagery: U.S. Geological Survey)

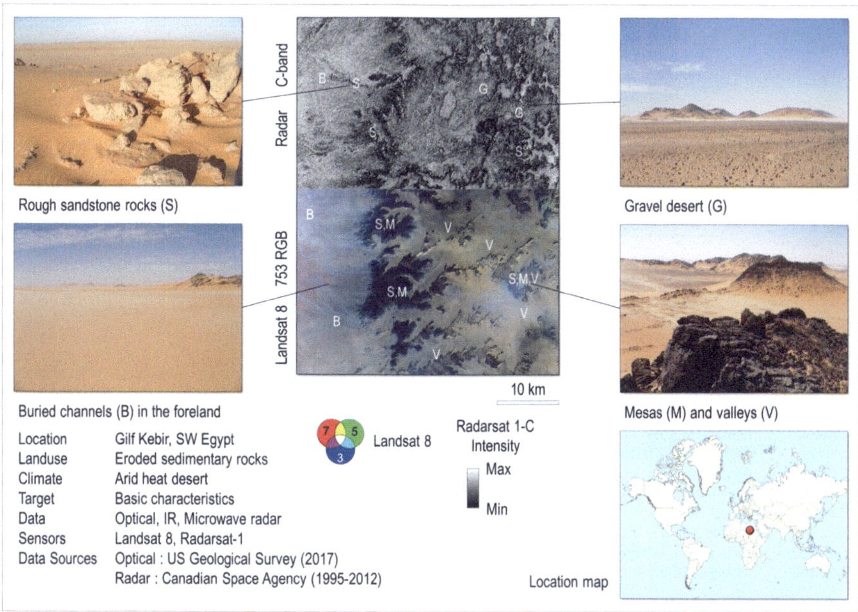

Fig. 2.3 Southwest Egypt case study: Landsat 8 false color 753 RGB with photographic examples of different surface classes. (Source of Landsat imagery: U.S. Geological Survey)

sharp lines in the radar image. When they are heat-weathered, they build a dark crust—desert varnish—like in the foreground of the photo showing mesas and valleys. Gravels form an intermediate state of erosion between solid rock mesas and sands and are found mainly on the highland of the Gilf Kebir plateau next to solid sandstone mesas.

Eolian erosion grinds down gravel to fine sand and which is blown into the valleys and out into the foreland. Widespread channel systems that existed during the humid period of the Pleistocene have been buried under sand (B) and can today only be detected by radar that penetrates the sand.

The shallow desert coast of the Abu Dhabi features in the last example (Fig. 2.4). The elevation of the coastal desert does in parts not exceed mean sea level, which results in temporal inundation with seawater. Upon evaporation in the hot arid climate, salt, gypsum and anhydrite are left over in salt pan deserts. This environment between sandy desert, salt pans and tidal shoreface is called coastal sabkha (S). Sandy beaches (B) line most of the very shallow lagoons with a water depth of less than 10 m. In the tidal flats, algae grow in the saltwater flooded sand leaving dark patches behind. Where sufficient nutrients are available, mangroves (M) grow along the shoreline. Since the focus of this study is the mapping into the seawater, the coastal, blue and green bands have been chosen for the representative multiband combination 321 RGB. The onshore coastal sabkha would be best imaged with a band combination of shortwave infrared and visible bands such as the 753 RGB multiband combination used in the previous example.

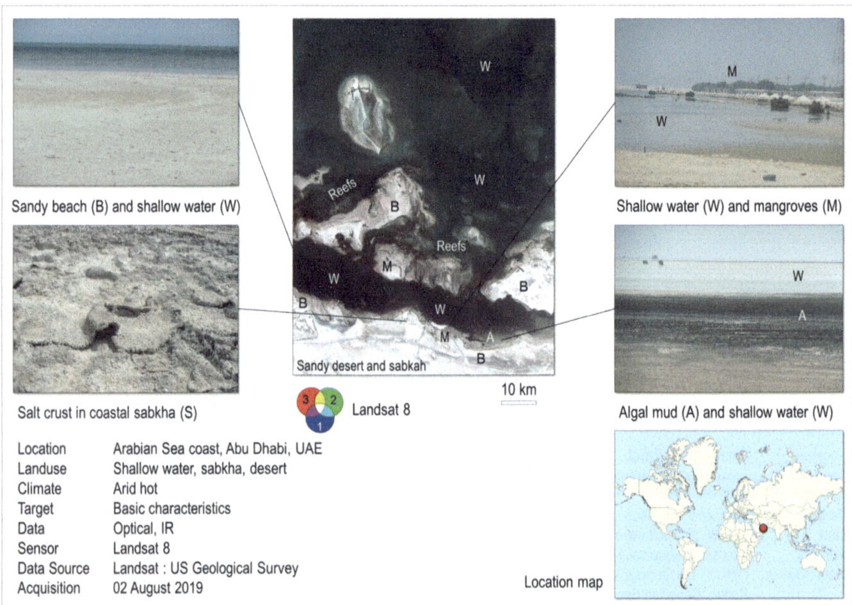

Fig. 2.4 Abu Dhabi coastal case study: Landsat 8 false color 321 RGB image with photographic examples of different surface classes. (Source of Landsat imagery: U.S. Geological Survey)

2.2 Optical Bands

The optical bands correspond to the spectrum visible to the human eye, which comprises the wavelength from 400 to 700 μm, i.e. the spectral colors blue to red. At Landsat 8, this part of the wavelength spectrum is represented by four bands: coastal blue, blue, green and red. A panchromatic band across part of the blue to red spectrum allows panchromatic sharpening to double resolution. The coastal blue band has been added at the design of Landsat 8 to enhance the color mapping in shallow water and to increase the detection of aerosols (USGS, 2019).

Optical bands are sensitive to water, vegetation and colors of the surface rock. In desert environment, where water and vegetation are absent, optical bands do not show significant differences between the four color bands (Fig. 2.5). Only the coastal blue band (a) has an overall lower response. Consequently, the panchromatic band (e) well represents the overall response in the optical spectrum, resulting in less spectral distortion in panchromatic sharpening (see Sect. 3.2).

The multiband color composite of blue, green and red bands is dominated by creme and black tones (f) with the black tones representing heat-weathered sandstone and the crème tones indicating soft sand.

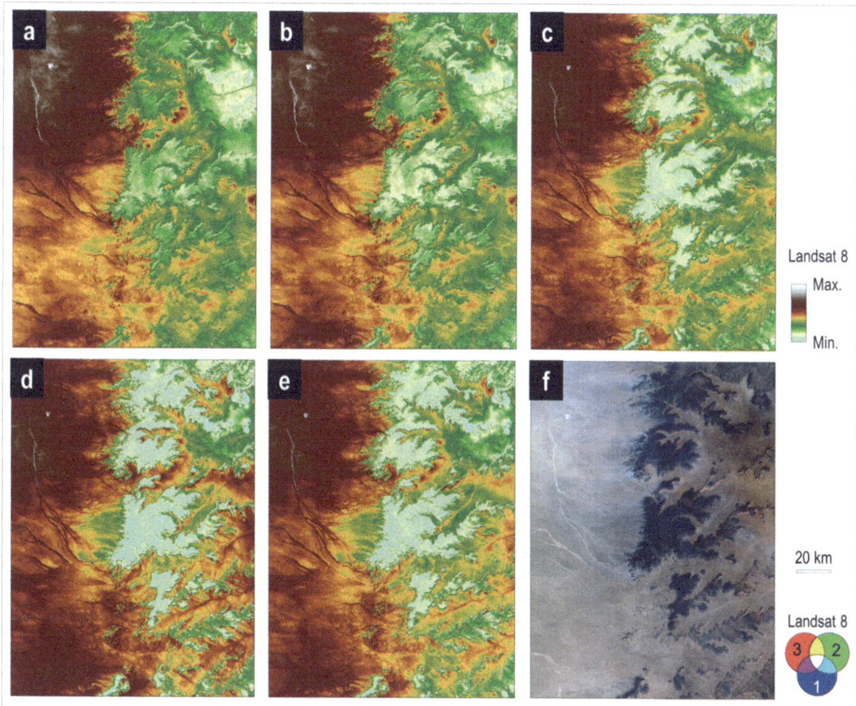

Fig. 2.5 Response of optical bands to surface cover in desert environment, Southwest Egypt. (**a**) coastal band 1, (**b**) blue band 2, (**c**) green band 3, (**d**) red band 4, (**e**) panchromatic band 8, (**f**) true color image LS8 321 RGB. (Source of Landsat imagery: U.S. Geological Survey)

The spectral response in the optical bands is markedly different in the Hawaii case study (see Fig. 2.6), where the surface cover ranges from tropical rain forest (V), lava flows (L) to human infrastructure (I) and seawater. The coastal blue (a) and blue bands (b) show a very similar response clearly differentiating between infrastructure on the one hand and vegetation and lava flows on the other but slightly deviate in their offshore signature. The green (c) and red (d) bands distinguish clearly between vegetation and lava flows. The panchromatic band (e) represents mostly the spectral characteristics of the green and red bands. Panchromatic sharpening (see Sect. 3.2) should, therefore, only be applied to multiband combinations using either the blue, green and red or the green, red and near infrared bands. Applying the panchromatic sharpening to other band combinations will result in a degradation of the color quality. The different response of the blue, green and red bands to the different types of surface cover results in a clear color representation when combined in a multiband composite (f).

In the case of shallow water and coastal sabkha, the response of the optical bands is remarkably different (see Fig. 2.7). On the maps for the individual bands the 10 m isobath is marked to give an idea about the water depth. The coastal band (a) shows the most detail in the range below 10 m water depth. The blue (b), green (c) and red

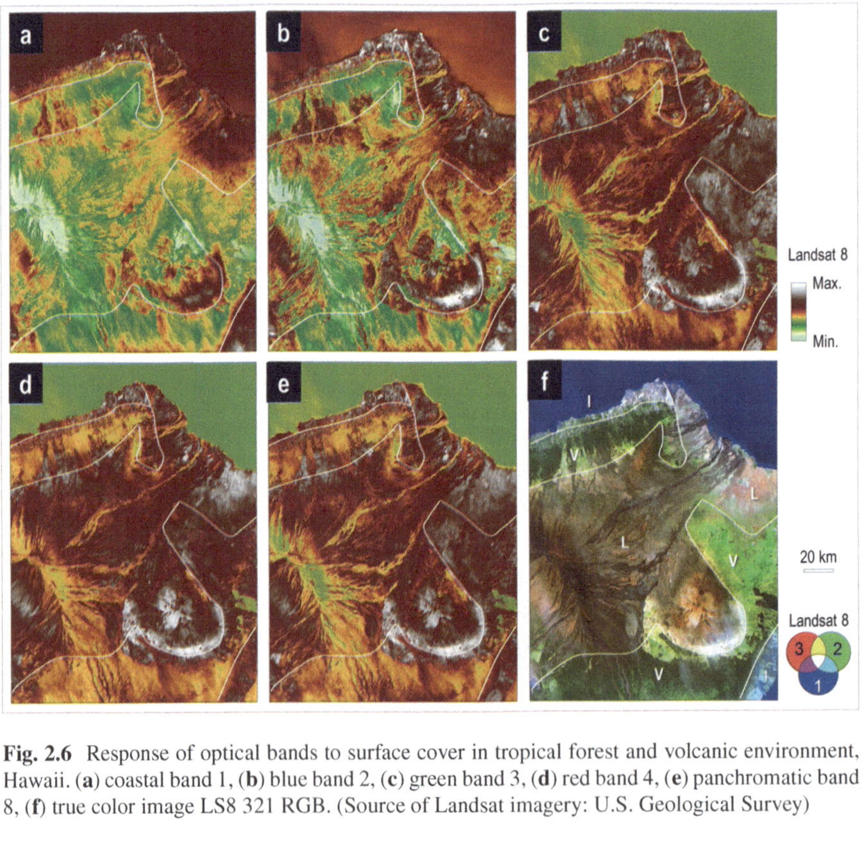

Fig. 2.6 Response of optical bands to surface cover in tropical forest and volcanic environment, Hawaii. (**a**) coastal band 1, (**b**) blue band 2, (**c**) green band 3, (**d**) red band 4, (**e**) panchromatic band 8, (**f**) true color image LS8 321 RGB. (Source of Landsat imagery: U.S. Geological Survey)

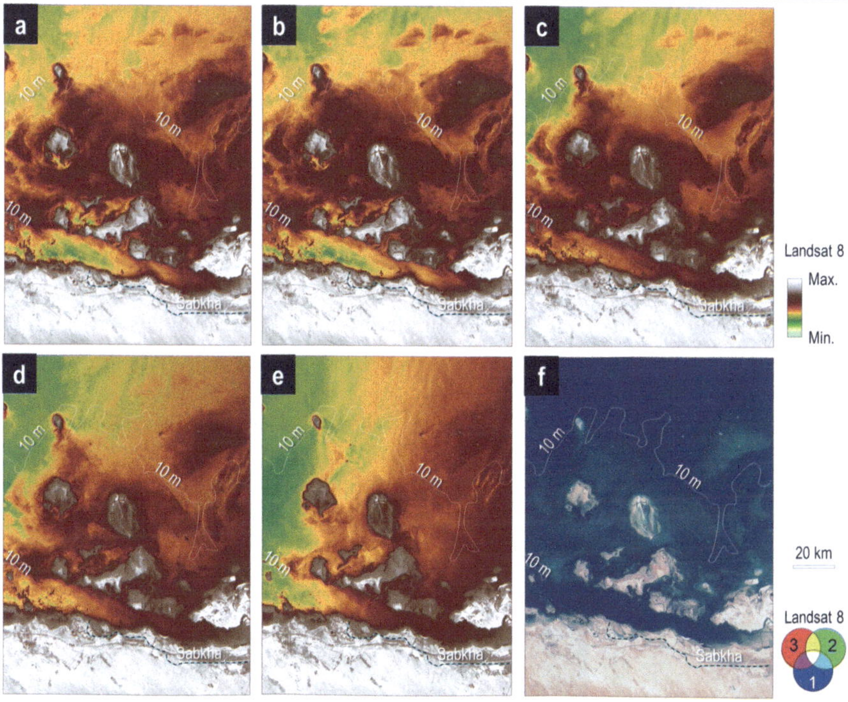

Fig. 2.7 Response of optical bands to surface cover in shallow water and coastal sabkha environment, Abu Dhabi coast. (**a**) coastal band 1, (**b**) blue band 2, (**c**) green band 3, (**d**) red band 4, (**e**) panchromatic band 8, (**f**) true color image LS8 321 RGB. (Source of Landsat imagery: U.S. Geological Survey)

(d) bands show decreasingly less information but still provide underwater information in very shallow water. The panchromatic band (e) is dominated by the reduced penetration of the red and green bands and consequently does not represent the deeper information provided by the blue and coastal bands. We will see in Sect. 3.2 the degrading effect of the pansharpening the shallow water multiband combination 321 RGB which is shown in (f). This band combination, however, only poorly discriminates the lithological facies of the sabkha onshore due to their overall bright response in all visible bands.

2.3 Infrared Bands

Remote infrared measurements are carried out in the near and shortwave reflected infrared and in the thermal infrared spectrum. Landsat 8 provides a near infrared (NIR) band that is sensitive to chlorophyll as well as two short wave infrared (SWIR) bands that are used for mineral spectroscopy. The thermal infrared is covered by two thermal infrared (TIR) bands that are sensitive to surface temperature as well as thermal radiation from surface rocks that convert the sunlight into heat radiation.

In the arid hot desert of SW Egypt, the near infrared bands produce a very different result than the thermal infrared. The desert varnish on the mesas that are showing as dark patches on the true color composite image (Fig. 2.8f) have a distinctly different mineral composition than the wadis in between and the sandy plains outside the mesas (light colors in Fig. 2.8f).

Their high iron contents yields a very low response for the mesas in all three near infrared bands (bright green), whereas the wadis and sands outside the mesas are composed mainly of quartz and show a higher response. The channels leading into the lowlands can be nicely traced because of the difference in lithology between the iron-rich channels and the quartzose sands.

The TIR bands, however, yield high response where the surface has a higher temperature. The reason for this is the high level of conversion of energy in the visible bands into infrared both on the mesas as well as in the sands outside the mesas. The mineral contents does not play as important a role as for the SWIR bands.

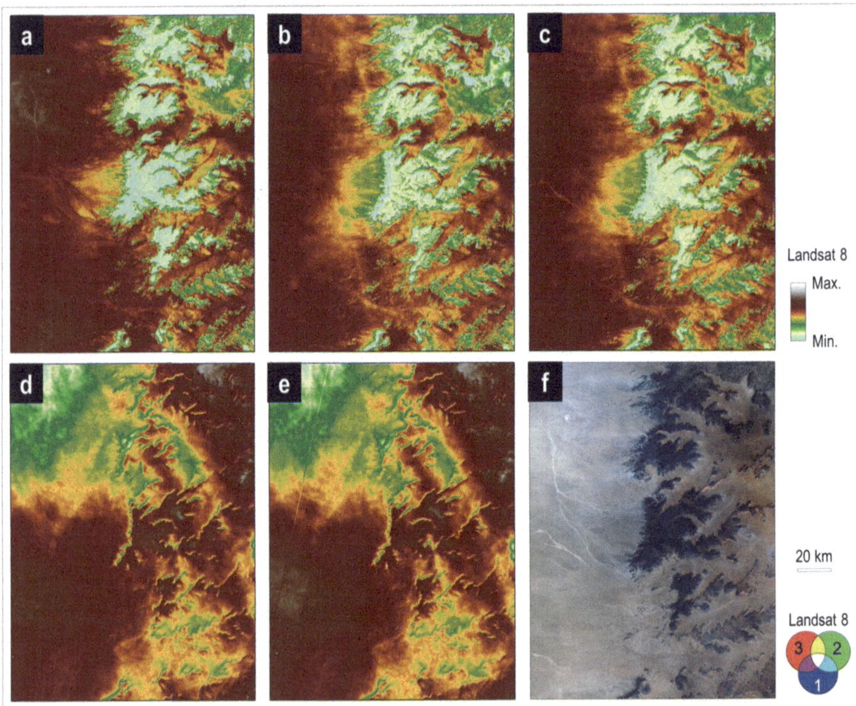

Fig. 2.8 Response of infrared bands to surface cover in arid desert environment, SW Egypt. (**a**) NIR band 5, (**b**) SWIR1 band 6, (**c**) SWIR2 band 7, (**d**) TIR1 band 10, (**e**) TIR2 band 11, (**f**) true color image LS 432 RGB. (Source of Landsat imagery: U.S. Geological Survey)

In the case of tropical rainforest and volcanics from Hawaii, the difference between near-infrared and TIR is strong as well. Figure 2.9f is labelled with the dominant surface cover as shown also in Fig. 2.1 in Sect. 2.1 of this chapter. The NIR band (a) shows a strong response in areas with dense vegetation and low signature over lava fields, whereas the SWIR bands (b and c) show high response for lava. The TIR bands, in contrast to this, show low response over vegetated areas and in the cold lava of the summit crater. Young lava, however, shows high to very response. These properties make them useful for the combination of NIR, SWIR and TIR bands in multispectral composite images (see Sect. 3.1).

The infrared spectrum is strongly absorbed in water resulting in a response that is not related to the water depth or the character of the seafloor. The true color composite of the visible coastal, blue and green bands (see Fig. 2.10f) reveals texture offshore that can be related to seafloor and water depth. The offshore signature of infrared bands, however, appears not related to water depth or seafloor texture, but rather to the surface temperature of the water and to currents that leave the imprint at the surface.

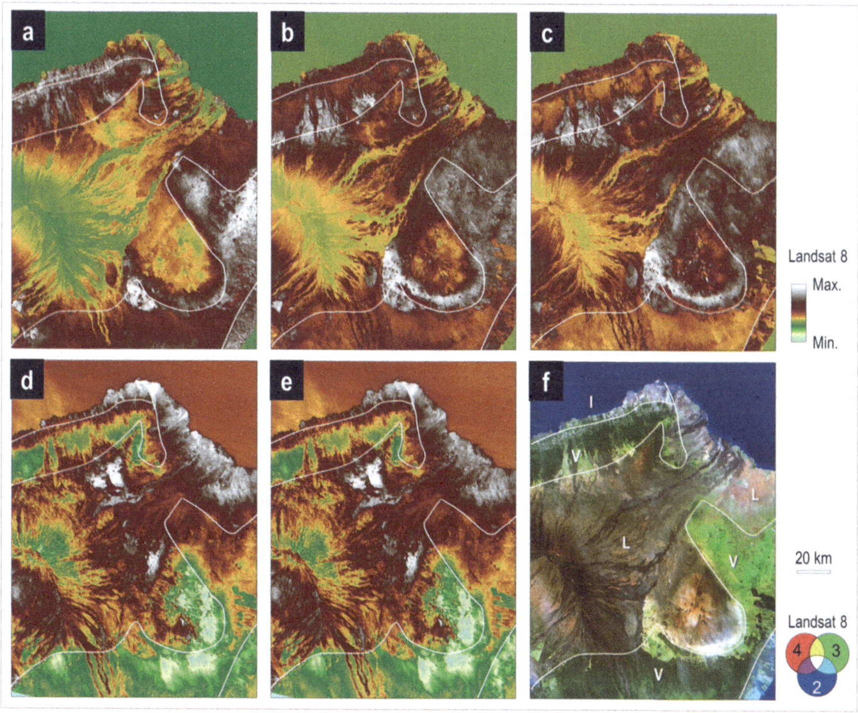

Fig. 2.9 Response of infrared bands to surface cover in tropical forest and volcanic environment, Hawaii. (**a**) NIR band 5, (**b**) SWIR1 band 6, (**c**) SWIR2 band 7, (**d**) TIR1 band 10, (**e**) TIR2 band 11, (**f**) true color image LS 432 RGB. *L* lava, *V* vegetation, *I* infrastructure. (Source of Landsat imagery: U.S. Geological Survey)

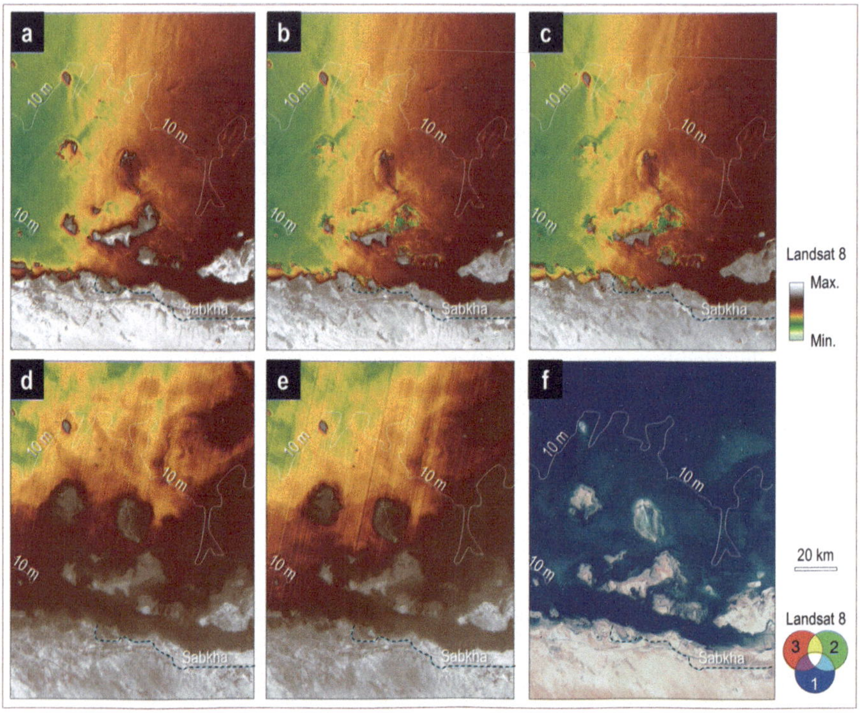

Fig. 2.10 Response of infrared bands to surface cover in shallow water and coastal sabkha environment, Abu Dhabi coast. (**a**) VNIR band 5, (**b**) SWIR1 band 6, (**c**) SWIR2 band 7, (**d**) TIR1 band 10, (**e**) TIR2 band 11, (**f**) true color image LS 432 RGB. (Source of Landsat imagery: U.S. Geological Survey)

A specialty for Landsat 8 is the new cirrus band 9, a very narrow shortwave infrared band with focus specifically on the detection of high cirrus clouds that often mask parts of the spectrum and are not always easily detected. Figure 2.11 shows the cirrus band together with the natural color LS8 432 RGB image for the case studies in SW Egypt (a, c) and Hawaii (b, d). In the hot arid desert, cirrus clouds (CC) are mostly spotted as white areas, specifically where the surface rocks are dark (a). Band 9 (c), however, reveals the full extent of the contamination with absorption by the cirrus clouds, which can now be corrected.

Over rainforest and volcanic areas, however, cirrus clouds are much more difficult to detect on a multispectral composite due to the vivid change in the surface cover (b). This setting is what the SWIR band 9 has been designed for. The cirrus band (d) shows the large extent of cirrus cloud absorption over almost half the extent of the visible color image.

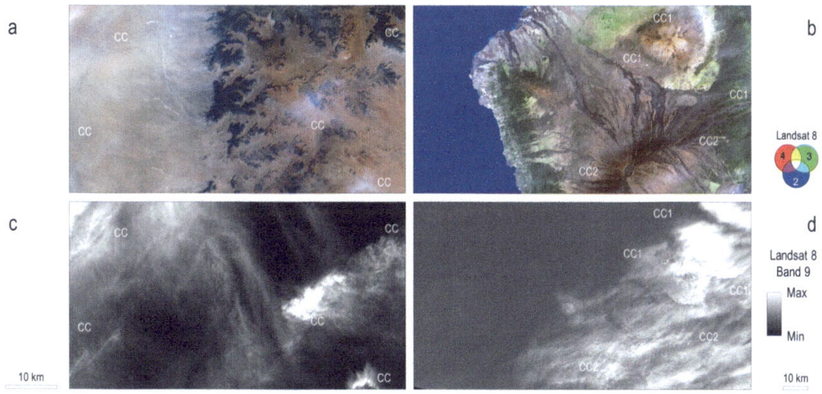

Fig. 2.11 Detection of cirrus clouds with Landsat 8 band 9 in dry arid desert and tropical rain forest environment. (**a**) Landsat 8 432 RGB southwest Egypt, (**b**) Landsat 8 432 RGB Hawaii, (**c**) Landsat 8 band 9 southwest Egypt, (**d**) Landsat 8 band 9 Hawaii. (Source of Landsat imagery: U.S. Geological Survey)

2.4 Microwave Radar

Another method to obtain information of the earth surface in the presence of clouds is microwave radar. The atmospheric absorption in the microwave band is significantly lower than in the visible and infrared regions (see Sect. 2.3). Clouds of water vapor, therefore, do not obstruct the transmission of microwaves and allow the structural mapping of the earth surface even in the presence of clouds. The comparison of a visible band multiband composite 432 RGB (Fig. 2.12a) with a C-band microwave radar image (b) clearly shows this advantage. Where the optical image from Landsat 8 reveals the lithological differences between the heat-weathered sandstones (black), the gravel deserts (reddish-brown) and the sands in the valleys and in the foreland (light colors), the radar image only distinguishes between hard and soft surfaces.

Hard surfaces backscatter the microwaves right at the earth surface with strong amplitudes (dark tones), whereas soft surfaces such as sands are associated with low backscatter amplitudes. Gravels yield intermediate backscatter amplitudes. Microwave radar is best suited for textural mapping of the earth surface. The foreland shows faint imprints of buried channels which will be discussed in more detail in Sects. 2.6, 9.5.1 and 9.5.2.

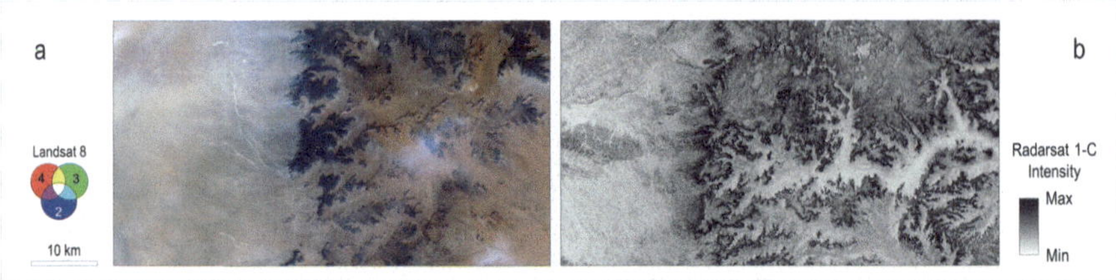

Fig. 2.12 Microwave C-band radar in SW Egypt. (**a**) true color image LS8 432 RGB, (**b**) radar backscatter intensity image. (Source of Landsat imagery: U.S. Geological Survey. Radar imagery contains information licensed under the Open Government License—Canada)

2.5 Spatial Resolution

The spatial resolution of the optical and the infrared is determined by their wavelength ranges. Visible bands range from 0.4 to 0.7 μm, shortwave infrared bands span from 0.7 to 3.0 μm and thermal infrared range at around 10 μm. The spatial resolution of the visible and shortwave infrared bands implemented in Landsat 8 is 30 m, the resolution of the thermal infrared bands is 100 m. Panchromatic sharpening using the panchromatic band 8, which offers 15 m spatial resolution, can achieve 15 m resolution also for the visible bands. Figure 2.13 demonstrates the impact of spatial resolution on the imaging capabilities at the example of a volcanic crated in the north of Hawaii's Big Island. The panchromatic band (a) images the crater of about 1 km diameter with details of erosional furrows in the crater rim when sampled at 15 m. The 30 m sampling of visible band 4 (b) and shortwave infrared band 7 (c) reveal only a blurred image of these furrows, whereas the 100 m resolution of the thermal infrared band 11 (d) shows only a faint outline of the crater. The thermal infrared image, however, is enhanced in contrast compared to the visible and shortwave infrared bands due to the higher sensitivity to the difference between illuminated parts of the crater and shadows.

2.6 Penetration into Soil and Water

The penetration of visible, infrared and microwave regions of the electromagnetic spectrum is studied at the images of a sandy plain east of Gilf Kebir in southwest Egypt (Fig. 2.14). Microwave C-band radar penetrates the sand for several meters and allows detecting fossil channel systems buried below the sand. The texture of patches with low radar backscatter reveal such channels as being part of a regional fossil drainage system (b). The extent of the inferred fossil drainage system of presently buried channels is indicated by dashed lines on the satellite images. Surface photos from three locations north of the channel area (1), in the channel area (2) and south of the channel area (3) do not reveal any indication of the fossil channels.

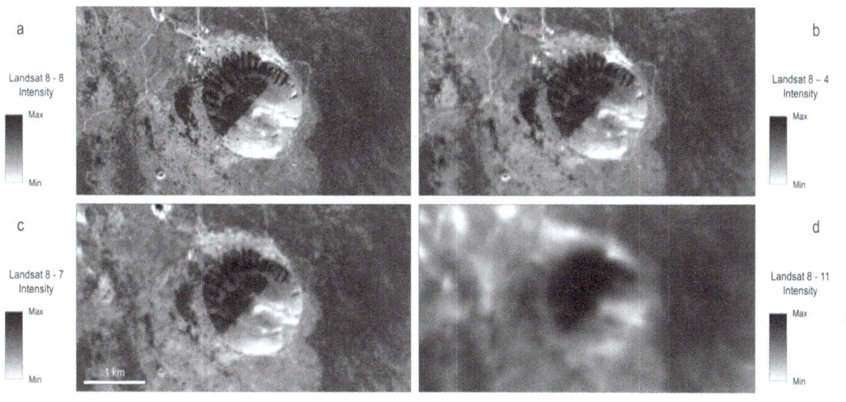

Fig. 2.13 Spatial resolution of Landsat8 bands at a volcanic crater on North Hawaii Island. (**a**) panchromatic band 8 at 15 m, (**b**) visible red band 4 at 30 m, (**c**) SWIR band 7 at 30 m, (**d**) TIR band 11 at 100 m resampled to 30 m. (Source of Landsat imagery: U.S. Geological Survey)

Fig. 2.14 Comparison of shortwave infrared and thermal infrared bands with microwave radar C-band images for a sand dune desert east of Gilf Kebir in southwest Egypt. (**a**) Landsat 8 band 7, (**b**) Radarsat-1 C-band, (**c**) Landsat 8 432 RGB, (**d**) Landsat 8 654 RGB, (**e**) field photos. The area of the buried channel system is outlined by dashed lines in the satellite images. (Source of Landsat imagery: U.S. Geological Survey. Radar imagery contains information licensed under the Open Government License—Canada)

Infrared bands are capable of mapping only the surface expression where the buried channels still outcrop due to erosion of the overlying deposits as indicated by faint high amplitude values (blue) in the Landsat 7 image.

The braided southern portion of the channel system is covered by sand and hence not visible to shortwave infrared waves that do not penetrate the sand. We make the same observation with the multiband composite Landsat 8 432 RGB. Not even the outcropping channels are visible due to their minimal distinction in the visible bands. Building the multiband composite out of infrared and visible red bands of Landsat 8 654 RGB yields similar indications for the outcropping parts of the buried channel system. But no evidence for the buried southern part of the braided channel system is evident. The calibration of the penetration of microwave radar waves and their interpretation in a larger geological context is described in Sects. 9.5.2 and 10.2.

The penetration of electromagnetic waves into seawater allows the mapping of the shallow water seafloor texture, provided the attenuation of the waves is sufficiently low. Figures 2.15 and 2.16 demonstrate the mapping capabilities of the coastal band 1, the blue band 2 and the green band 3 as well as the red band 4, the near infrared band 5 and the shortwave infrared band 6, respectively. The dashed lines indicate the 10 m isobath.

The multiband composite Landsat 8 321 RGB in Fig. 2.15d provides a detailed mapping of the shallow seafloor texture comprising reefs, some of which formed around islands and shallow channels. Channel 1 stands out from the other channels as its bottom shows significantly darker color due to algal mats giving it the appearance of much greater depth. This observation indicates that it is necessary to

Fig. 2.15 Mapping of shallow seafloor texture offshore Abu Dhabi. (**a**) Landsat 8 band 1, (**b**) Landsat 8 band 2, (**c**) Landsat 8 band 3, (**d**) multiband composite Landsat 8 321 RGB. (Source of Landsat imagery: U.S. Geological Survey)

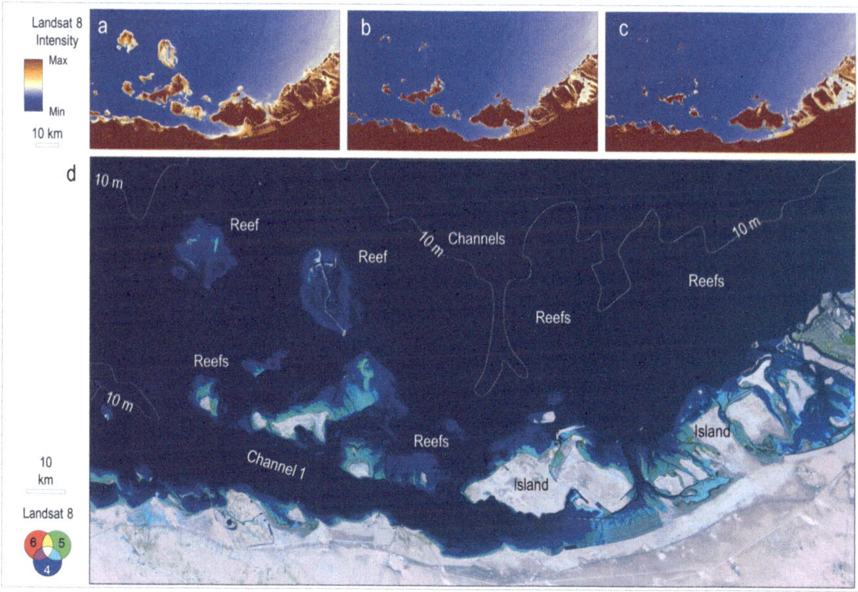

Fig. 2.16 Mapping of shallow seafloor texture offshore Abu Dhabi. (**a**) Landsat 8 band 4, (**b**) Landsat 8 band 5, (**c**) Landsat 8 band 6, (**d**) multiband composite Landsat 8 654 RGB. (Source of Landsat imagery: U.S. Geological Survey)

include site calibration when trying to derive the water depth from the reflection intensity of optical bands.

The red band (Fig. 2.16a) does not penetrate significantly into seawater. The texture of the underwater areas indicates that the penetration is most likely less than 1 m.

The green band 5 (b) and near infrared band (c) do not penetrate water at all. Hence the multiband composite Landsat 8 654 RGB shows only faint shadows of the underwater texture in the blue color. The cyan areas around the islands are tidal zones with only little water coverage.

References

Landsat

USGS. (2019). *Landsat 8 (LS8) Data users handbook, USGS LSDS-1573 Version 4.0*. USGS EROS.

Radarsat

Raney, R. K., Luscombe, A. P., Langham, E. J., & Ahmed, S. (1991). RADARSAT. *Proceedings of the IEEE, 79*(6) (Reprint).

Chapter 3
Foundations of Multi-Band Processing of Satellite Images

Abstract The joint processing, evaluation and interpretation aims at integrating the information otherwise obtained in separate processes. Geologic features at the earth surface leave their imprint on remote sensing data in various ways. We will study multi-band RGB images from two or more spectral images. For two band combinations band difference and band ratio images will be studied. The multi-band processing of three or more spectral bands merges the intensity images of different spectral bands into one multi-color image that renders the impact of the earth surface electromagnetic properties in an integrated way. Amplitude inversion of single or multi-band images may highlight different features than the original image. Pansharpening of multi-band images with high-resolution panchromatic images is studied for both, benefits and limitations. For all aspects, applications in onshore and offshore settings as well as across different climate zones are studied.

Keywords Multi-band composite image · Pansharpening · Band difference · Band ratio · Band inversion

3.1 Introduction

Combing the information of multiple satellite images obtained at different wavelength bands enhances the insight into the characteristics of the earth surface. In this section the following processing methods are described and illustrated with examples from different environments (Fig. 3.1).

Multi-band RGB composites (a): this approach reproduces the color perception of the human visual reception. Three images from different wavelength bands are allocated to the red, green and blue color channels and co-rendered as RGB composite image. Thereby the dynamic range of the human perception is increased substantially. The human eye can only detect approximately 256 shades of gray but more than ten million shades of colors in the RGB color scheme. This means that multi-band RGB composites can convey significantly more information at finer levels for human perception. Typical RGB composites are true color images that

A. Laake, *Remote Sensing for Hydrocarbon Exploration*, Springer Remote Sensing/Photogrammetry, https://doi.org/10.1007/978-3-030-73319-3_3

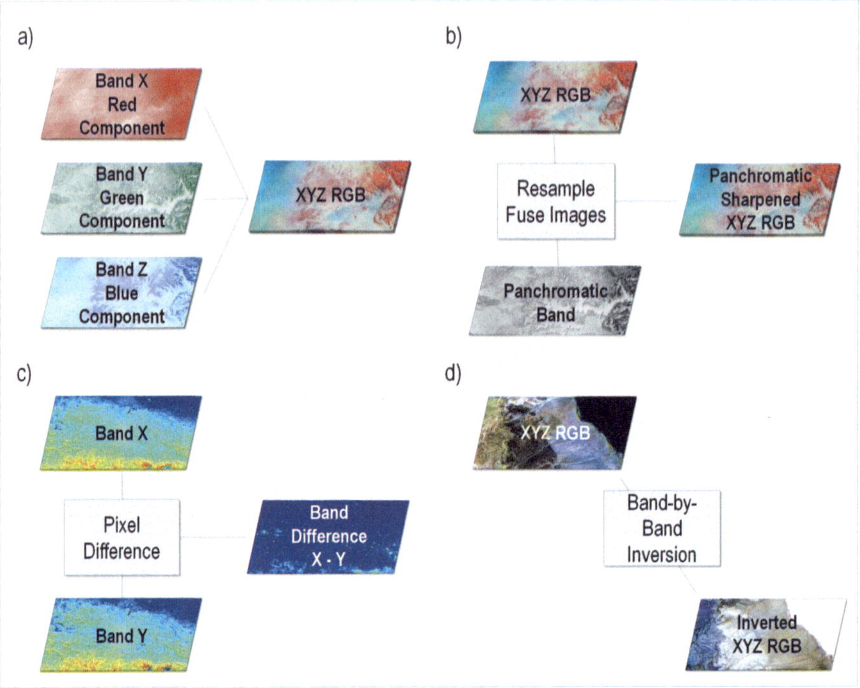

Fig. 3.1 Methods for processing multi-band satellite images. (**a**) Multi-ban RGB composites, (**b**) Panchromatic sharpening, (**c**) Band difference (**d**) Band inversion

resemble visible photographic colors in a combination familiar to the human perception or false color images that highlight certain properties of the earth surface in color combinations other than naturally looking colors.

Panchromatic sharpening (b): the spatial resolution of satellite images measured in a single wavelength band is given by the design of the sensor. To increase the spatial resolution some satellites include in their sensor a broader wavelength band at higher resolution, the panchromatic band. This band is used to increase the sharpness of lower resolution bands through a process called panchromatic sharpening, where the lower resolution bands are resampled at the spatial sampling rate of the panchromatic band and fused with the panchromatic band. Using this method a lower resolution multi-band RGB image can be sharpened to resemble the resolution of the panchromatic band as a RGB color image.

Band difference and band ratio (c): executing an algorithm between pairs of wavelength bands enhances subtle differences between these bands. The reflectance of the earth surface shows specific wavelength characteristics depending on the composition of the earth surface. Consequently, building the difference or the ratio between two bands of different wavelength enhances the signature of areas where the wavelength spectra of the surface composition is very different. One application is the highlighting clay-filled karst holes in carbonate rock at the earth surface using the band ratio of a near infrared and the red band due to the different reflectivity in these two bands.

Band inversion (d): if the rendering of a single band or a multi-band composite does not reveal sufficient contrast for a specific feature, then the inverted image may be an option to improve the contrast. To achieve the inversion the reflectance maxima and minima are swapped. This approach proves useful for example when mapping subtle channel patterns in a floodplain.

3.2 Multi-Band Composite Images

The most frequently used multi-band processing methods for satellite images are the multi-band composite images because these offer the largest dynamic range through co-rendering of different spectral bands as a RGB color image. A selection of commonly used RGB images and their designed mapping focus is listed in Table 3.1 for the examples shown in Fig. 3.2 at the example of Big Island of Hawaii. The natural color image 432 RGB in Fig. 3.2a serves as an orientation for the characteristic features present in this environment: seawater in blue, lava flows in reddish to black and grey-brown tones for fresh to weathered lava. Vegetation is shown in shades of dark to light green for pine forest, jungle and meadows. Infrastructure such as buildings, roads and airports appear in white to grey tones depending on the contents of limestone, gypsum and tar. Not all band combinations are equally useful in remote sensing for hydrocarbon exploration. The most frequently used band combinations are 432 for natural colors, 321 for seafloor mapping in shallow water, 654 for vegetation mapping and 753 for landuse mapping of vegetation, rocks and water. In deserts the combination 754 is useful for lithology mapping. The combination 11 7 5 adds a thermal component to the shortwave infrared lithology, which is specifically useful for the detection desert varnish.

To cover a desert example, the commonly used multi-band combinations are shown in Fig. 3.3 for an example in the Gilf Kebir area in southwest Egypt. The natural color image (a) reveals the overall character of the landscape that is composed of smooth sand dunes (yellow-beige), sandstone mesas (blue-black) and beds of temporary rivers (brownish-beige). Since no water is present, the combination 321 (b) does not provide any additional information. The combinations of shortwave infrared and visible 543 (c) and 753 (e) as well as the shortwave infrared combination 654 (d) highlight the difference between heat-weathered mesas and soft sands in the wadis and the sand dune foreland and appear interchangeable. The similarity of combination 654 is due to the absence of vegetation. The combination of thermal and shortwave infrared with red 11 7 5 enhances surface areas with strong conversion of sunlight to heat radiation, which occurs preferably in dark surfaces for example in sandstone or limestone where the surface is weathered in

Table 3.1 Commonly used Landsat 8 multi-band combinations and their applications. Most frequently used combinations are highlighted in bold

Landsat 8 bands	Description	Mapping focus	Figure 3.2
4 3 2	**Natural color**	**Natural "true color"**	**a**
3 2 1	**Shallow water**	**Seafloor in shallow water**	**b**
5 4 3	Color Infrared	Vegetation	c
5 6 2	Healthy vegetation	Vegetation	d
6 5 2	Agriculture	Vegetation	f
6 5 4	**Vegetation analysis**	**Vegetation**	**g**
5 6 4	Land/water	Land-water boundary	e
7 5 3	**Natural with atmospheric removal**	**Landuse**	**h**
7 6 5	Atmospheric penetration	Landuse	j
7 6 4	False Color (urban)	Landuse, infrastructure	k
7 5 4	**Shortwave infrared**	**Lithology**	**i**
11 7 5	**Infrared and visible**	**Lithology and weathering**	**l**

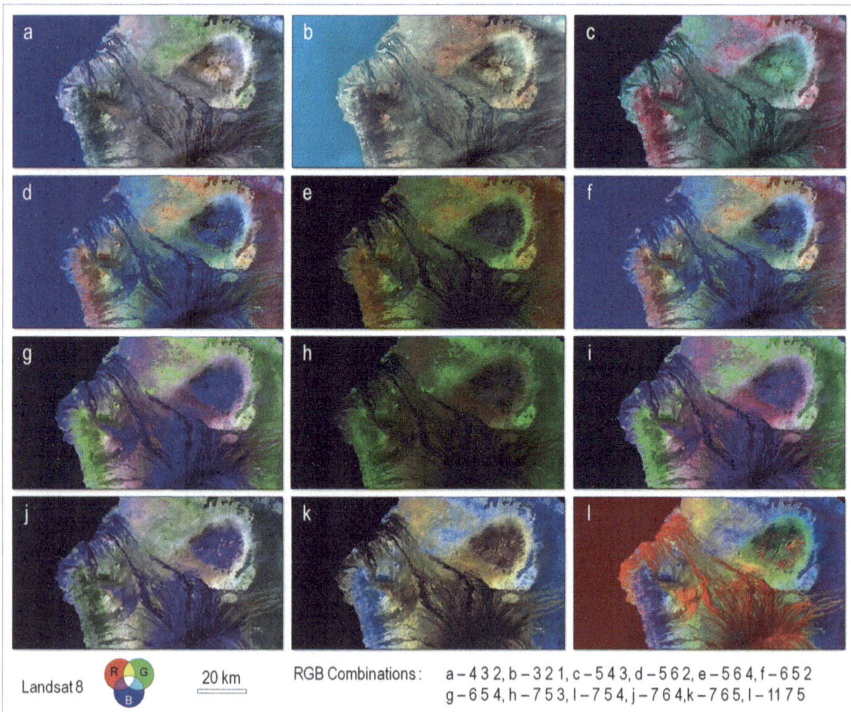

Fig. 3.2 **(a–l)** Examples for commonly used multi-band composite images, Hawaii Big Island. (Source of Landsat imagery: U.S. Geological Survey)

Fig. 3.3 Examples for common multiband RGB composite images of Landsat 8 bands, Gilf Kebir, southwest Egypt. (**a**) 4 3 2 RGB, (**b**) 3 2 1 RGB, (**c**) 5 4 3 RGB, (**d**) 6 5 4 RGB, (**e**) 7 5 3 RGB, (**f**) 11 5 3 RGB. (Source of Landsat imagery: U.S. Geological Survey)

the heat to form black desert varnish. Desert varnish appears in this band combination as bright red which stands in sharp contrast to the blueish tones of the sands of the sand dune foreland where surface reflectance and backscatter dominate the spectrum of the signal returning from the earth's surface.

3.3 Panchromatic Sharpening

The color-coded information about the earth surface that is provided by multi-band composites often does not provide the spatial resolution desirable for the design of geophysical surveys or infrastructure planning. An option to provide higher resolution color imagery from a data set that offers a panchromatic image is panchromatic sharpening. In the case of Landsat 8, the natural color combination of the visible bands 4, 3 and 2 has 30 m spatial resolution whereas the panchromatic band 8 is sampled at 15 m. To achieve pan-sharpening of the visible bands, these are resampled to 15 m. The composition of the multi-band image is guided by the contrast texture of band 8 in a process called image fusion. Different methods for pan-sharpening based on different fusion algorithms are offered by image processing software (Belfiore et al., 2016; DadrasJavan et al., 2019; Nikolakopoulos & Oikonomidis, 2015). Figure 3.4 demonstrates the improvement of spatial resolution

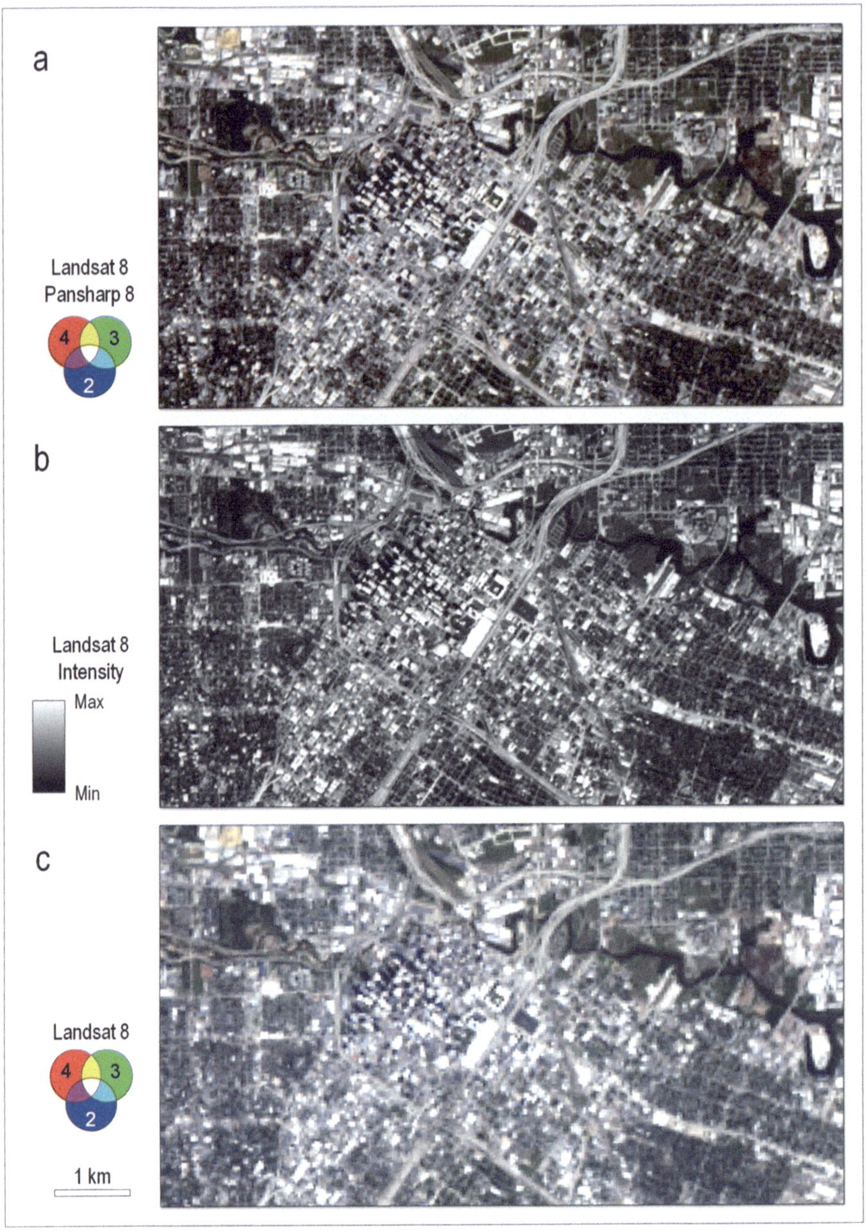

Fig. 3.4 Urban environment example for panchromatic sharpening, Houston TX. (**a**) 4 3 2 RGB pansharpened with band 8, (**b**) panchromatic band 8, (**c**) 4 3 2 RGB multiband composite image. (Source of Landsat imagery: U.S. Geological Survey)

of the Landsat 8 432 RGB composite (a) achieved by a factor of 2 through panchromatic sharpening of a natural color image (c) using the panchromatic band (b) at the example of the metropolitan area of Houston, Texas U.S.A.

Another successful case is the pan-sharpening of the visible band combination for an arid desert environment in southwest Egypt (Fig. 3.5). In the natural color combination 432, the course of the wadi is enhanced through the deposition of clay and evaporite minerals, that show as white in (c). They stand out in stark contrast to eolian sands (beige) and heat weathered sandstone mesas (black). The zoom clearly shows the 30 m resolution limit of the 432 RGB image. The grayscale panchromatic band (b) provides a 15 m resolution guide for the fusion process used in the panchromatic sharpening, which leads to the spatially enhance image (a). Since the panchromatic band covers mainly the green and red bands and only partially the blue band, a slight discoloring of the composite image is observed, which appears acceptable.

The pan-sharpening has improved the spatial resolution allowing for better interpretation capabilities for lithology and depositional features whilst retaining largely the colors of the multi-band composite image.

However, when applying panchromatic sharpening to multi-band composites comprising wavelength bands that are not covered by the wavelength range of the panchromatic band, distorted colors are observed. The composite of thermal infrared, near infrared and green 11 5 3 (Fig. 3.6c) shows vivid red colors for thermally weathered sandstone on the sandstone mesas, which become less prominent in the panchromatic sharpened image (a). Furthermore, the thermal properties of the valley filled with eolian sands and comprising the wadi also looses its detail after application of panchromatic sharpening. On the other hand, the improvement of the spatial sampling for the thermal infrared band is by almost a factor of 7. Pan-sharpening in this specific case requires to decide whether to retain the analytical properties of the multi-band combination or to substantially improve the spatial sampling of the image.

In the case of the coastal and shallow water environment from the Abu Dhabi coast the band combination of coastal band, blue and green bands 321 has been chosen to enabling mapping the texture of the seafloor below shallow water down to approximately 20 m water depth (Fig. 3.7c). Details of reefs and channels are mapped at the same level of detail as the dry ground of the islands and the sabkha of the onshore coastal plain. The panchromatic band 8 (b) only covers red and green as well as part of the blue visible bands but not the coastal blue band. Hence the information about the seafloor texture that comes mainly from the coastal blue band is suppressed which results in the loss of the seafloor texture in the panchromatic sharpened image (a).

We conclude that panchromatic sharpening has the advantage of improving the lateral spatial resolution through fusion of the input image with the panchromatic sharpened image provided that the bands used in the multi-band input image are close to the wavelength band that is covered by the panchromatic band. If bands with a wavelength outside the wavelength covered by the panchromatic band are used, then the information provided by the bands outside the panchromatic range is

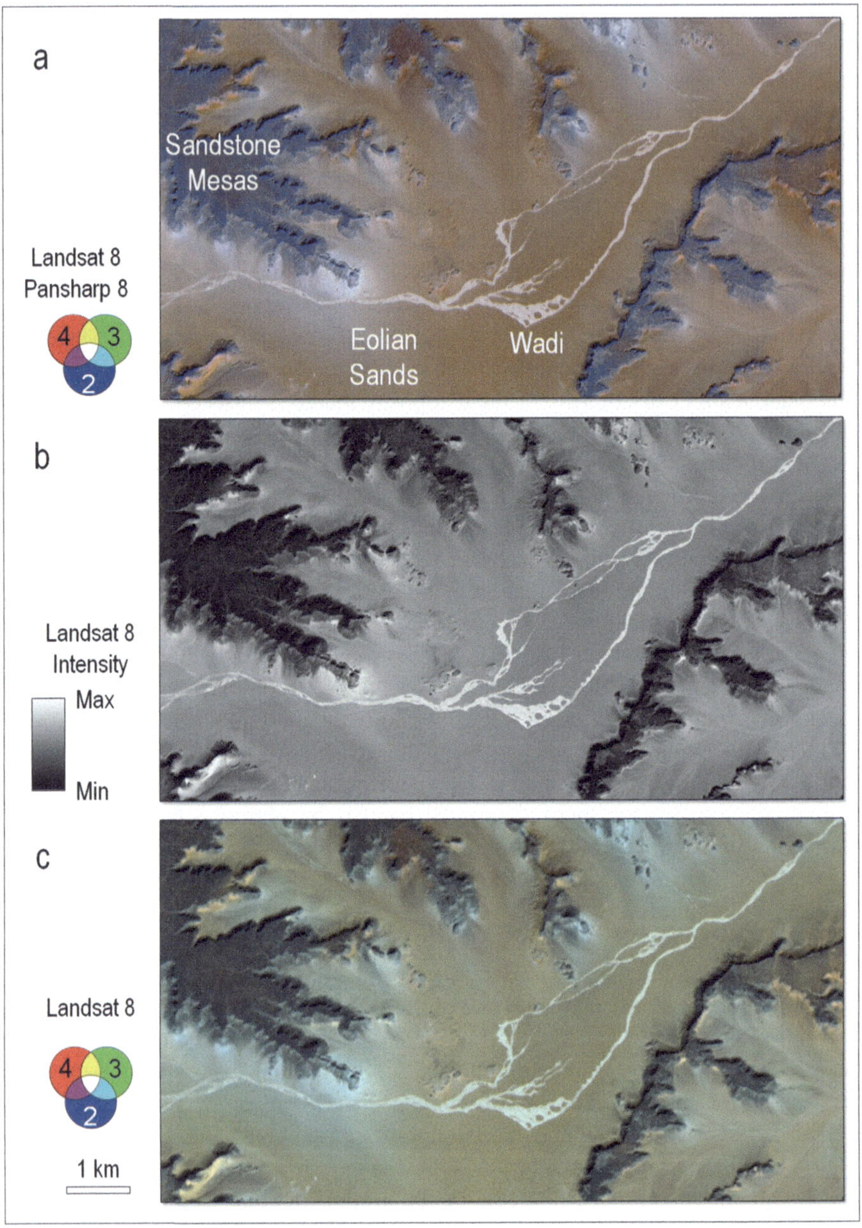

Fig. 3.5 Arid desert environment example for image resolution improvement of Landsat image through panchromatic sharpening, Gilf Kebir, Southwest Egypt. (**a**) 4 3 2 RGB pansharpened with band 8, (**b**) panchromatic band 8, (**c**) 4 3 2 RGB multiband composite image. (Source of Landsat imagery: U.S. Geological Survey)

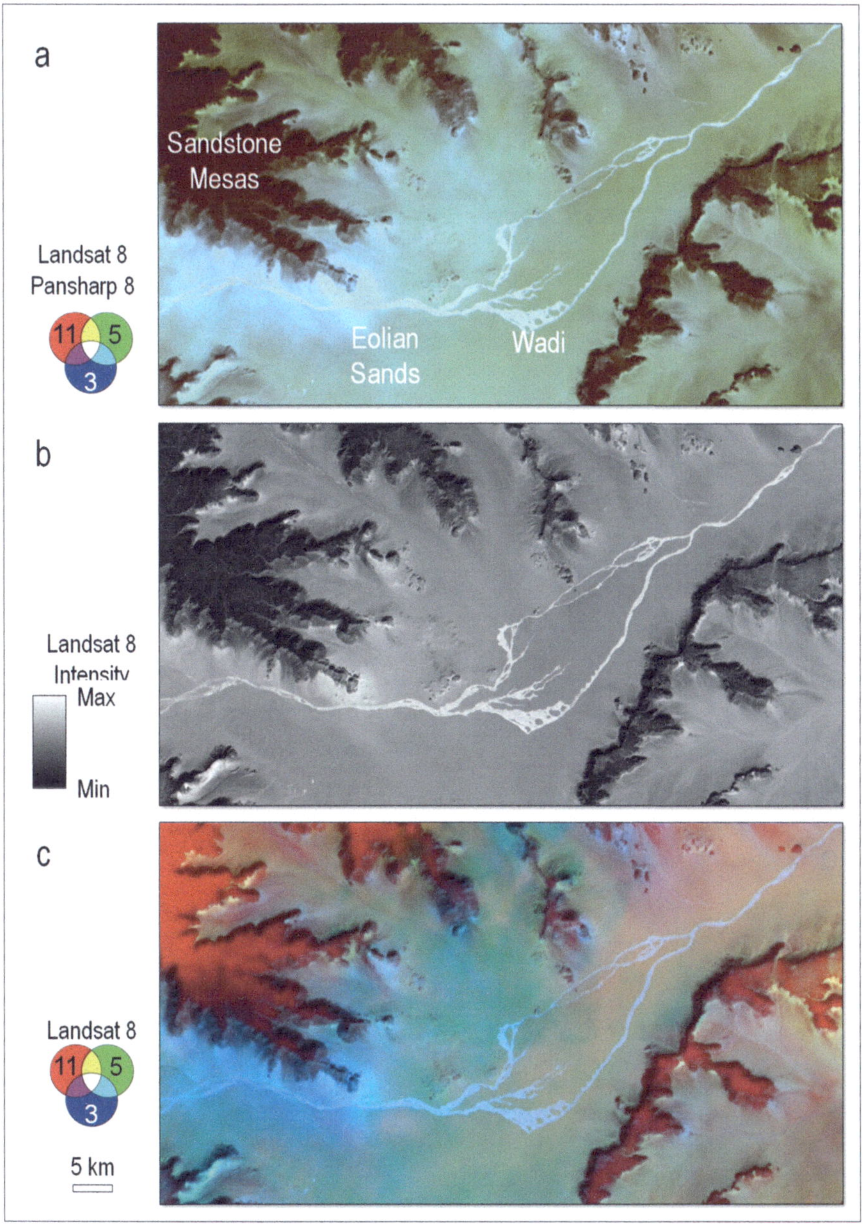

Fig. 3.6 Arid desert environment example for failure of image resolution improvement of Landsat image through panchromatic sharpening, Gilf Kebir, Southwest Egypt. (**a**) 11-5-3 RGB pansharpened with band 8, (**b**) panchromatic band 8, (**c**) 11-5-3 RGB multiband composite image. (Source of Landsat imagery: U.S. Geological Survey)

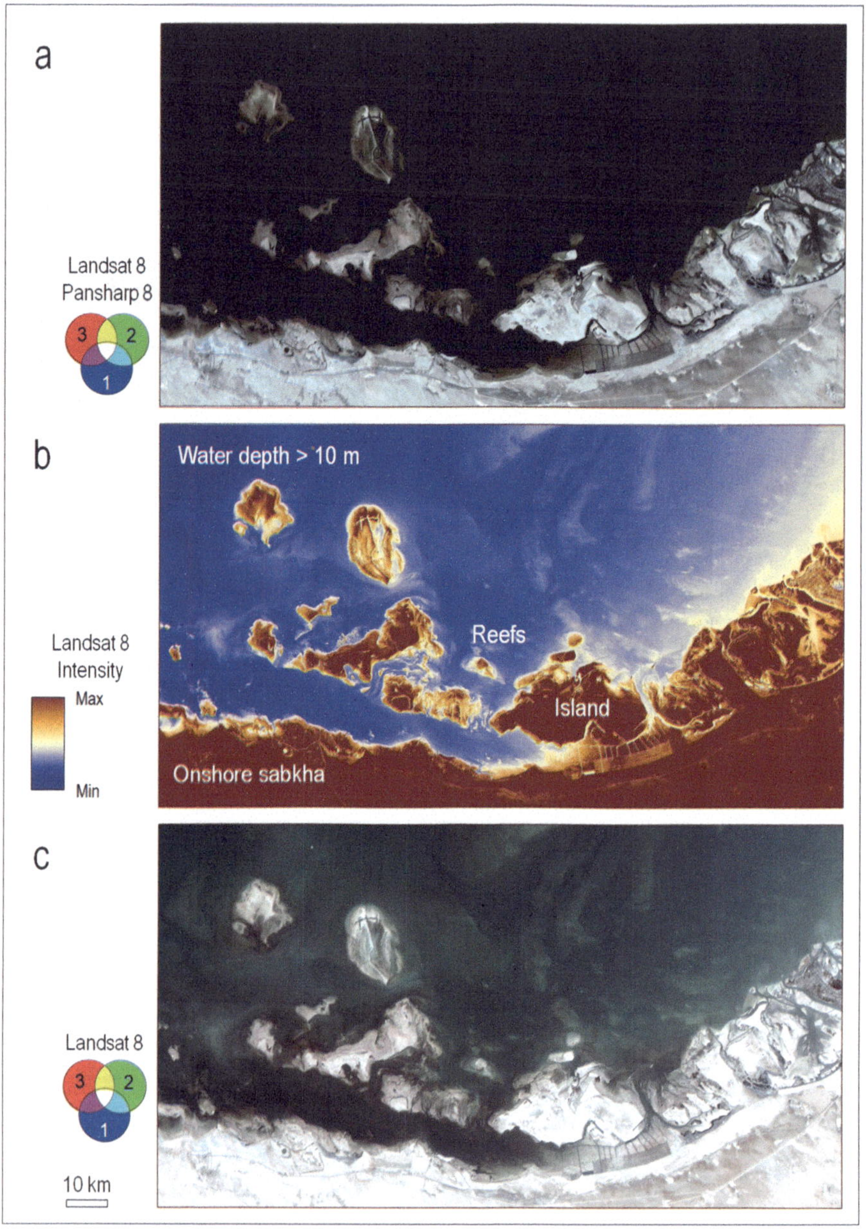

Fig. 3.7 Coastal environment example for impact of panchromatic sharpening on Landsat image, Abu Dhabi coast. (**a**) 3 2 1 RGB pansharpened with band 8, (**b**) panchromatic band 8, (**c**) 3 2 1 RGB multiband composite image. (Source of Landsat imagery: U.S. Geological Survey)

suppressed which leads to a change in the image color from discoloring to complete absence of analytical information. The satellite image interpreter, therefore, needs to choose carefully between improved resolution and retention of the analytical properties of the multi-band combination chosen for the image.

3.4 Band Difference and Band Ratio Images

Band difference and band ratio images are frequently used in mineral exploration especially on hyperspectral data where the reflection spectra for different minerals are sampled more finely than with multispectral sensors like Landsat (Crosta & Souza Filho, 2000; Papp & Cudahy, 2002; Martinez et al., 2005). Using band ratios, Landsat imagery can also be used for regional mapping minerals (Sultan et al., 1987). For applications to hydrocarbon exploration, surface mineral mapping plays an important role in geological modeling of the surface and near-surface layers for logistic and infrastructure planning as well as for velocity model building in seismic modeling (Laake & Insley, 2004; Laake et al., 2009; Laake & Zaghloul, 2009).

Band difference images are constructed between two spectral bands for which the targeted mineral shows a distinguishable reflectance difference (Fig. 3.8). A typical case is the delineation of clay-filled pans in limestone, which is typically the surface expression of subsurface karsting, a severe logistic hazard. The example is located in the surface limestone karst of the Western Desert of Egypt. We choose the appropriate band combination from the reflectance spectra, which are available from the U.S. Geological Survey (Kokaly et al., 2017). These are shown in Fig. 3.8e.

The reflectance difference is strongest in the visible and near to short-wave infrared. We choose bands 2 and 7 which show for clay minerals and gypsum a decrease in reflectance by 15% and 40%, respectively. For limestone the reflectance increases from band 2 to band 7 by about 25%.

When we build the difference between bands 2 and 7 (Fig. 3.9), we anticipate negative values for clay and gypsum and positive values for limestone. The reflectance maps are shown in (a) for band 2 and in (b) for band 7. Red to yellow colors indicated positive differences that are associated with limestone (c). They are related to escarpments of limestone, which is much harder than clay and gypsum. Dark blue colors indicate negative values which we associate with clay and gypsum that fills the karst holes in the limestone. To closer study this observation we zoom into the area with numerous dark blue areas (d) and compare the band difference map with the multi-band composite for the same area. The composite image (f) confirms that the reddish lineaments are associated with escarpments whereas as the blue spots can be located in so-called clay pans.

Band ratio images are created from two spectral bands by computing the pixel-by-pixel ratio from the two input images. Since the band ratio is sensitive to even subtle deviations in the reflectance, also fine differences in the reflectance can be detected. To demonstrate this, we select bands 4 and 5 for which clay and gypsum show almost identical reflectance whereas limestone has a slightly increased

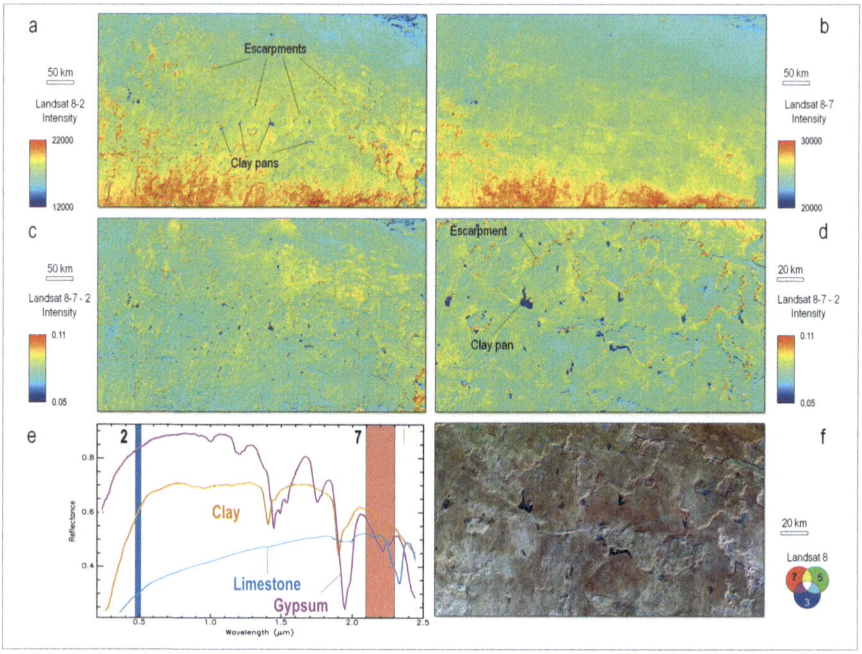

Fig. 3.8 Example for Landsat 8 band difference image, Western Desert, Egypt. (**a**) band 2, (**b**) band 7, (**c**) band difference 7-2, (**d**) band difference 7-2 zoom, (**e**) reflectance spectra, (**f**) 753 RGB zoom. (Source of Landsat imagery and mineral reflectance data: U.S. Geological Survey)

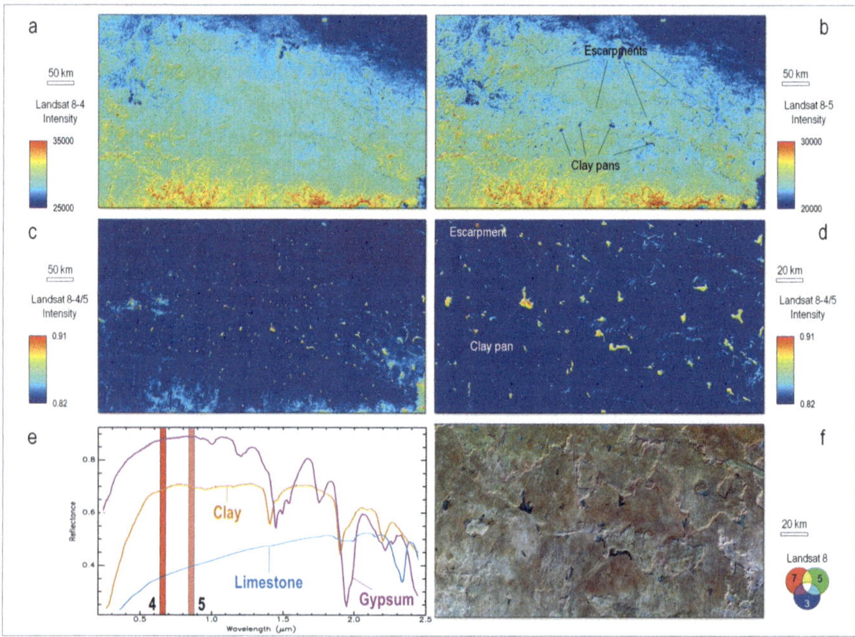

Fig. 3.9 Example for Landsat 8 band ratio image, Western Desert, Egypt. (**a**) band 4, (**b**) band 5, (**c**) band ratio 4/5, (**d**) band ratio 4/5 zoom, (**e**) 753 RGB, (**f**) 753 RGB zoom. (Source of Landsat imagery and mineral reflectance data: U.S. Geological Survey)

reflectance from band 4 to 5 (Fig. 3.9e). The related reflectance maps are shown in (a) for band 4 and in (b) for band 5. The band ratio image (c) clearly highlights the clay pans whereas the limestone escarpments are not distinguishable. The zoom into the same area (d) which we have just studied in Fig. 3.8d confirms the mapping of clay pans as does the multi-band composite image (e).

3.5 Inverted Colors

If neither of the previously discussed method produces sufficient discrimination between the objects of study and the background geology, then using the method of inversion may be considered. The inversion of a single band image is achieved by reversing the colormap associated with the color-coded amplitude. For an example from the Gulf of Suez floodplain in Egypt, Fig. 3.10a displays the original band 7 reflectance image with blue tones indicating low reflectance values and red to yellow tones representing high reflectance values. The seawater of the Gulf of Suez and the dark Paleozoic rocks of the Red Sea Hills show low reflectance.

The Gulf of Suez floodplain is crossed by numerous wadi channels that deposited eroded rocks across the floodplain into the Gulf of Suez. The network of these channels is known to give indications about the subsurface structural framework

Fig. 3.10 Examples of color inverted Landsat 8 images, Gulf of Suez, Egypt. (**a**) band 7, (**b**) band 7 inverted, (**c**) 321 RGB, (**d**) 321 RGB inverted, (**e**) 753 RGB, (**f**) 753 RGB inverted. (Source of Landsat imagery: U.S. Geological Survey)

(Laake et al., 2011a, b). The light yellow-brownish tones do not provide strong distinction between the channels and the host rock. The inversion of the band 7 (b) converts the yellow tones to blueish tones which better highlight the channel paths. Interestingly, the fault network crisscrossing the Red Sea Hills that is clearly distinguishable in the original image has become less clear in the inverted image.

The same approach can be applied to multi-band composite images. The natural color composite (c) allows to clearly distinguish the Red Sea Hills, the floodplain and the Gulf of Suez. The depositional environments of the wadi channels in the floodplain as well as the faults in the Red Sea Hills are poorly distinguished, however. The inverted image (d) preserves the distinction between the three major environments and adds clarity for the channels in the floodplain. When choosing the multi-band combination of shortwave infrared, near infrared and green (e), the Red Sea Hills stand out from the floodplain. This band combination is even capable of discriminating the dark Paleozoic basement rock from a younger granite intrusion. When inverting the image (f) the channels in the floodplain as well as the fault network in the Red Sea Hills are well delineated. The distinction of the basement rock and the granite, however, is no longer possible.

In summary, inversion of single band images and multi-band composites can improve the contrast for selected features, but it may reduce the contrast for other features.

3.6 Multiband RGB Difference Composite

We close the multiband analysis with a special case, the multiband RGB difference image. So far we have considered composite images of maximum three bands which was inspired by the concept of true and false color RGB images. For the multiband difference image in Fig. 3.11b we create two RGB composites of very similar bands,

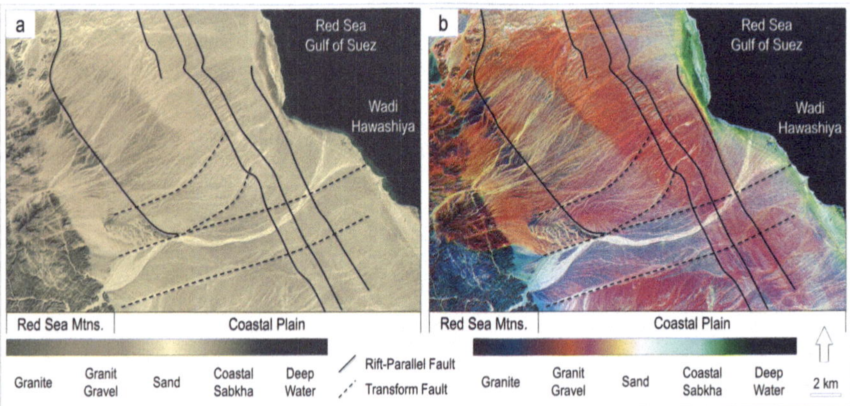

Fig. 3.11 Comparison of true image and multiband RGB difference composite image. (**a**) Landsat 5 321 RGB, (**b**) difference Landsat 5 (531-675) RGB. (Source of Landsat imagery: U.S. Geological Survey)

in this case the very near infrared—visible composite 531 RGB and a thermal and near-infrared composite 675 RGB. Each image has it's own focus with 531 RGB discriminating lithology, vegetation and water, whereas 675 RGB focuses on lithology. The RGB difference of these two images, however, renders the band differences in RGB. The result is a high-discrimination landuse and lithology image.

The comparison with the true color 321 RGB composite (a) shows a significant improvement not only in the discrimination of different lithological units such as granite (dark blue-green and dark red depending on mineral composition), sand (light orange to beige) and sabkha (green to cyan). It also highlights structural lineaments in the coastal plain that are related to the Gulf of Suez rift tectonics. Where rift-parallel fault outcrop above tilted fault blocks in the subsurface, they influence the deposition of gravel and sand which can be detected from colour boundaries in the multiband composite image. Similarly, transform faults that are generally perpendicular to the rift-parallel faults, structure the coastal plain the SW-NE direction. One prominent structure is the horst-graben feature that guides the course of Wadi Hawashiya. The wadi bed itself is highlighted by non-granitic sands that originate from the area west of the Red Sea Mountains outside the area covered in the figure. Due to the very shallow dip of the transform fault-driven valleys in the Red Sea Mountains, the flashfloods can wash the sands through the Red Sea Mountains across the coastal plain into the Gulf of Suez, where they formed sand reservoir rocks in the subsurface.

References

Multi-Band Composite Images

Laake, A., Sheneshen, M. S., Strobbia, C., Velasco, L., & Cutts, A. (2011a). Integration of surface/subsurface techniques reveals faults in Gulf of Suez oilfields. *Petroleum Geoscience, 17*, 165–179. https://doi.org/10.1144/1354-079310-014

Panchromatic Sharpening

Belfiore, O. R., Mengehini, C., Parente, C., & Santamaria, R. (2016). Application of different pan-sharpening methods on Worldview-3 images. *ARPN Journal of Engineering and Applied Sciences, 11*(1), 490–496.

DadrasJavan, F., Samadzadegan, F., Mehravar, S., & Toosi, A. (2019). A review of spatial quality assessment methods for evaluation of pan-sharpened satellite imagery. In *The International Archives of the Photogrammetry, Remote Sensing and Spatial Information Sciences, Volume XLII-4/W18, 2019, GeoSpatial Conference 2019 – Joint Conferences of SMPR and GI Research, 12–14 October 2019, Karaj, Iran* (pp. 255–261).

Nikolakopoulos, K., & Oikonomidis, D. (2015). Quality assessment of then fusion techniques applied to Worldview-2. *European Journal of Remote Sensing, 48*, 141–167. https://doi.org/10.5721/EuJRS20154809

Band Difference and Band Ratio Images

Crosta, A. P., & Souza Filho, C. R. (2000). Hyperspectral remote sensing for mineral mapping: A case-study at Alto Paraiso de Goias, Central Brazil. *Revista Brasileira de Geosciencas, 20*(3), 551–554.

Martinez, Y., Khan, S., Link, P., & Glenn, N. (2005). *Mapping geology and structure using multi-spectral and hyperspectral data and evaluating topographic correction methods: Case study, Salmon River Mountains of East-Central Idaho, SEG Houston 2005 Annual Meeting, paper NSE P1.5* (pp. 1204–1207).

Papp, E., & Cudahy, T. (2002). Hyperspectral remote sensing. In E. Papp (Ed.), *Geophysical and remote sensing methods for regolith exploration, CRCLEME open file report 144* (pp. 13–21). CRCLEME.

Sultan, M., Arvidson, R. E., Sturchio, N. C., & Guiness, E. (1987). Lithological mapping in arid regions with Landsat thematic mapper data: Meatiq dome, Egypt. *Geological Society of America Bulletin, 99*, 748–762.

Multiband Differences and Ratios

Kokaly, R. F., Clark, R. N., Swayze, G. A., Livo, K. E., Hoefen, T. M., Pearson, N. C., Wise, R. A., Benzel, W. M., Lowers, H. A., Driscoll, R. L., & Klein, A. J. (2017). *USGS Spectral library version 7: U.S. Geological Survey data series* (Vol. 1035). USGS. https://doi.org/10.3133/ds1035. 61 p.

Laake, A., Sheneshen, M. S., Strobbia, C., Velasco, L., & Cutts, A. (2011b). Integration of surface/subsurface techniques reveals faults in Gulf of Suez oilfields. *Petroleum Geoscience, 17*, 165–179. https://doi.org/10.1144/1354-079310-014

Introduction to Seismic Planning, Acquisition and Data Processing

Laake, A., Zaghloul, A. & Strobbia, C. (2009). Geomorphology—Understanding the near-surface impact on seismic data. In *71st EAGE Conference & Exhibition Amsterdam, 2009.*

Seismic Logistics Case Studies

Laake, A., & Insley, M. (2004). Applications of satellite imagery to seismic survey design. *The Leading Edge, 23*(10), 1062–1064.

Remote Sensing Case Study Qattara Depression and Western Desert, Egypt

Laake, A., & Zaghloul, A. (2009). Estimation of static corrections from geologic and remote-sensing data. *The Leading Edge, 28*, 192–196.

Chapter 4
Digital Relief Models

Abstract Digital relief models provide a digital elevation and/or bathymetry representation of the earth surface onto which data from remote sensing satellites can be projected to show these data in a virtual 3D rendering. Prior to satellite altimetry mapping, the only way to obtain such maps was from surface mapping or photogrammetry from stereo photos from aircraft for onshore and from onboard echolot measurements of the water depth. Remote sensing has changed the spatial coverage and the accuracy of digital relief models completely. We provide a summary on the different types of digital relief models, show their global coverage and list the characteristic parameters before showing applications from local to global scale.

Keywords Digital relief model · Digital elevation model · Digital surface model · Digital bathymetry model · SRTM DEM · ASTER GDEM · ETOPO · GTOPO · GMTED2010 · ArcticDEM · LiDAR

4.1 Introduction

In the previous sections we had studied the information electromagnetic waves and their reflection and backscatter from the surface of the earth can provide about the earth's surface itself. The shape of the earth surface gives clues for the rocks and their layering as well. In this section we will study in detail the different digital earth relief models that remote sensing offers. We distinguish the following types of relief models the data for which are mostly measured by remote sensors on satellites, airplanes or helicopters:

- **Digital elevation model (DEM)**: digital model of the surface of the earth excluding vegetation, buildings, bridge and other man-made objects. The DEM also includes the surface of water bodies. This model is sometimes referred to as **bedrock DEM**.

A. Laake, *Remote Sensing for Hydrocarbon Exploration*, Springer Remote Sensing/Photogrammetry, https://doi.org/10.1007/978-3-030-73319-3_4

- **Digital surface model (DSM)**: digital model of the earth's surface including vegetation and all fixed man-made objects. A derivative of the DSM is the **ice surface DEM** that includes the large ice bodies in Antarctica, Greenland and other places whereas the **bed rock surface DEM** excludes the ice bodies. The correction to bed rock, however, originates from surface point measurements.
- **Digital bathymetry model (DBM)**: digital model of the seafloor of the oceans. Since direct remote sensing methods using electromagnetic waves do not penetrate water to any significant extent, indirect methods based on gravity data and geoid models are used to obtain the seafloor bathymetry. The computed model is calibrated with shipborne measurements.

The first global mapping of the earth's topography was a compilation of surveying data that was started by Schreier of the DLR (Hastings & Dunbar, 1999). The initiative resulted in the global digital terrain models TerrainBase at 5 km resolution in 1994 (Wilson & Gallant, 2000; Row et al., 1995) and the improved GLOBE at 1 km resolution in 1999 (Hasting & Dunbar, 1999). Stereo-photogrammetry from photos taken from airplanes was used to compute digital elevation models. Some of these models had a very good quality locally, but the overall quality varied substantially (Farr et al., 2007). This approach is also used in the creation of the ArcticDEM (Porter et al., 2018) that covers the area of 60° N latitude to the North Pole, where it fills in for SRTM DEM V3.

The first remote sensing mission dedicated to the mapping of the global topography was the Shuttle Radar Topographic Mission (SRTM) in 2000 (NASA, 2000; Smith & Sandwell, 2003) that improved the resolution by a factor 10–90 m. A global DEM based on the processing of stereo-optical imagery from the ASTER satellite improved specifically the imaging of steeply dipping mountainous terrain (Lee et al., 2014). Due to limitations in the orbits of the satellite used to create this DEM, the coverage is limited from 60° N to 56° S.

Today the most frequently used global digital elevation models are the latest versions of the radar-based SRTM DEM V3 and the stereo optic-based ASTER GDEM V3. The SRTM DEM has the advantage of using active radar, which makes it independent from cloud cover and illumination by the sun, but it is sensitive to very steep terrain dips. The ASTER DEM provides better results for steep dips but does not work in frequently clouded areas due the absorption of visible and infrared wavelength bands.

Figure 4.1 gives an overview of the main techniques for creating DEMs, Table 4.1 summarize their key characteristics and Fig. 4.2 shows the global coverage of the different digital elevation and digital relief models.

4.2 Generation of Relief Models

We will now have a brief look into the data and methods that are used to generate the different types of relief models. Active electromagnetic methods use a source mounted on a flying platform to generate electromagnetic waves (Fig. 4.3). The

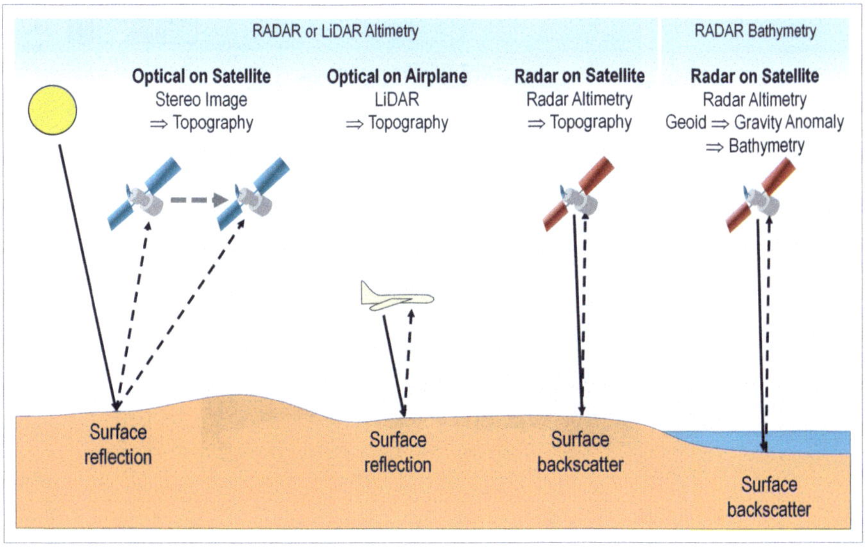

Fig. 4.1 Methods for remote relief mapping

Table 4.1 Overview of frequently used digital relief models

DEM	SRTM DEM 3	ASTER GDEM 3	GTOPO30	ETOPO1[a]	ArcticDEM V3	LiDAR
Method	Radar altimetry	Optical stereoscopy	Radar altimetry	Radar altimetry	Optical stereoscopy	Laser altimetry
Coverage	60° N	83° N	90° N	90° N	60° N –90° N	n.a.
	56° S	83° S	90° S	90° S		n.a.
Lateral resolution	30/90 m 1/3 arcsec	30 m 1 arcsec	1000 m 30 arcsec	2000 m 1 arcmin	2 m	dm–m
Vertical accuracy	10 m	7–14 m	30 m	30 m	<4 m	cm–dm
Missing data	Steep mountains	Clouded areas	None	None	Void areas	n.a.

[a]Onshore part of ETOPO1

travel time from the source down to the surface point on the earth and back to the sensor on the platform is measured and the path length computed. For calibration, a network of global position system (GPS) stations on the ground provides points with known elevation to correct for the distortion of the traveltime measurement due to roll and tilt of the platform and atmospheric effects. The high vertical accuracy of this method is the consequence of using interferometry.

This method is employed by satellite and airplane-based radar measurements such the SRTM DEM (NASA, 2000; Smith & Sandwell, 2003; Berry et al., 2007; Farr et al., 2007; Lu & Kiang, 2013) and airplane-based LiDAR measurements such

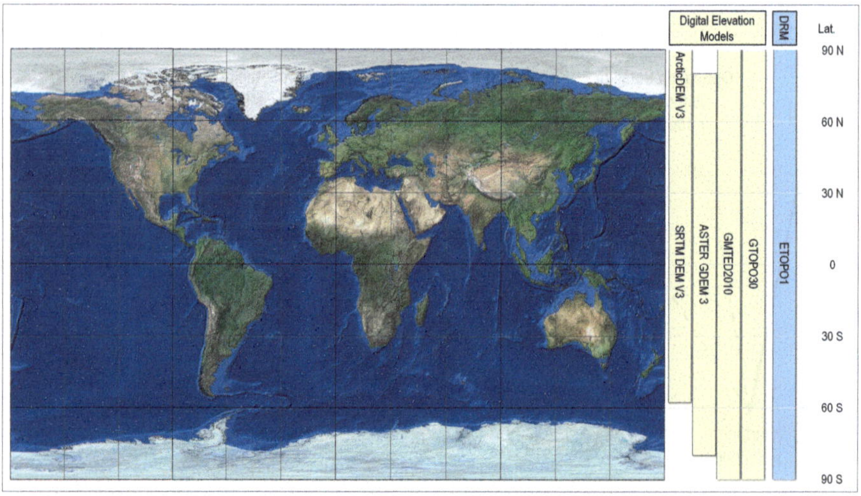

Fig. 4.2 Global coverage of common digital elevation and digital relief models

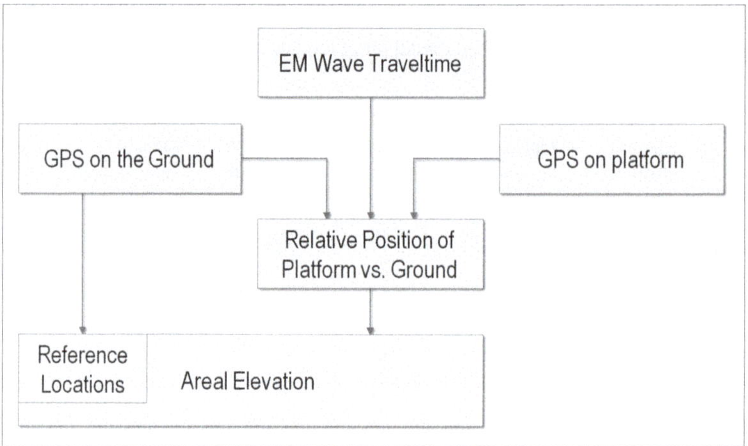

Fig. 4.3 Principle of active electromagnetic RADAR (radar altimetry) and LiDAR digital relief mapping

as the LiDAR DEM (Heidemann, 2018; Bartolini et al., 2000; NOAA, 2012; Wozencraft & Irish, 2000). A variant of this method has been developed for seafloor mapping in shallow water (Maslov et al., 2000; Harsdorf et al., 2000).

Bathymetry mapping of the oceans cannot use direct remote sensing methods because of the strong attenuation of electromagnetic waves in water (Sandwell & Smith, 2009). Radar-based bathymetry mapping uses the radar altimetry as shown in Fig. 4.4 to measure the distance of the satellite to the sea surface. Anomalies in

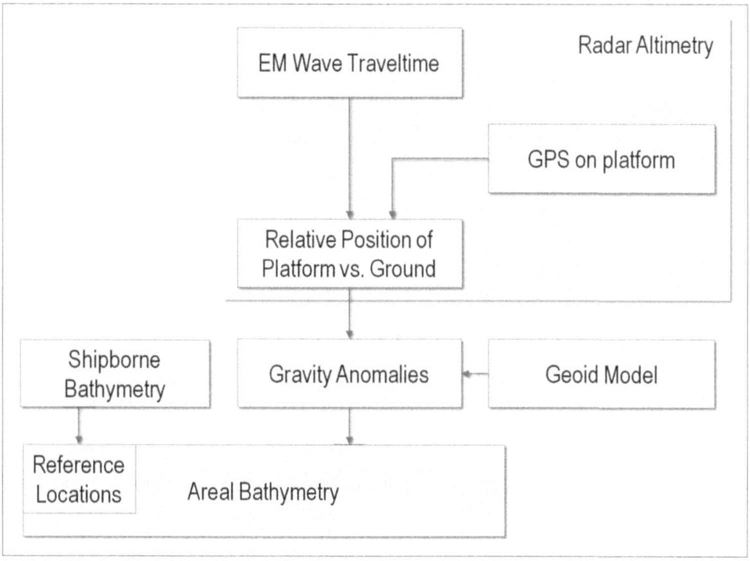

Fig. 4.4 Principle of active electromagnetic RADAR bathymetry mapping

this measurement are attributed to gravity anomalies that deviate the satellite's orbit which is measured by the satellite's GPS system. In combination with a geoid model the gravity anomalies are converted to bathymetry, which is calibrated at reference locations with shipborne bathymetry measurements. The marine relief to the ETOPO1 global relief model (Amante & Eakins, 2009), however, is entirely based on shipborne bathymetry and sonar measurements.

The stereo-optic method uses stereo images in the optical wavelength range to compute the surface elevation (Raggam, 2006; Abrams & Crippen, 2019; ASTER GDEM Validation Team, 2009). The ASTER satellite acquires two images at a known angle along its flight path and computes the distance to the ground from stereo-processing. Since permanent cloud coverage disables this method in certain areas, other digital elevation models are used to fill the missing areas (Fig. 4.5).

4.3 Methods for Rendering Digital Relief Models

The rendering digital relief models can highlight topographic features for specific applications. Figure 4.6 illustrates three typical methods at the example of the White River bend at Guion in Arkansas. The flat map (a) shows low lying areas in the White River valley in light blue to yellow tones whereas the Ozark Mountains are associated with brown tones. The satellite image (b) gives the photographic

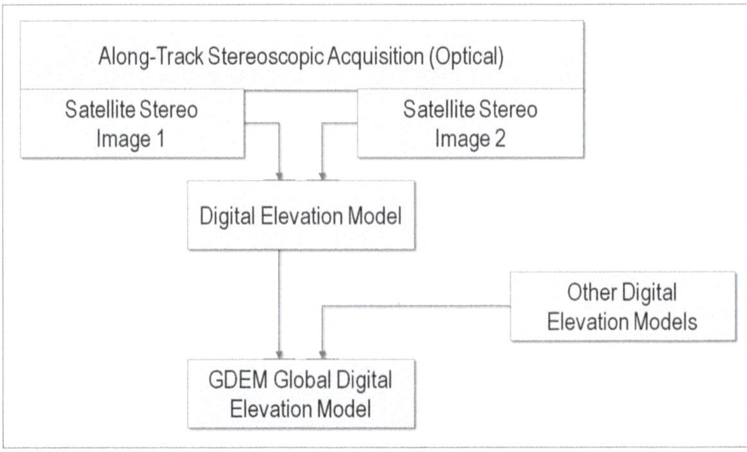

Fig. 4.5 Principle of passive stereo-optic digital relief mapping

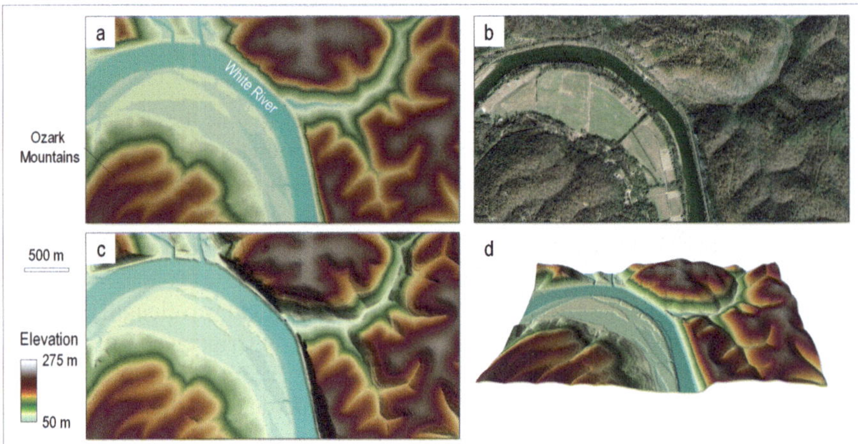

Fig. 4.6 Methods for rendering of digital relief models. (**a**) Flat map, (**b**) satellite image of study area, (**c**) map view in two vertical exaggeration and virtual slant illumination, (**d**) 3D view in 2× vertical exaggeration and virtual slant illumination. (LiDAR courtesy of DSM Arkansas Government, satellite imagery courtesy of U.S. Geological Survey)

context. When adding vertical exaggeration and virtual slant illumination (c), the map view image gives a virtual 3D impression whilst avoiding warping of the topographic features as seen in the 3D or birds-eye view (d). The 3D gives the most realistic impression of the surface features but cannot be used for measurements since the scale of the image changes from front to back as a result of the perspective.

When mapping floodplains, historic river paths can be important to determine whether the terrain may flood, whether the ground is stable or how ground water flow patterns are connected. The lower White River watershed exposes meanders typical for low flow velocities in very shallow floodplains (Fig. 4.7). The satellite image (a) shows the present course of White River (green for water and bright yellow for shoals) with some indications for historic courses with their clear delineation obstructed by farming (brownish tones). We have marked hints for fossil oxbows and lakes that can be interpreted as oxbow lakes with cyan dots. Slight changes in the topography are detected by LiDAR and become visible in the LiDAR DSM, which are even more enhanced when introducing 5× vertical exaggeration and slant illumination from two directions (c). The color coding of the subtle elevation differences in the otherwise virtually flat Mississippi floodplain allows detecting continuous historic river courses. This effect is particularly strong at the confluence of White River and Blackwater.

The area comprising historic channels marks the White River floodplain. At high water levels of White River the floodplain is more likely to flood than the higher grounds outside. Floodplains typically contain more silts than the areas outside. Both aspects can be used for infrastructure planning. Overlaying satellite image and color-coded LiDAR DSM (b) combines the information depth of each individual image. The dark blue colors are related to present river channels and open water, pastel blues and pink represent historic channels and greenish colors indicate marshes and wetlands. All features are confined to the floodplain delimited by the light yellow dashed lines. The satellite image contributes the landuse information, i.e. fields for the areas outside the floodplain.

4.4 Resolution of Digital Relief Models

The spatial and vertical resolution determines how well represented the earth's topography is by the digital elevation model. We increase the area from the previous section and study how frequently used digital relief models represent the topography (Fig. 4.8). We compare the LiDAR DSM of Arkansas in the decimation to 5 m resolution, ASTER GDEM V3 at 30 m resolution and GMTED at 1 km resolution. The Lidar DEM (a) resolves the floodplain of White River with the river-bed in cyan blue as well as the dendritic valley pattern that is enhanced by the grayscale rendering (b). ASTER GDEM (c) and SRTM DEM (d) resolve the floodplain but not the contours of the riverbed with marginal difference between the two models. The GMTED model (e) clearly shows the impact of reduced resolution. The White River valley and the texture of the mountains with the dendritic valley patterns is not resolved at all. The overlay of the LiDAR DSM texture on GMTED (f) reveals the loss of information as a result of insufficient sampling.

In 3D view (Fig. 4.9) this loss of textural resolution becomes even more evident. The LiDAR DSM (a) shows a realistic representation of the mountains, dendritic valleys and the details of the White River valley. ASTER GDEM (b) and

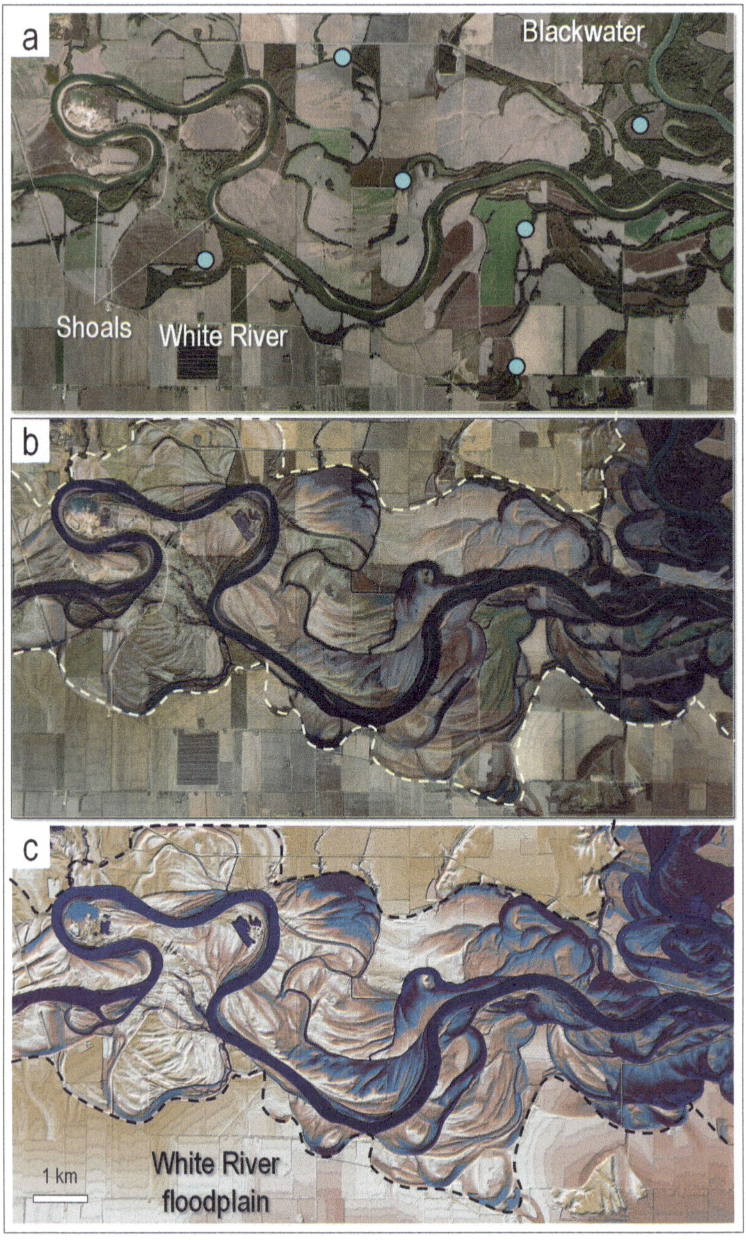

Fig. 4.7 Mapping of floodplain from optical satellite imagery (**a**), and overlay of satellite image and LiDAR DSM. (**b**) and LiDAR DSM with 5× vertical exaggeration (**c**). (LiDAR courtesy of DSM Arkansas Government, satellite imagery courtesy of U.S. Geological Survey)

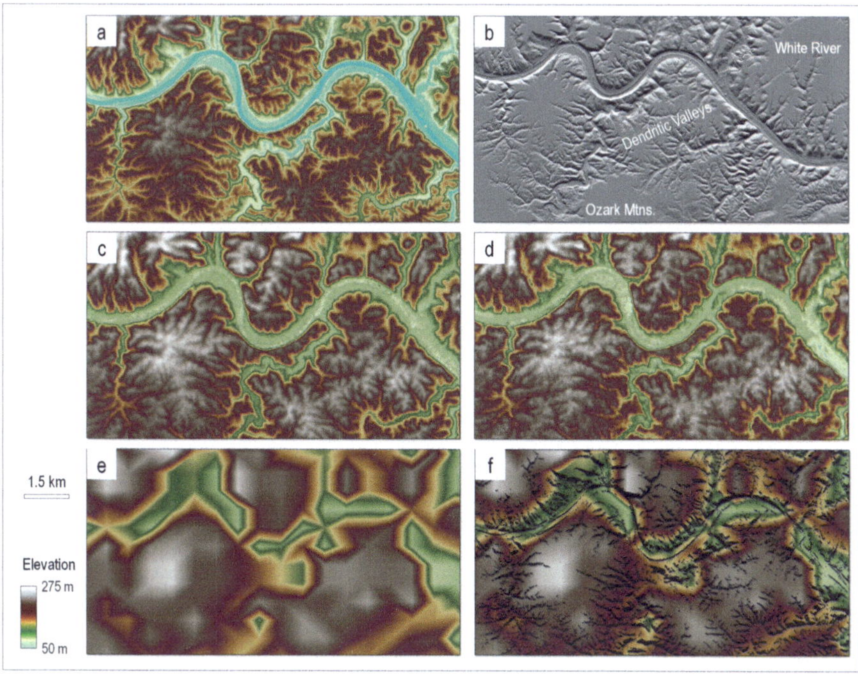

Fig. 4.8 Resolution of frequently used digital relief models in map view. (**a**) LiDAR DSM of Arkansas (5 m), (**b**) grayscale relief of (**a**), (**c**) ASTER GDEM V3 (30 m), (**d**) SRTM DEM V3 (30 m), (**e**) GMTED, 1 km resolution, (**f**) LiDAR DEM overlay on GMTED (1 km). (LiDAR courtesy of DSM Arkansas Government, SRTM DEM courtesy U.S. Geological Survey, ASTER GDEM courtesy METI)

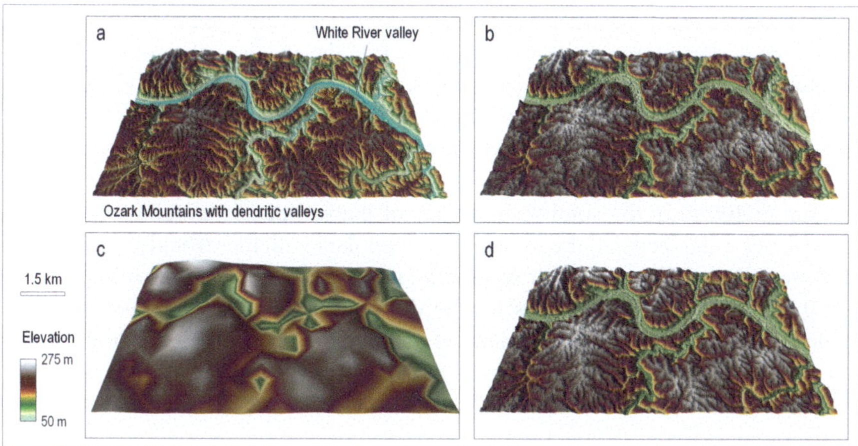

Fig. 4.9 Resolution of frequently used digital relief models in 3D view. (**a**) LiDAR DSM of Arkansas 5 m resolution, (**b**) ASTER GDEM V3, 30 m resolution, (**c**) GMTED, 1 km resolution, (**d**) SRTM DEM V3 (SRTM DEM Ver 3 courtesy of U.S. Geological Survey) 30 m resolution. (LiDAR courtesy of DSM Arkansas Government, SRTM DEM courtesy U.S. Geological Survey, ASTER GDEM courtesy METI)

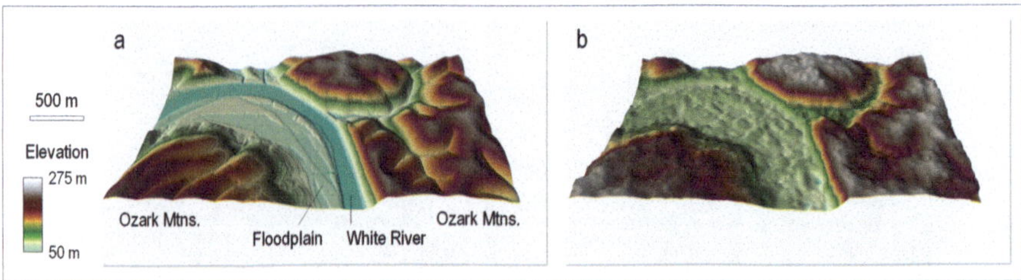

Fig. 4.10 Resolution of DEM for White River bend near Guion, (**a**) Arkansas LiDAR DSM 1 m resolution, (**b**) ASTER GDEM V3 30 m resolution. (LiDAR courtesy of DSM Arkansas Government, SRTM DEM courtesy U.S. Geological Survey, ASTER GDEM courtesy METI)

SRTM DEM (c) represent well the texture of the Ozark Mountains with the dendritic valleys, but the White River valley is insufficiently represented. GMTED (d) appears not useful for the appropriate representation of features of this level of detail. The zoom into the river bend (Fig. 4.10) highlights again the importance of appropriate spatial and vertical sampling for the interpretation of digital relief models.

4.5 Assessment of Absolute Elevation in Digital Relief Models

The absolute elevation above sea level is important for studies related to flooding in coastal areas. Therefore, we now have a look into the absolute elevations given by the most frequently used digital elevation models. Typically, LiDAR models give the most reliable elevation when calibrated properly with ground GPS stations. For the assessment we zoom out a bit further and compare the different digital elevation models (Fig. 4.11a). The terrain is comprised of the Ozark Mountains with elevations up to 400 m in the southwest, a zone of marked dendritic valleys along the White River watershed and the Mississippi floodplain with the strongly meandering Blackwater River. We start with setting the range for the elevation to the interval 0–500 m and compare the colors for the different models in the region with the dendritic valleys (b, c) and in the marshes of the Mississippi floodplain at the junction of White River and Blackwater (d–g). Similar scaling to the interval 0–500 m produces similar colors for ASTER GDEM (d) and SRTM DEM (e) but different colors for the LiDAR DSM (g).

When shifting the LiDAR DSM by approximately 30 m down (f), we obtain the same colors as for ASTER GDEM and SRTM DEM. We observe the same shift in the area with the dendritic valleys, i.e. the LiDAR DSM (c) needs to be shifted down by 30 m to match the ASTER GDEM. We conclude: when working in areas where

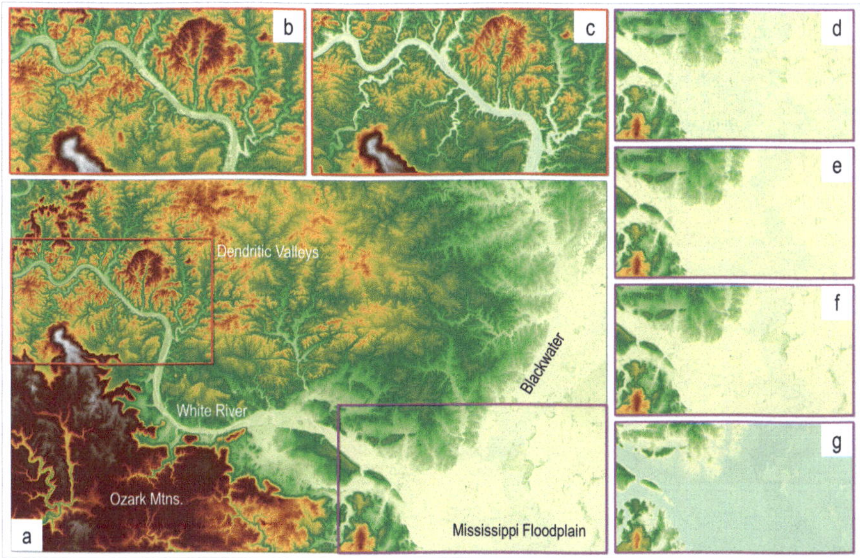

Fig. 4.11 Comparison of absolute relief in digital elevation models. (**a**, **f**) LiDAR DSM original, scale 0–500 m, (**c**, **g**) DSM −30 m, scale −30 to 470 m, (**b**, **d**) ASTER GDEM V2, scale 0–500 m, (**e**) SRTM DEM V3 0–500 m. (LiDAR courtesy of DSM Arkansas Government, SRTM DEM courtesy U.S. Geological Survey, ASTER GDEM courtesy METI)

absolute elevation values are required, then ASTER GDEM and SRTM DEM should be checked at least against a locally available LiDAR DSM and ideally also against ground GPS stations.

4.6 Applications for Different Digital Relief Models

Concluding this section, we will provide some typical use cases for the different relief models (Fig. 4.12). Detailed mapping of river beds, watersheds and mountain texture in areas of a few sq. km requires high resolution LiDAR at 1 m (a) resolution. If a larger area of a few 100 sq. km is studied, then LiDAR at 5 m resolution (b) is more economic because the data size shrinks by a factor of 25. Regional mapping of areas of 1000s to 10,000s of sq. km is best done with ASTER GDEM (c) and SRTM DEM. Continental (d) to global mapping (e) is sufficient at km resolution, i.e. GMTED and ETOPO1 are suitable for such tasks.

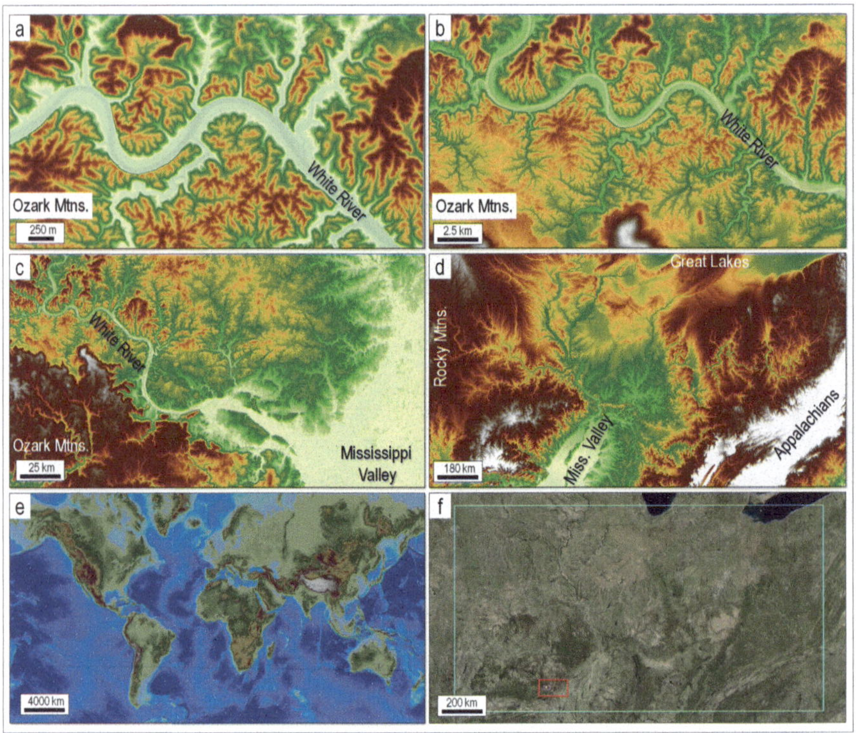

Fig. 4.12 Applications for digital relief models of different resolution. (**a**) LiDAR DSM 1 m, (**b**) LiDAR DSM 5 m, (**c**) ASTER GDEM V3 30 m, (**d**) GMTED 1 km, (**e**) ETOPO1 2 km, (**f**) overview map for areas shown in (**a–d**). (LiDAR courtesy of DSM Arkansas Government, SRTM DEM and GMTED courtesy U.S. Geological Survey, ASTER GDEM courtesy METI)

References

Digital Elevation Models General

Wilson, J. P., & Gallant, J. C. (2000). Chapter 1. In J. P. Wilson & J. C. Gallant (Eds.), *Terrain analysis: Principles and applications* (pp. 1–27). Wiley. ISBN 978-0-471-32188-0.

ASTER GDEM V3

Abrams, M., & Crippen, R. (2019). *ASTER GDEM V3 (ASTER global DEM), User Guide Version 1, Japan's Ministry of Economy, Trade and Industry (METI)*. National Aeronautics and Space Administration (NASA), Jet Propulsion Laboratory/California Institute of Technology.

ASTER GDEM Validation Team. (2009). *ASTER global DEM validation, summary report*. METI/ERSDAC, NASA/LPDAAC, USGS/EROS.

Lee, C., Oh, J., Hong, C., & Youn, J. (2014). Automated generation of a digital elevation model over steep terrain in Antarctica from high-resolution satellite imagery. *IEEE Transactions on Geoscience and Remote Sensing, 53*(3), 1186–1194.

Porter, C., Morin, P., Howat, I., Noh, M.-J., Bates, B., Peterman, K., Keesey, S., Schlenk, M., Gardiner, J., Tomko, K., Willis, M., Kelleher, C., Clouties, M., Husby, E., Foga, S., Nakamura, H., Platson, M., Wethington, M. Jr., Williamson, C., Bauer, G., Enos, J., Galen, K., Kramer, W., Becker, P., Doshi, A., D'Souza, C., Cummens, P., Laurier, F. & Bojesen, M. (2018). ArcticDEM, Harvard Dataverse, V1, https://doi.org/10.7910/DVN/OHHUKH, [accessed 15 Feb 2021]

ETOPO1

Amante, C., & Eakins, B. W. (2009). *ETOPO1 1 arc-minute global relief model: Procedures, data sources and analysis, NOAA Technical Memorandum NESDIS NGDC-24*. National Geophysical Data Center, Marine Geology and Geophysics Division.

GLOBE

Hastings, D. A., & Dunbar, P. K. (1999). *Global land one-kilometer base elevation (GLOBE) digital elevation model, documentation, Volume 1.0* (Key to geophysical records documentation (KGRD) 34). National Oceanic and Atmospheric Administration, National Geophysical Data Center.

LiDAR DEM

Bartolini, L., Bordone, A., Fantoni, R., Ferri de Collibus, M., Fornetti, G., Moriconi, C., & Poggi, C. (2000). Development of laser range finder for the Antarctic Plateau. In *Proceedings of EARSeL-SIG-Workshop LIDAR, Dresden/FRG, June 16–17, 2000* (pp. 148–156).

Harsdorf, S., Tönebön, S., Reuter, R., & Wachowicz, B. (2000). Submersible Lidar for seafloor inspection. In *Proceedings of EARSeL-SIG-Workshop LIDAR, Dresden/FRG, June 16–17, 2000*.

Heidemann, H. K. (2018). Lidar base specification, Chapter 4 of Section B, U.S. Geological Survey Standards Book 11, Collection and Delineation of Spatial Data, Version 1.3.

Maslov, D. V., Fadeev, V. V., & Lyashenko, A. I. (2000). A shore-based Lidar for coastal seawater monitoring. In *Proceedings of EARSeL-SIG-Workshop LIDAR, Dresden/FRG, June 16–17, 2000* (pp. 46–52).

NOAA. (2012). *Lidar 101: An introduction to Lidar technology, data and applications, revised*. National Oceanic and Atmospheric Administration (NOAA), Coastal Service Center.

Wozencraft, J. M., & Irish, J. (2000). Airborne Lidar surveys and regional sediment management. In *Proceedings of EARSeL-SIG-Workshop LIDAR, Dresden/FRG, June 16–17, 2000* (pp. 28–38).

Radar-Based Bathymetry

Sandwell, D. T., & Smith, W. H. (2009). Global marine gravity from retracked Geosat and ERS-1 altimetry: Ridge segmentation versus spreading rate. *Journal of Geophysical Research, 114*(B01411), 1–18. https://doi.org/10.1029/2008JB006008

SRTM DEM

Berry, P. A. M., Garlick, J. D., & Smith, R. G. (2007). Near-global validation of the SRTM DEM using satellite radar altimetry. *Remote Sensing of Environment, 106*(1), 17–27. https://doi.org/10.1016/j.rse.2006.07.011

Farr, T. G., Rosen, P. A., Care, E., Crippen, R., Duren, R., Hensley, S., Kobrick, M., Paller, M., Rodriguez, E., Roth, L., Seal, D., Shaffer, S., Shimada, J., Umland, J., Werner, M., Oskin, M., Burbank, D., & Alsdorf, D. (2007). The Shuttle Radar Topography Mission. *Reviews of Geophysics, 45*(2), RG2004. https://doi.org/10.1029/2005RG000183

Lu, K.-Y., & Kiang, J.-F. (2013). Terrain height estimation using a stereo-SAR technique aided by a reference point. *Progress in Electomagnetics Research M, 31*, 1–11.

NASA. (2000). *Shuttle Radar Tomography Mission, first shuttle flight of the new millennium.* Shuttle Press Kit, NASA.

Smith, B., & Sandwell, D. (2003). Accuracy and resolution of shuttle radar topography mission. *Geophysical Research Letters, 30*(9), 20.1–20.4.

Stereo Image-Based DEM

Raggam, H. (2006). Surface mapping using image triplets: Case studies and benefit assessment in comparison to stereo image processing. *Photogrammetric Engineering and Remote Sensing, 72*(5), 551–563.

TerrainBase

Row, L. W., Hastings, D. A., & Dunbar, P. K. (1995). *TerrainBase worldwide digital terrain data—Documentation manual, CD-ROM Release 1.0.* NOAA National Geophysical Data Center.

Part II
Primary Applications for Geosciences

Chapter 5
Primary Applications for Geosciences

Abstract After introducing the principles of remote sensing for hydrocarbon exploration, the data provided by remote sensing sensors and methods for processing the data, we will now turn to the primary applications to geosciences including rendering and interpreting remote sensing data. We will study how the earth surface texture is related to the properties and configuration of the subsurface rock layers using the concept of geomorphology.

In this section we show how the rendering of remote sensing data can reveal properties of individual remote data sets as well as the correlation between multiple data sets. Equipped with the data, the processing and the rendering, we will then study three geological settings that are important for various aspects of exploration. We start with the mapping of structural features, i.e. folds and faults, and then study depositional and erosional features in arctic and hot desert environment using both direct methods mapping features at the earth surface as well as using indirect methods to infer geomorphologic properties from observations of vegetation and water features.

Keywords Geomorphology: topography, lithology, depositional environment, surface dip · Rendering: flat maps, vertical exaggeration, slant illumination, colormaps, co-rendering, image overlay, surface attributes, image draping · Structural mapping: fold, anticline, fault, thrust, compressional tectonics, transpressional tectonics, structural front, incised valley, fluvial erosion, dendritic drainage system, vegetation · Mapping arctic depositional and erosional features: digital elevation model at high latitude, ArcticDEM, GMTED2010, moraines, ground moraines, terminal moraines, glacier terminus, floodplain · Mapping hot arid desert depositional and erosional features: lithology mapping, mesa, plateau, canyon, fault-guided erosion, surface gradient, multi-spectral co-rendering

A. Laake, *Remote Sensing for Hydrocarbon Exploration*, Springer Remote Sensing/Photogrammetry, https://doi.org/10.1007/978-3-030-73319-3_5

5.1 Introduction

The interpretation of remote sensing data typically follows two routes, the study of the geometry and texture of the earth surface which gives clues for structural, depositional and erosional aspects and the spectral characteristics of the surface reflectance which provides information about lithology, vegetation, water features and landuse. For both approaches we distinguish direct methods that use the remote sensing directly for interpretation and indirect methods where processing of the remote sensing data precedes the interpretation of the remote sensing data thus providing interpretation based on inferred and modeled subsurface structures and properties. The combination of the two approaches yields the most comprehensive information obtainable from remote sensing data (Table 5.1).

The approaches and methods are introduced using geological settings where individual features and processes can be isolated. They form the basis for the targeted workflows we will study in Part III.

An overview of the remote sensing data sets used in Part II is given in Table 5.2, Fig. 5.1 shows the data sets on a map.

Table 5.1 Approaches and methods for the interpretation of remote sensing data for geosciences

Approach	Method
Surface geometry and texture	Direct: Digital elevation model, surface gradient Detect lineaments from edges Delineate drainage patterns
	Indirect: Regional approach for detection of trends 3D rendering and optical correlation
Spectral reflectance	Direct: Lithology and landuse mapping from spectral reflectance Indirect: Regional trends for terrain filtering to detect overprints Classification and feature extraction
Combined approach (Geomorphology)	Interpretation of features for "hidden" influence from the subsurface processes: Incised valleys, anomalous river courses

Table 5.2 Overview of remote examples used in Part II

Section/ Chapter	Geologic setting	Area of interest	Remote sensing input for
5.2	Folded and faulted strata	Dinosaur NM, U.S.A.	Explanation of concept of geomorphology
Chapter 6	Regional glacial overprint	Great Lakes, North America	Demonstration of rendering options
7.2	Fold-and-thrust belt	Appalachian Mtns, U.S.A.	Extraction of structural geologic features
7.3	Regional glacial overprint	Pechora Lowlands, Russia	Extraction of deposition and erosional features in arctic environment
7.4	Erosion into carbonate plateau	Colorado Canyon and Colorado Plateau, U.S.A.	Extraction of erosional features in hot arid environment

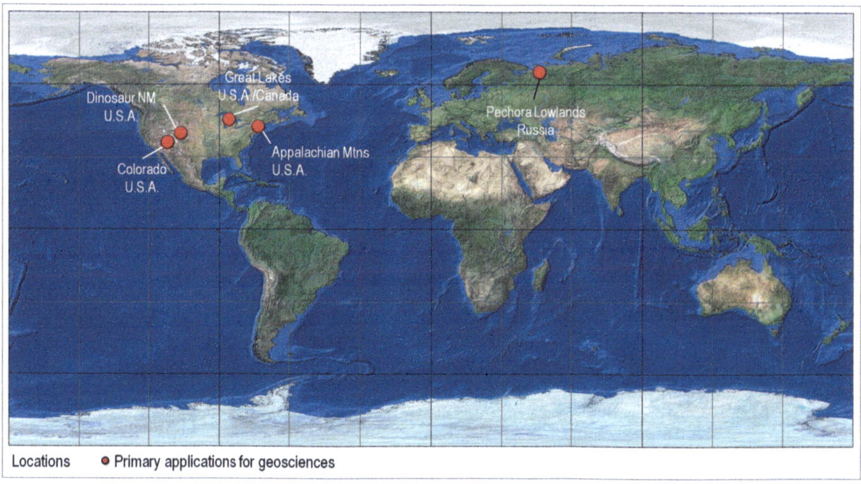

Fig. 5.1 Location of the remote sensing examples used in Part II

5.2 The Concept of Geomorphology

The concept of geomorphology aims at understanding topographic and bathymetric features at the earth's surface by combining surface and subsurface observations and interpretations of geological processes (Goudie, 2004). Geomorphology has also been employed as a tool for interpreting seismic data (Zeng, 2002). Recently, dynamic aspects have been included in the previously static concept (Orme, 2017). Geomorphology has been one of the early targets of remote sensing (Short & Blair, 1986).

We will introduce the concept of geomorphology using satellite remote sensing and surface geological data from the area around the Mitten Park Fault in the Dinosaur National Monument in Utah and Colorado, U.S.A. (Bradley, 1936; Brown, 1996; Sears, 1962). The view from the observation point near the confluence of the Green and Yampa rivers exhibits the impact subsurface tectonic processes and the differential erosion of the exposed rock strata have on the surface topography.

The view from observation point at Harpers Corner (Fig. 5.2) comprises the complete inventory of processes that contribute to geomorphology: in the northeast-southeast running Mitten Park Monocline the rock layers are warped almost vertically. The Mitten Park Fault marks the apex of the almost 90° bend where the Warm Springs Monocline forms the continuation towards the east. We will study the setting in map view in more detail. The tectonic warping exposes hard sandstone and limestone as well as soft shales to weathering which leaves a rough topography in the area to the right of the Mitten Park fault.

Hard rocks like limestone and dense sandstone weather significantly less than shales. Consequently, cliffs and escarpments are often topped by hard rocks whereas the slopes are determined by the amount of soft rocks present in the lithologic column. Tectonic processes warp sedimentary rock strata that are horizontal and flat at deposition into folds that expose different rock layers to the erosional forces at the surface.

Fig. 5.2 Correlation of topography, differential erosion and subsurface formation dip at the observation point at Harpers Corner near the confluence of Green River and Yampa River, Dinosaur NM, U.S.A.

In the background, the formations did not undergo tectonic deformation and remained horizontal and flat layered. The erosion of the Green River into these rock layers created the Canyon of Lodore, which exposes the rock layers of different hardness below a smooth undulating surface.

Remote sensing data can assist the understanding of the geological setting of the area of Mitten Park. The digital elevation model provides the topography whereas multispectral satellite imagery gives clues for the rock types exposed at the surface. We study the remote sensing data first in a flat map representation to compare the expression of the geomorphology in a topographic and a lithologic map (Fig. 5.3). The digital elevation model (a) shows the highs of the Yampa Plateau, Wild, Douglas and Blue Mountains. Lows are formed by the valleys eroded into the highlands by the Green and Yampa Rivers. A striking anomaly is Split Mountain into which the Green River has carved a canyon. The visible band Landsat 8 432 RGB composite (b) shows larger geologic units associated with the topography. The highlands are mostly composed of sandstones and conglomerates (brown-gray). Along its edges, softer lighter sandstone forms cliffs (light shades). The lower parts of the slopes towards the rivers are formed by soft shales the fluvial erosion of which does not

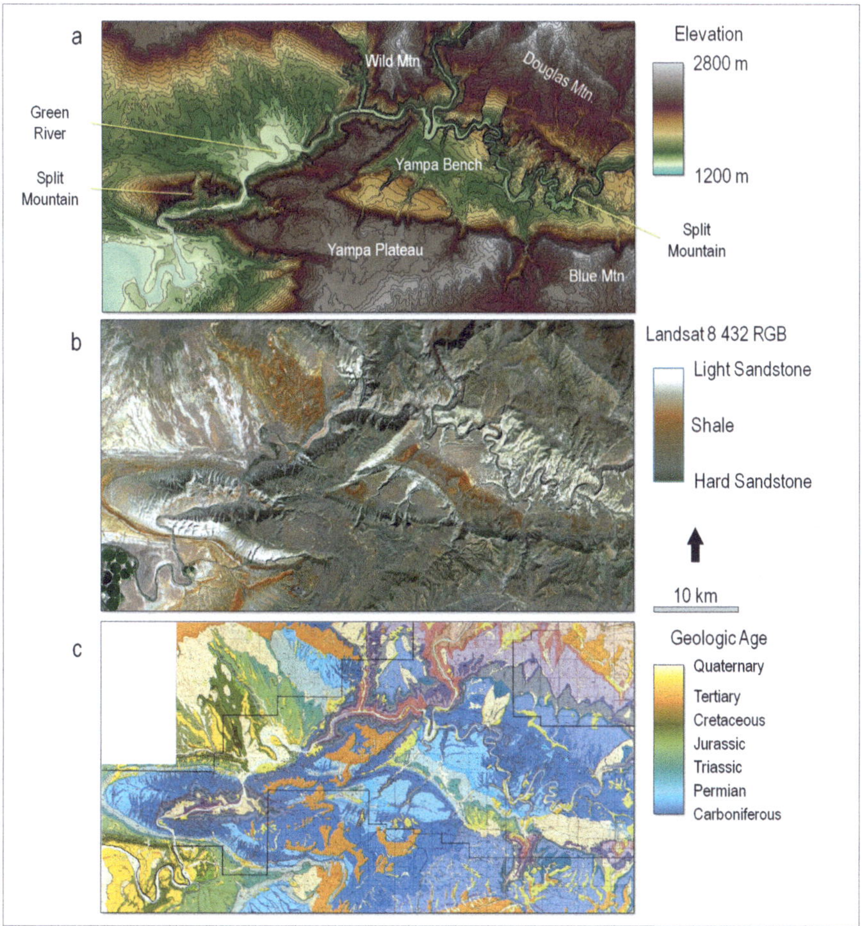

Fig. 5.3 Contributions to geomorphology. (**a**) Digital elevation model, (**b**) satellite image Landsat 8432 RGB pansharpened with band 8, (**c**) geological map. (SRTM DEM V3, satellite imagery and geological map courtesy of U.S. Geological Survey)

form canyons. The geological map (c) (Hansen et al., 1991) allows the correlation between the structural features observed at the surface model with their lithologic composition and geologic age. Most of the highlands are of Carboniferous to Permian age. A band of Jurassic rocks is exposed on the shallow slopes in a band running from northwest to southeast. Portions of the Yampa Plateau are covered in very coarse Tertiary conglomerate.

The 3D rendering of the same data (Fig. 5.4) reveals how the rocks of different composition resist erosion. The digital elevation model (a) provides the topography whereas the Landsat 8 432 RGB composite gives clues for the rocks and the geological map correlates rocks and topography with geological age and rock types. Highlands are mainly covered in hard sandstone of Permo-Carboniferous age

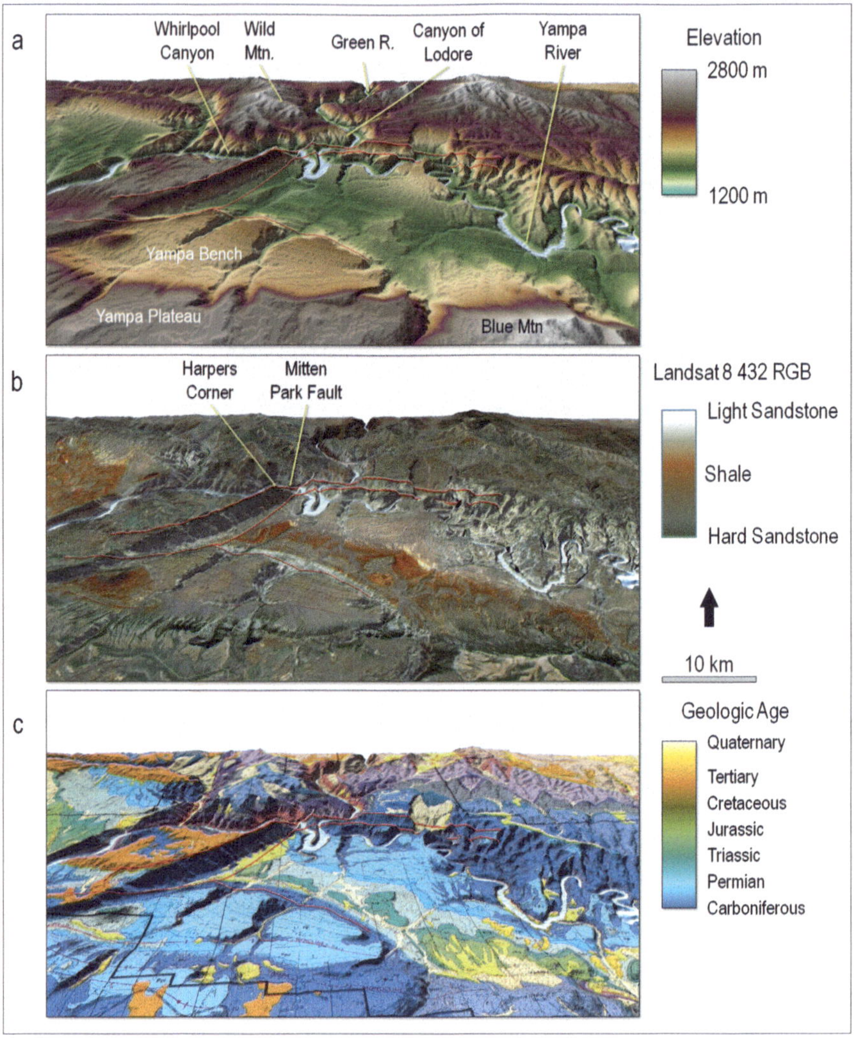

Fig. 5.4 Rendering elements of geomorphology in 3D with structural interpretation. (**a**) Digital elevation model, (**b**) satellite image Landsat 8432 RGB pansharpened with band 8, (**c**) geological map. (SRTM DEM V3, satellite imagery and geological map courtesy of U.S. Geological Survey)

showing gray-brown in the satellite image and the smooth slopes below the Yampa Bench are formed by soft shales of Cretaceous-Jurassic age. We conclude that the topography is directly correlated with the rock type exposed at the surface.

To study the influence of the layering of the subsurface strata on the topography we focus now on the bend of the Green River at its junction with the Yampa River where the Mitten Park Fault is exposed in the cliff face above the Green River (Fig. 5.5). We overlay the geologic map on the satellite image and drape the combined image on the digital elevation model (a). The geological map shows the Mitten Park Monocline approaching the focus area from the southwest.

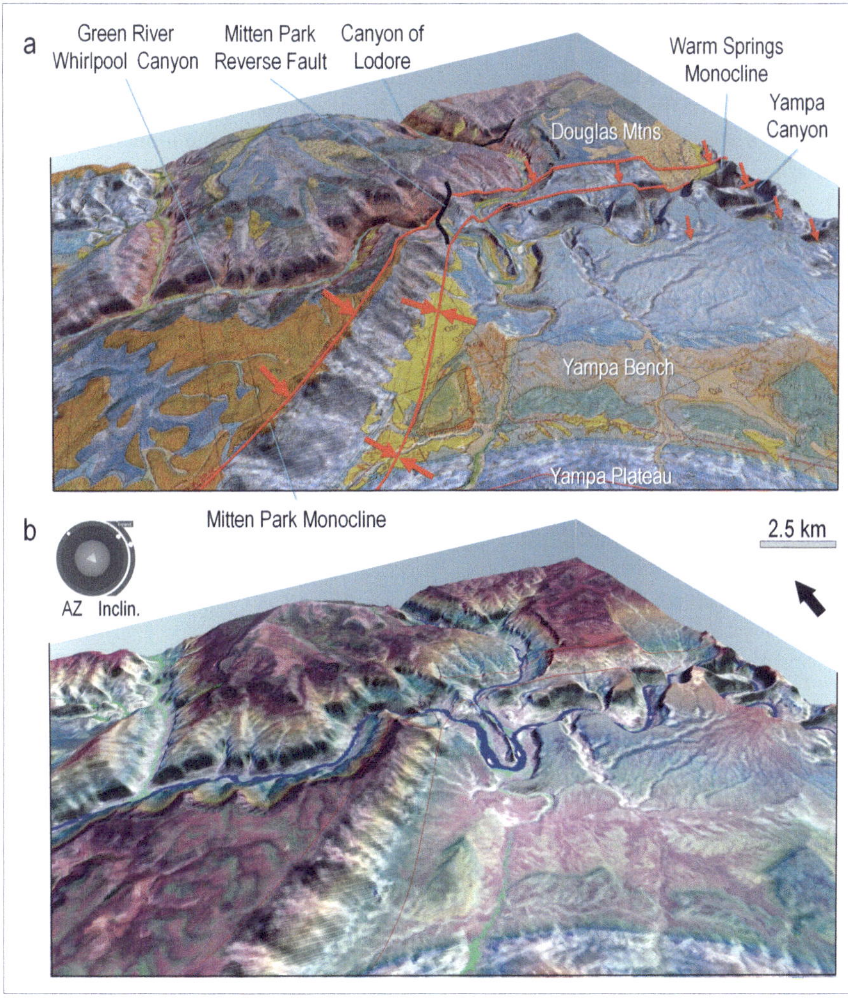

Fig. 5.5 Photo-geologic and geomorphological maps draped on digital elevation model in 3D rendering, 1× vertical exaggeration and slant illumination. (**a**) Photo-geologic map composed of geological map and Landsat 8432 RGB composite, (**b**) geomorphologic map composed of digital elevation model an Landsat 8432 RGB composite. (SRTM DEM V3, satellite imagery and geological map courtesy of U.S. Geological Survey)

Where Mitten Park Fault is exposed by the canyon carved out by the Green River, the Mitten Park Monocline bends towards the east and continues in the Warm Springs Monocline. The bend itself represents a complex pattern of steeply dipping strata of widely varying hardness that are exposed to the fluvial erosion in the Green Canyon creating the geological setting we have seen in the photo in Fig. 5.2. Immediately outside the Monoclines, the formations are rather flat layered, which creates the Yampa Plateau in the south and the smooth mountains in the north. The Yampa River has eroded its canyon into the southern flank of the Warm Springs

Monocline. The geomorphologic map that is composed of the overlay of the satellite image on the color-coded digital elevation model confirms the correlation of the surface topography with the lithology of the rock strata exposed at the surface.

The erosion of the Green River and Yampa River canyons follows both trends of weak exposed rocks as well as weak structural zones related to the monoclines. The most complex geological setting is created where fold and fault tectonics, weak rocks and fluvial erosion are acting in combination. To visualize this situation, we render the data at 5× vertical exaggeration with slant illumination from the northwest and the northeast at a low inclination angle. This approach highlights the steep slopes that dominate the complex erosional setting.

References

Remote Sensing Case Study Uinta Mountain and Dinosaur NM, Utah U.S.A.

Bradley, W. H. (1936). *Geomorphology of the north flank of the Uinta Mountains*, USGS Professional Paper 185-I, U.S. Geological Survey, 52 p.

Brown, C. M. (1996). *Structural analysis of the Mitten Park reverse fault and related deformation in Dinosaur National Monument, Northwestern Colorado and Northeastern Utah,* MSc dissertation, Utah State University, 127 p.

Hansen, W. R., Rowley, P. D., & Carrara, P. E. (1991). *Geologic map of Dinosaur National Monument and vicinity, Utah and Colorado*, USGS Miscellaneous Investigations Series Map I-1407. https://ngmdb.usgs.gov/Prodesc/proddesc_9190.htm

Sears, J. D. (1962). *Yampa Canyon in the Uinta Mountains, Colorado, A study of some unusual features and the possible origin and development of the canyon*, USGS Geological Survey Paper 374-I, U.S. Geological Survey. https://www.nps.gov/parkhistory/online_books/geology/publications/pp/374-I/

Geomorphology

Goudie, A. S. (Ed.). (2004). *The encyclopedia of geomorphology*. Routledge. ISBN 1134482752.

Orme, A. R. (2017). Dynamic geomorphology: Historical convergence towards modern practice. In W. Mayer, R. M. Clary, L. F. Azuela, T. S. Mota, & S. Wołkowicz (Eds.), *History of geoscience: Celebrating 50 years of INHIGEO* (Vol. 442, pp. 141–154). Geological Society, London, Special Publications. https://doi.org/10.1144/SP442.34

Short, N. M., & Blair, R. W., Jr. (1986). *Geomorphology from space, special publication 486.* National Aeronautics and Space Administration, 717 p.

Zeng, H. (2002). Geomorphology-based, automated seismic facies analysis. In *Expanded Abstracts, SEG International Exposition and 72nd Annual Meeting, Salt Lake City*. https://doi.org/10.1190/1.1817327

Chapter 6
Rendering of Remote Sensing Data

Abstract Rendering is an essential task to view remote sensing data for interpretation and modeling. We will study three different realizations of spatial rendering:

Flat maps without vertical exaggeration and perpendicular illumination.

With this rendering method, the remote sensing data are represented like maps with the advantage that accurate measurements can be taken for geospatial mapping and planning. This approach is widely used in geographic information systems (GIS).

Flat maps with vertical exaggeration and slant illumination.

To realize this method, a digital relief model is required onto which the remote sensing data are mapped or projected. We call this process draping the image. The vertical exaggeration then offers to the option to introduce illumination from the side to create the visual impression of a 3D effect, whilst the data still remain in map view. This approach is beneficial for structural analysis when 3D distortion from parallax is not acceptable. One version with pre-defined vertical exaggeration and slant illumination is implemented in GIS as hill-shading, for example in topographic maps.

3D rendering with vertical exaggeration and slant illumination.

This approach also requires a digital relief model on which to drape the remote sensing data. Vertical exaggeration and slant illumination create a virtual reality impression that lets interpreters experience the data as if they were immersed in the landscape. The use of slant illumination with multiple light sources enhances the virtual reality impression because it creates the impression of diffuse light which creates a very real impression.

A second important aspect of rendering is the use of color to emphasize features for interpretation. For single data sets, colormaps assist in making certain geologic features accessible to the interpretation. For multiple co-located data sets such as multi-spectral remote sensing data the co-rendering of three bands in RGB is a powerful method because it allows to merge the expression of features on different spectral bands whilst preserving specific properties in color.

Keywords Rendering: flat maps, vertical exaggeration, slant illumination, colormaps, co-rendering, image overlay, surface attributes, image draping

A. Laake, *Remote Sensing for Hydrocarbon Exploration*, Springer Remote Sensing/Photogrammetry, https://doi.org/10.1007/978-3-030-73319-3_6

6.1 Introduction

The rendering of remote sensing data aims at displaying the data in a way that highlights features for interpretation. The joint rendering of multiple data sets focuses on the detection and documentation of the correlation between different data sets. Figure 6.1 shows the digital elevation model for the area of the Great Lakes in North America a sequence of rendering techniques that visualize specific features: the grayscale image shows the texture of the digital elevation model. Grayscale images are frequently used to transmit data from satellites to ground stations and transfer processed satellite images from image providers to end-users.

For the analysis of the digital elevation data topographic or GIS colormaps are frequently used to convert the data into a map for easier interpretation. Vertical exaggeration and slant illumination support the visual impression and further ease the interpretation. Finally, the draping of a satellite image on to the digital elevation model and rendering the combined data with vertical exaggeration and slant illumination provides a near-realistic impression of the study area.

We will now study the different rendering methods in more detail using the digital elevation model and satellite imagery of the Great Lakes area and the upper Mississippi and Minnesota River valleys as an example. In Fig. 6.2 we collected an overview of the main methods with descriptions and examples.

2D flat map rendering is the simplest method. The data values for each point are coded with a color and plotted as a map. The relation between the data value and the color value is called a colormap. In the example shown here we have used a

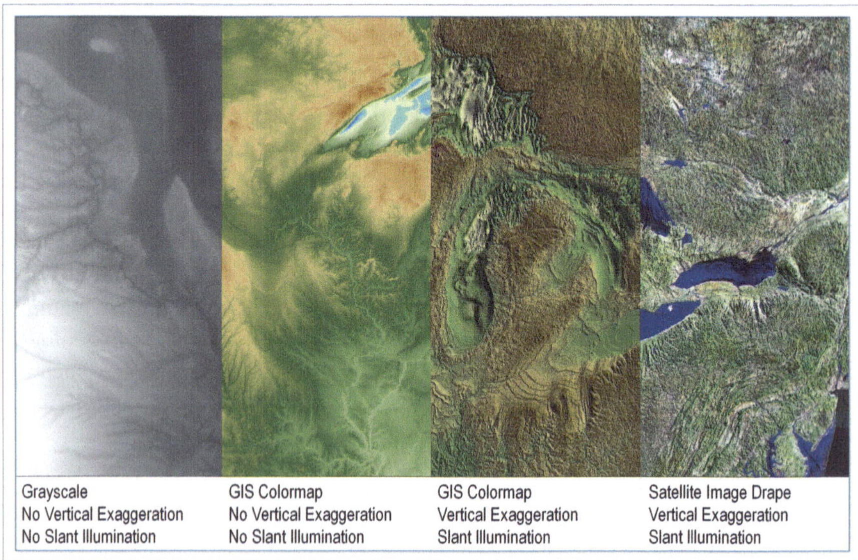

Fig. 6.1 Impact of rendering on the interpretation of remote sensing data illustrated at rendering of DEM and satellite image. (DEM and satellite imagery courtesy of U.S. Geological Survey)

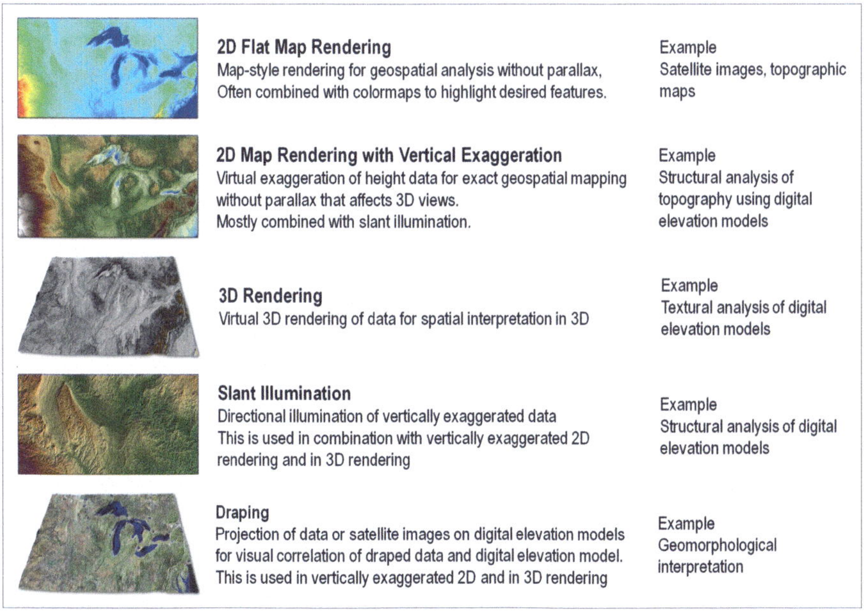

Fig. 6.2 Methods for rendering remote sensing data. (DEM and satellite data courtesy of U.S. Geological Survey)

so-called rainbow colormap from blue tones indicating low values to red tones indicating high values. 2D flat map rendering has the advantage that the data are represented in a geospatially correct way without parallax. Therefore, geospatial measurements such as distances and areas are possible with this rendering.

2D map rendering with vertical exaggeration is employed when the data are studied for structure and texture whilst preserving the geospatially correct representation. Vertical exaggeration gives the illusion of 3D effects without parallax distortion. Slant illumination is often combined with vertical exaggeration to even enhance the virtual 3D impression.

3D rendering is the most realistic spatial rendering method. It is preferred for qualitative studies of structural and depositional features. However, when details and measurements are required, the interpreter will often return to the 2D rendering because of parallax and perspective effects, which enhance the understanding of features in 3D but disable exact measurements that are required for quantitative geospatial analysis.

Slant illumination is a virtual technique to simulate illumination from the side with the goal of enhancing directional structural features in 3D surfaces. When studying this technique in more detail we will cover the effects of multiple light sources and different azimuthal illumination. Slant illumination works best with vertically exaggerated data.

The method of **draping** data on to a digital elevation model is best described as throwing a cloth over a structured surface. Typically, either attributes highlighting aspects of the surface such as surface gradient or data independent of the surface such as satellite images from the same area are draped on the digital elevation model to study the correlation of the draped data with the elevation model for example using the concept of geomorphology (Sect. 5.2).

6.2 2D Rendering and Colormaps

2D rendering of remote sensing data represents the data values as a map (Fig. 6.3). A linear grayscale colormap renders the data elevation values as shades of gray with linear steps. For very low-profile terrain a logarithmic colormap may be useful (b). In our example the outline of the Great Lakes is better distinguished by this colormap. Rendering the data with color (c) assists further in distinguishing different features.

When choosing the colormap the maximum and minimum values for the rendering need to be considered. For too low values for the maximum and too high values for the minimum of the display range, clipping occurs and features with amplitudes larger than the maximum and/or smaller than the minimum are not represented correctly (d). However, clipping may be desirable when focusing only on the areas where the rendered data are not clipped.

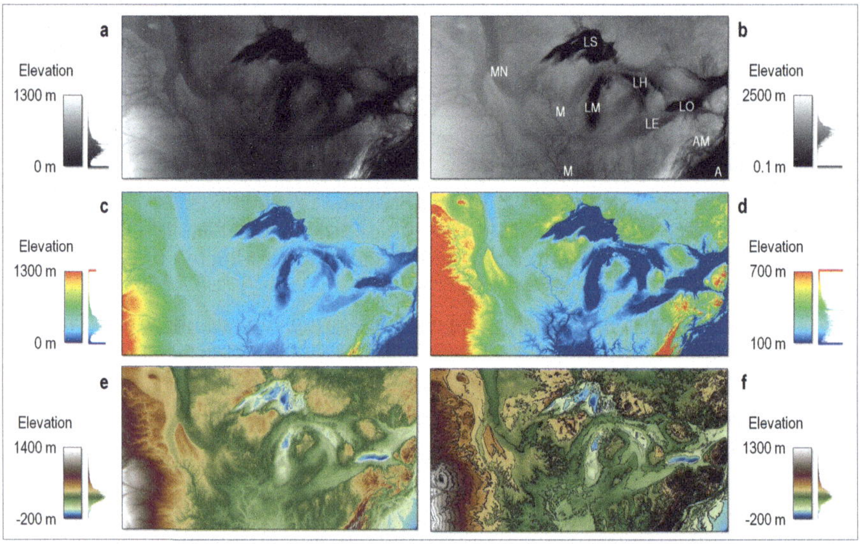

Fig. 6.3 Rendering of remote sensing data in flat 2D map view. GMTED DEM for the Great Lakes area, North America. (**a**) Linear grayscale, (**b**) logarithmic grayscale, (**c**) linear rainbow colormap, (**d**) linear rainbow colormap with clipping, (**e**) topographic colormap, (**f**) topographic colormap with elevation contours. *M* Mississippi River, *MN* Minnesota River, *LS* Lake Superior, *LM* Lake Michigan, *LH* Lake Huron, *LE* Lake Erie, *LO* Lake Ontario, *AM* Appalachian Mountains, *A* Atlantic. (GMTED DEM courtesy of U.S. Geological Survey)

A frequently used colormap for digital elevation models is the topographic or GIS colormap that renders water in blue tones, low coastal areas in light green and mountains in brown tones (e). Using this method, the regional glacial features such as moraines around the Great Lakes, the incised valleys of Mississippi and Minnesota Rivers and the texture of the Appalachian Mountains become very obvious. When more accurate elevation information is required, elevation contours may be added (f).

6.3 Vertical Exaggeration and Slant Illumination

Vertical exaggeration provides a virtual 3D impression of the data, whilst the rendering is still in 2D. In Fig. 6.4 we show vertical exaggeration for factors of 0, 5, 50 and 500. Please note that slant illumination has been added to enhance the virtual 3D effect, but the illumination angles have been kept constant to isolate the effect of the vertical exaggeration. The flat image without vertical exaggeration (a) shows very little texture outside the mountain ranges.

In the glacial basins south of the Great Lakes and along the Minnesota and Mississippi Rivers only the color coding gives clues for the location of the moraines left by the melted Pleistocene glaciers. At 5× vertical exaggeration only the mountainous areas reveal more detail (b). 50× times vertical exaggeration helps understanding the erosional features on the highlands in the west and north of the Great Lakes (c). At 500x vertical exaggeration different stages of moraines south and west of Lake Erie and Lake Michigan become visible so that reconstructions of the

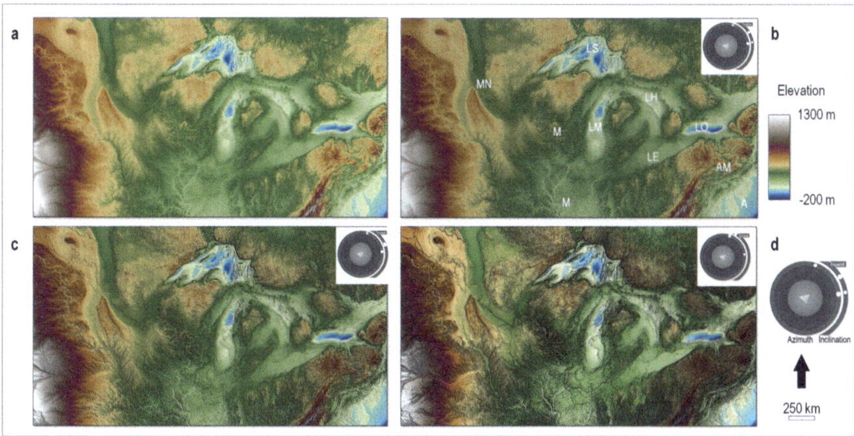

Fig. 6.4 Rendering of remote sensing data in 2D map view with vertical exaggeration and slant illumination. GMTED DEM for the Great Lakes area, North America. (**a**) No vertical exaggeration, (**b**) 5× vertical exaggeration, (**c**) 50× vertical exaggeration, (**d**) 500× vertical exaggeration, (**b–d**) with slant illumination. *M* Mississippi River, *MN* Minnesota River, *LS* Lake Superior, *LM* Lake Michigan, *LH* Lake Huron, *LE* Lake Erie, *LO* Lake Ontario, *AM* Appalachian Mountains, *A* Atlantic. (GMTED DEM courtesy of U.S. Geological Survey)

glaciation are possible (d). The choice of the amount of vertical exaggeration depends on the relative terrain gradient and the elevation expression of the studied features.

We study the impact of slant illumination on data that are vertical exaggerated 100× (Fig. 6.5). The orientation of the virtual light sources used for virtual illumination is shown by the dial indicating the azimuth on the inner circle and the angle of inclination on the outer half-circle. The illumination with a single light source (a) mimics the directional illumination by the sun. The edges of the Cote de Prairie ridge and the rough terrain into which the Mississippi River has incised are highlighted.

Adding a second light source at 45–90° azimuth simulates a more diffuse illumination which is beneficial for structural and textural interpretation of surfaces. In Fig. 6.5b–d we rotate the fixed combination of two light sources at about 50° relative azimuth and 30–40° inclination in steps of about 120°. The slant illumination from different directions enhances features that are perpendicular to the illumination whereas it virtually suppresses the signature of parallel features. For example, in (b) and (d) one flank of the Cote de Prairie ridge is highlighted whereas in (c) neither of the flanks receives direct illumination. This gives the impression that the ridge is not as prominent in (c) as if it were illuminated with the source configurations in (b) or (d). This effect can even lead to the impression that the ridge appears to be a depression.

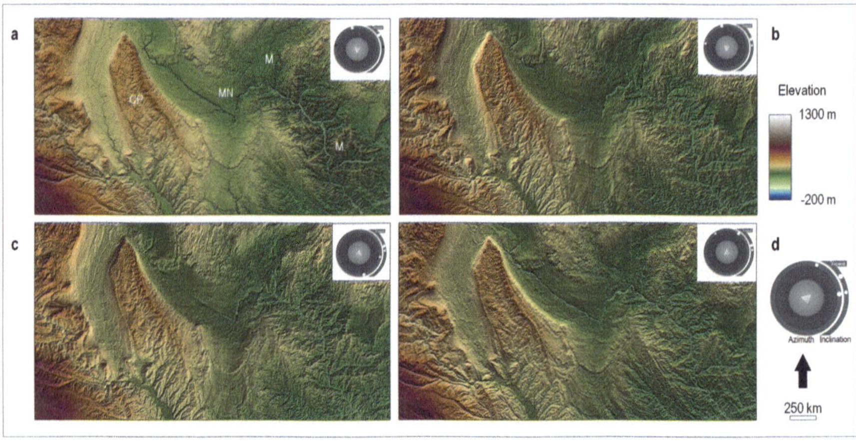

Fig. 6.5 Rendering of remote sensing data in 2D map view at 100× vertical exaggeration and slant illumination at different azimuths. GMTED DEM for the Great Lakes area, North America. (**a**) Single source illumination, (**b–d**) dual source illumination at different azimuths. *M* Mississippi River, *MN* Minnesota River, *CP* Cote de Prairie ridge. (GMTED DEM courtesy of U.S. Geological Survey)

6.4 3D Rendering and Draping

The most intuitive rendering method that delivers results that are easiest to interpret is the 3D rendering method, specifically when satellite images are draped over the surface rendered in 3D. In Fig. 6.6 we have collected the most frequently used 3D options: digital elevation model with topographic colormap (a). We have chosen 50× vertical exaggeration and slant illumination from the south and the east.

The rendering in 3D clearly reveals that the mountain ranges in the west and east of the Mississippi Valley provide the boundaries for the central basins that have been heavily overprinted by the Pleistocene glaciers. In the southeast, the sharp edge of the Appalachian Mountains towards the Atlantic coastal plain is clearly visible.

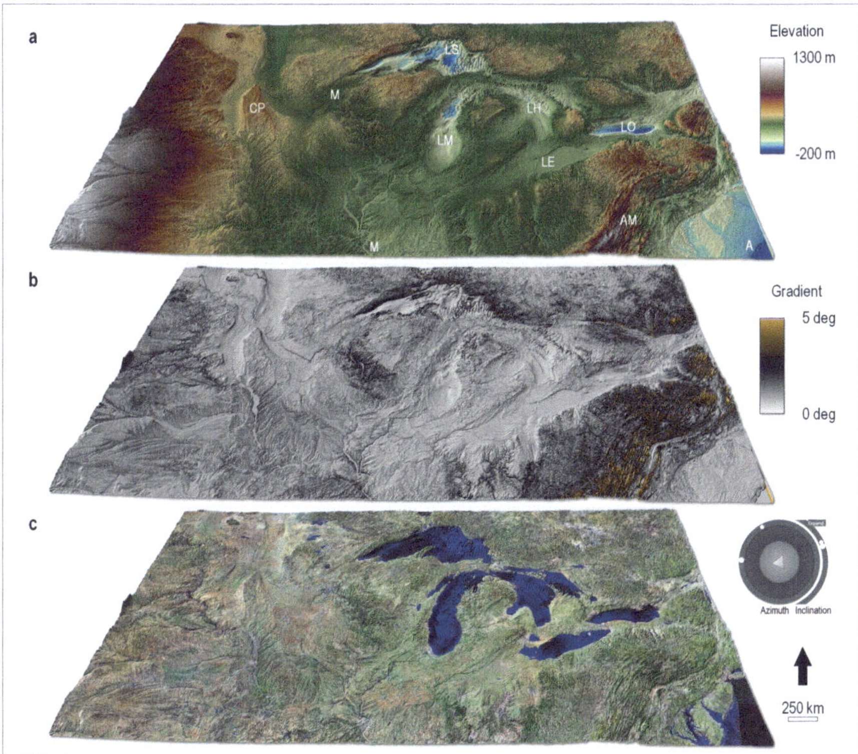

Fig. 6.6 Rendering of remote sensing data in 3D at 50× vertical exaggeration and slant illumination. GMTED DEM for the Great Lakes area, North America. (**a**) DEM, (**b**) surface gradient draped on DEM, (**c**) Landsat 7 742 RGB composite draped on DEM. *M* Mississippi River, *CP* Cote de Prairie, *LS* Lake Superior, *LM* Lake Michigan, *LH* Lake Huron, *LE* Lake Erie, *LO* Lake Ontario, *AM* Appalachian Mountains, *A* Atlantic. (GMTED DEM and satellite imagery courtesy of U.S. Geological Survey)

When draping the terrain gradient on to the digital elevation model (b), structural features that have a high gradient are emphasized. The strongest gradients occur in the Appalachian Mountains, however, incised rivers and terminal moraines are enhanced, too. Draping the Landsat 7 742 RGB composite image on the digital elevation model yields even more interpretation routes. The band combination used here is correlated with the landuse in a color setting that the human brain intuitively correlates with real landscape. The vegetation depends strongly on the elevation, which is reflected in the different shades of green seen on the Appalachian Mountains and in around the Great Lakes. The drier prairies west of the Mississippi, however show more shades of brown which is only loosely correlated with the elevation.

Chapter 7
Geologic Feature Extraction from Remote Sensing Data

Abstract In this section we will use remote sensing data to extract structural, depositional and erosional features from digital relief models and satellite imagery, jointly interpret them using the concept of geomorphology with goal of building geologic models to explain the geological setting of the study area. This process benefits from including existing information by other experts familiar with the target area and with the processes that shaped the geologic setting. Therefore, we have provided references related to the geologic processes on a global and regional scale, historic studies that provided insights into the geological setting and maps that can be used to correlate and calibrate the interpretation obtained from the remote sensing data.

Keywords Structural mapping: fold, anticline, fault, thrust, compressional tectonics, transpressional tectonics, structural front, incised valley, fluvial erosion, dendritic drainage system, vegetation · Mapping arctic depositional and erosional features: digital elevation model at high latitude, ArcticDEM, GMTED2010, moraines, ground moraines, terminal moraines, glacier terminus, floodplain · Mapping hot arid desert depositional and erosional features: lithology mapping, mesa, plateau, canyon, fault-guided erosion, surface gradient, multi-spectral co-rendering

7.1 Introduction

For the demonstration of the capabilities of remote sensing mapping, we have selected three target areas for which describe an as broad as possible spectrum of geologic features that can be mapped from the interpretation of remote sensing data. We follow the most often used geologic assessment process used in exploration: start with the structural framework, continue with structural horizons and then study depositional and erosional features. We document this approach in three different areas:

A. Laake, *Remote Sensing for Hydrocarbon Exploration*, Springer Remote Sensing/Photogrammetry, https://doi.org/10.1007/978-3-030-73319-3_7

- Appalachian Mountains and Alleghany Plateau, U.S.A.: complex regional folding, thrusting and faulting in a moderate humid climate. Both rock outcrops as well as vegetation and water features are used for the interpretation of the structural framework.
- Pechora Lowlands and northern Ural Mountains, Russia: arctic lowlands with glacial overprint from multiple episodes that left glacial deposits which show a present-day erosional overprint. With the exception of the Ural Mountains, the area does not yield any bedrock outcrops which demands indirect interpretation methods to be used for the analysis. The interpretation of topographic features, water features and vegetation is combined to provide input for the interpretation of past and present geological processes.
- Colorado Plateau, U.S.A.: Grand Canyon that is incised into the Colorado Plateau with its extreme continental climate provides a wealth of geological interpretation options. The interpretation of spectral reflectance for lithology mapping can be directly correlated and calibrated with detailed geologic maps and descriptions from the field. The canyon depth of up to 1.6 km provides the unique opportunity to use remote sensing for both, lateral and vertical mapping. Small erosional and volcanic features on the plateau east of Grand Canyon contribute additional geologic features to the mix.

The observations and results are presented side-by-side with the scope to show which remote sensing data set or its derivative provides the most information or which combination of data sets to use to achieve the most comprehensive result.

7.2 Structural Elements: Folds and Faults Appalachian Mountains U.S.A.

The remote sensing of regional structural elements comprises the delineation of folded and faulted strata at the surface. Eastern Pennsylvania exhibits a suite of such regional structural elements in the northern Appalachian Mountains and the adjacent Alleghany Plateau (Fig. 7.1). In this area, we clearly distinguish two very different tectonic regimes, the heavily folded Appalachian Fold Belt that forms the Valley and Ridge landscape (Freedman et al., 1964) and the Appalachian Plateau Decollement Sheet, that forms the Alleghany Plateau with deeply incised valleys. The two regimes are separated sharply by the Alleghany Structural Front (Gillespie & Kampfer, 2017). The different tectonic appearance has been attributed to subsurface salt that allowed the Appalachian Plateau Sheet to decouple from the underlying strata and contribute to the specific fold-and-thrust tectonics of the Appalachian Mountains (Davis & Engelder, 1985). A more detailed study of the area revealed strike-slip faults that are evidence of transpressional tectonics (Valentino et al., 2004; Yeh & Bell, 2004).

We start the assessment how remote sensing data can assist understanding and delineating the complex structural development with the Landsat 7 742 RGB band

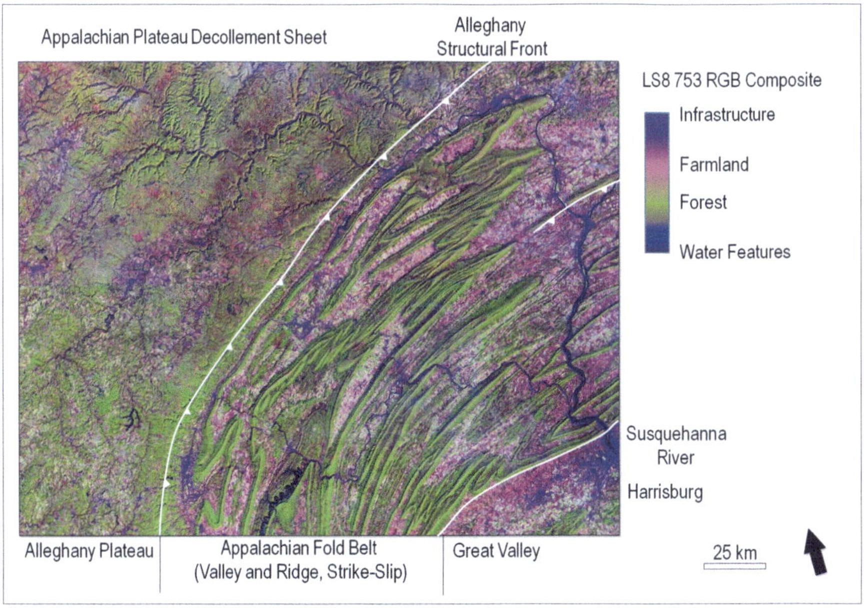

Fig. 7.1 Regional structural overview of Appalachian bend in Pennsylvania, Landsat 8753 RGB composite, regional structural interpretation after Gillespie and Kampfer (2017). (Satellite data courtesy of U.S. Geological Survey)

combination because the humid climate supports dense vegetation of the outcropping rocks thus disabling the lithological interpretation of the rock units. We notice that in the Valley and Ridge terrain of the Appalachian Fold Belt, the plains between the folds are used for farming (purple tones) whereas the folded ridges are covered in forest (light green). This highlights the shape of the folded ridges and indicates anticlinal structures which we will study in detail later. The Alleghany Plateau, in sharp contrast shows an overall smooth plateau with deeply incised narrow valleys in a dendritic pattern.

The digital elevation model of the area (Fig. 7.2a) confirms the observation of steeply folded ridges of the Appalachian Fold Belt we had already seen on the landuse interpretation of the satellite image, but also observe folded ridges on the Alleghany Plateau, the regional trends of which are indicated by dashed lines. However, these ridges are so gentle that they are not distinguishable on the elevation contour map (b) which rather highlights sharp terrain edges such as steeply folded ridges and dendritic incised valleys. The surface gradient (c) does not show the gentle ridges either. It rather provides insight into the regional structural patterns with sharper definition of steep folds and incised valleys, some of which show remarkably straight shape. In all three maps the Alleghany Structural Front is evident as the boundary between the Appalachian Fold Belt and the Alleghany Plateau.

The regional structural framework of folds and faults becomes evident when rendering the digital elevation model in 3D with vertical exaggeration and slant

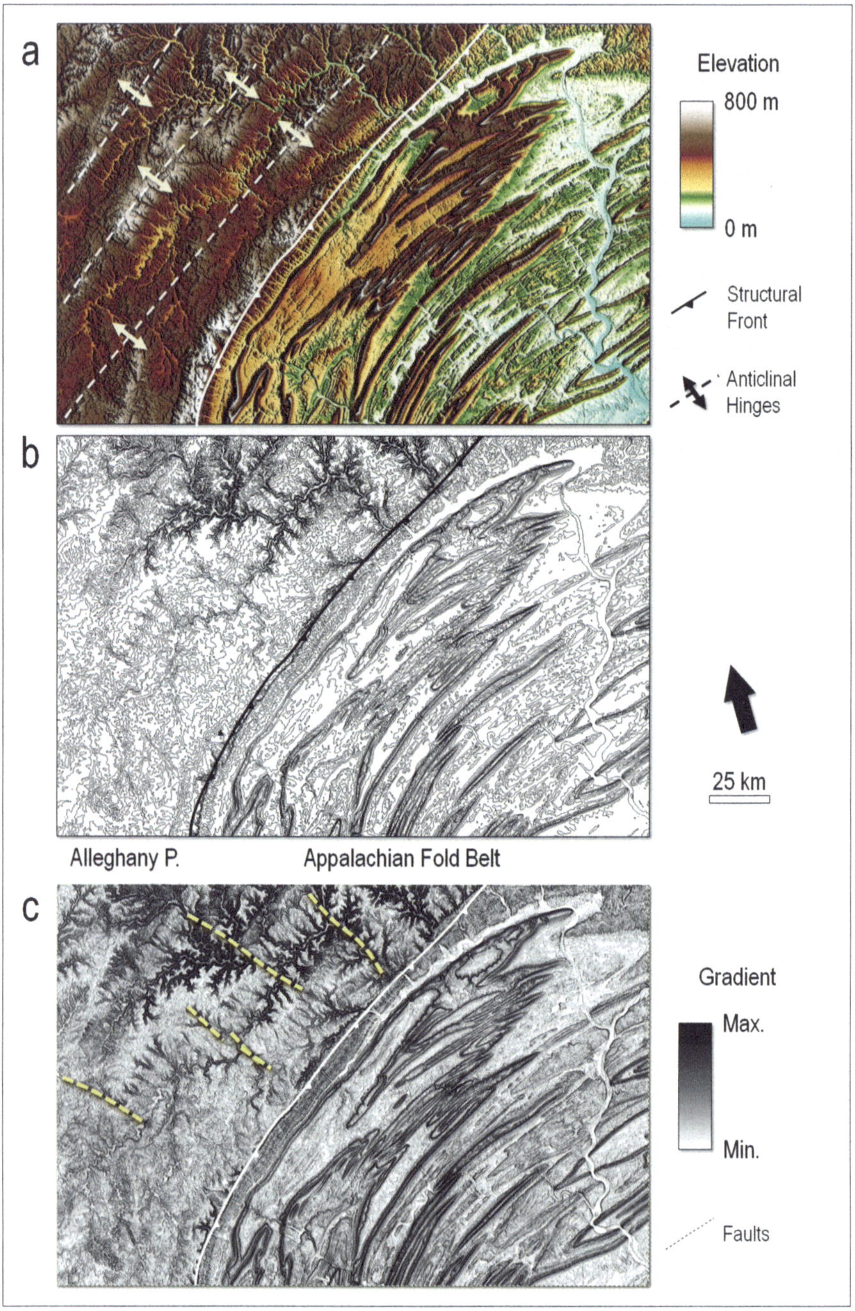

Fig. 7.2 Representation of regional structural framework in derivatives of the digital elevation model. (**a**) SRTM DEM, (**b**) DEM contours, (**c**) DEM surface gradient. (SRTM DEM data courtesy of U.S. Geological Survey)

illumination (Fig. 7.3). The structural zones become evident without the need for detailed interpretation: the gently folded Alleghany Plateau with deeply incised valleys, the intensely folded Appalachian Fold Belt with narrow steep ridges, the almost flat Great Valley and towards the coast the coastal plain. In the east the Susquehanna River has incised its valley across the ridges with its entrance at the Alleghany Structural Front and its exit into the Great Valley.

We will now study the structural features in detail at a smaller scale. The steep folds near the Alleghany Structural Front represent narrow anticlines (Fig. 7.4a). The larger anticlines can be recognized by the erosion of softer layers in their core (white arrows) leaving the dipping harder formations as a rim either side of the core of the anticline. Their strike is indicated by dashed lines, double arrows show the direction of the main dip. The anticlines are separated by right-lateral strike-slip faults (continuous lines) that offset adjacent faults by several kilometers.

The Landsat 8 753 RGB composite shows the anticlinal ridges as wooded green features that contrast sharply with the purple tones of the farmland in the valleys (b). The elevation contours (c) and the surface gradient (d) both delineate the folds as dark features, but the continuous color shades of the gradient map provides a smoother image that is easier to interpret.

We suggest using the gradient map as the preferred method to render terrain that shows intensely folding including thrust folding (Fig. 7.5). The flat gradient map without vertical exaggeration renders the flanks of anticlines as continuous features the steep dip of which contrast sharply with the smooth terrain in the anticline core and in the valleys. It also allows the identification of local perpendicular faults that break up the fold trends.

When we introduce 5 s vertical exaggeration and slant illumination we shift the attention to structural features. Figure 7.6b shows the gradient of the elevation model draped as a surface attribute on the digital elevation model. We use slant

Fig. 7.3 Regional structural framework in 3D, SRTM DEM at 5× vertical exaggeration and slant illumination. (SRTM DEM data courtesy of U.S. Geological Survey)

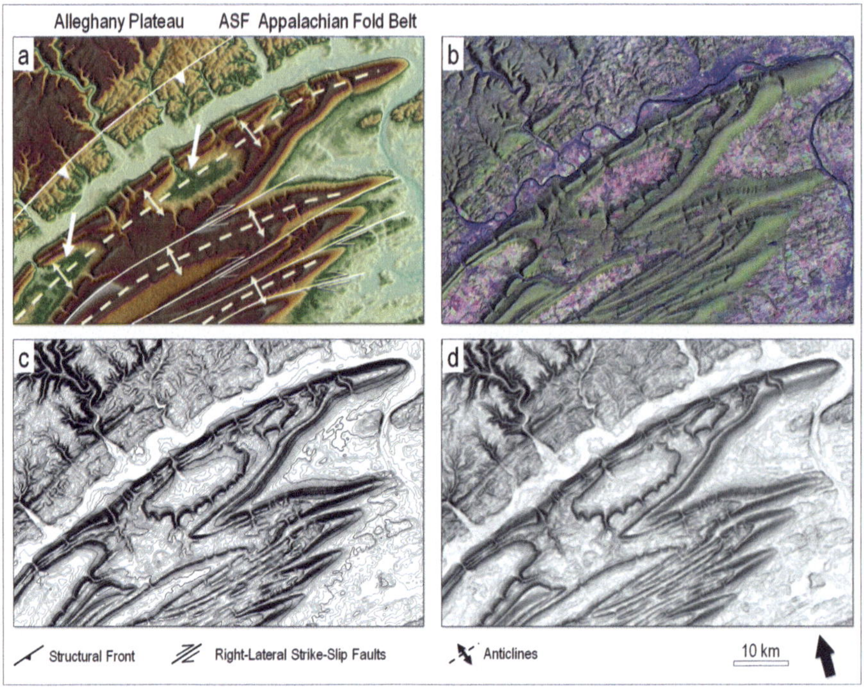

Fig. 7.4 Mapping of anticlinal folds in the strike-slip fold-fault zone of the Appalachian Fold Belt. (**a**) SRTM DEM, (**b**) Landsat 8 753 RGB composite, (**c**) contours of SRTM DEM, (**d**) gradient of SRTM DEM, *ASF* Alleghany Structural Front. (DEM data courtesy of U.S. Geological Survey)

illumination from two angles nearly perpendicular to the dominant orientation of the fault ridges. In addition to the anticlinal ridges with Brush Mountain and Loop Mountain as their most striking representatives, we can see now also a sequence of local faults that are perpendicular to the anticlinal ridges and the Alleghany Structural Front (yellow arrows, Fig. 7.6a). Besides these local faults we also observe a regional fault of about 50 km length that cuts through all fold ridges. The Little Juanita River follows this fault and eroded steep valleys into the folded ridges (white arrows).

Whilst the Appalachian Fold Belt shows only a few regional faults that guide the rivers, the fluvial erosion into the Alleghany Plateau forms a highly dendritic network of deeply incised valleys. The digital elevation model (Fig. 7.7a) shows an incision depth of up to 400 m into the plateau. It also reveals the gentle linear ridges that structure the plateau in the southwest-northeast direction. The surface gradient exaggerates the steep flanks of the incised valleys (b) and thereby highlights the valley network. To extract this dendritic network, we cut off the gradient data at a threshold of 5°. The resulting surface comprises only the valleys and their flanks and leaves the plateau empty (c). To enhance the impression of the valleys, we drape the satellite image that we used in Fig. 7.4b over this valley surface (c). The extracted

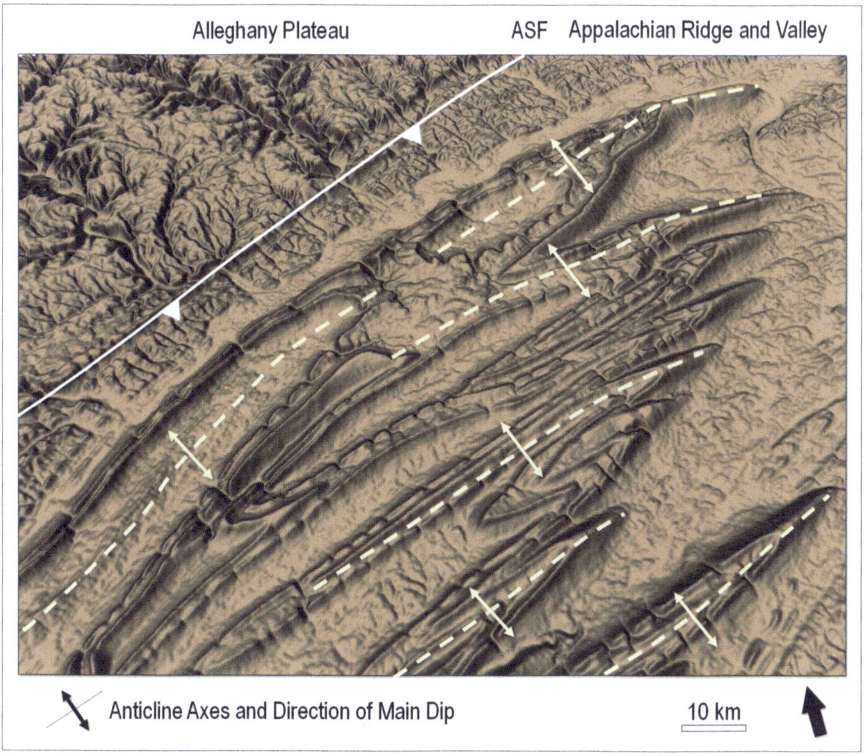

Fig. 7.5 Mapping of folds in the strike-slip tectonic zone of the Appalachian Ridge and Valley fold belt, gradient of SRTM DEM without vertical exaggeration. (DEM data courtesy of U.S. Geological Survey)

pattern of the dendritic valley network may be used as a geologic analog for dendritic river patterns that can be observed in the subsurface.

On the Alleghany Plateau we have observed both gentle folded ridges and incised valleys following major faults. We will now attempt to separate these two structural features. We return to the digital elevation model and render it with vertical exaggeration (Fig. 7.8a).

We adapt the topographic colormap such that it reveals the gentle ridges as dark brown features contrasting the lighter brown tones of the lower portions of the plateau. Spatial filtering a filter width of 50 km suppresses the signature of the dendritic valley network and extracts the long gentle ridges (b). To correlate the shape of the valley network with the texture of the digital elevation model, we merge the structural image we show in Fig. 7.7c with the digital elevation model (c). We interpret straight portions of the valley as generated by fluvial erosion along straight faults (white lineaments). The resulting fault network is perpendicular to the gentle ridges.

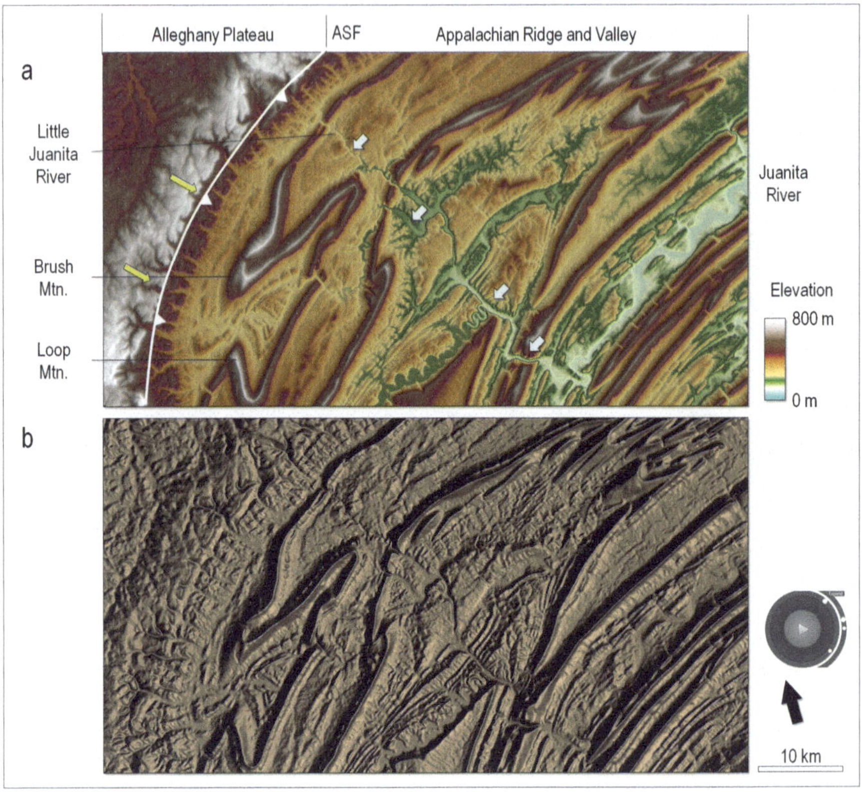

Fig. 7.6 River valleys incised along faults into the anticlines of the Appalachian Ridge and Valley. (**a**) SRTM DEM, (**b**) gradient of SRTM DEM at 5× vertical exaggeration. (DEM data courtesy of U.S. Geological Survey)

In the final step, we merge the satellite image with the digital elevation model and drape the resulting image on the digital elevation model (d). With this approach we obtain a 3D impression of the geomorphology. The gentle ridges are imprinted on the satellite image as darker patches, whereas the faults are enhanced by the shadows of the incised valleys where these face away from the slant illumination.

We conclude the structural analysis of folds and faults and its correlation with the landuse with a 3D rendering of the Landsat 8 753 RGB image that is draped on the digital elevation model (Fig. 7.9). Folded ridges, both steep ridges in the Appalachian Ridge and Valley province in the east as well as the gentle ridges of the Alleghany Plateau in the west appear as green patches representing the forest cover whereas valleys between the ridges are used as farmland that shows in purple colors. Normal faults are delineated by incised valleys from fluvial erosion along zones in which the rocks have been weakened by the faulting.

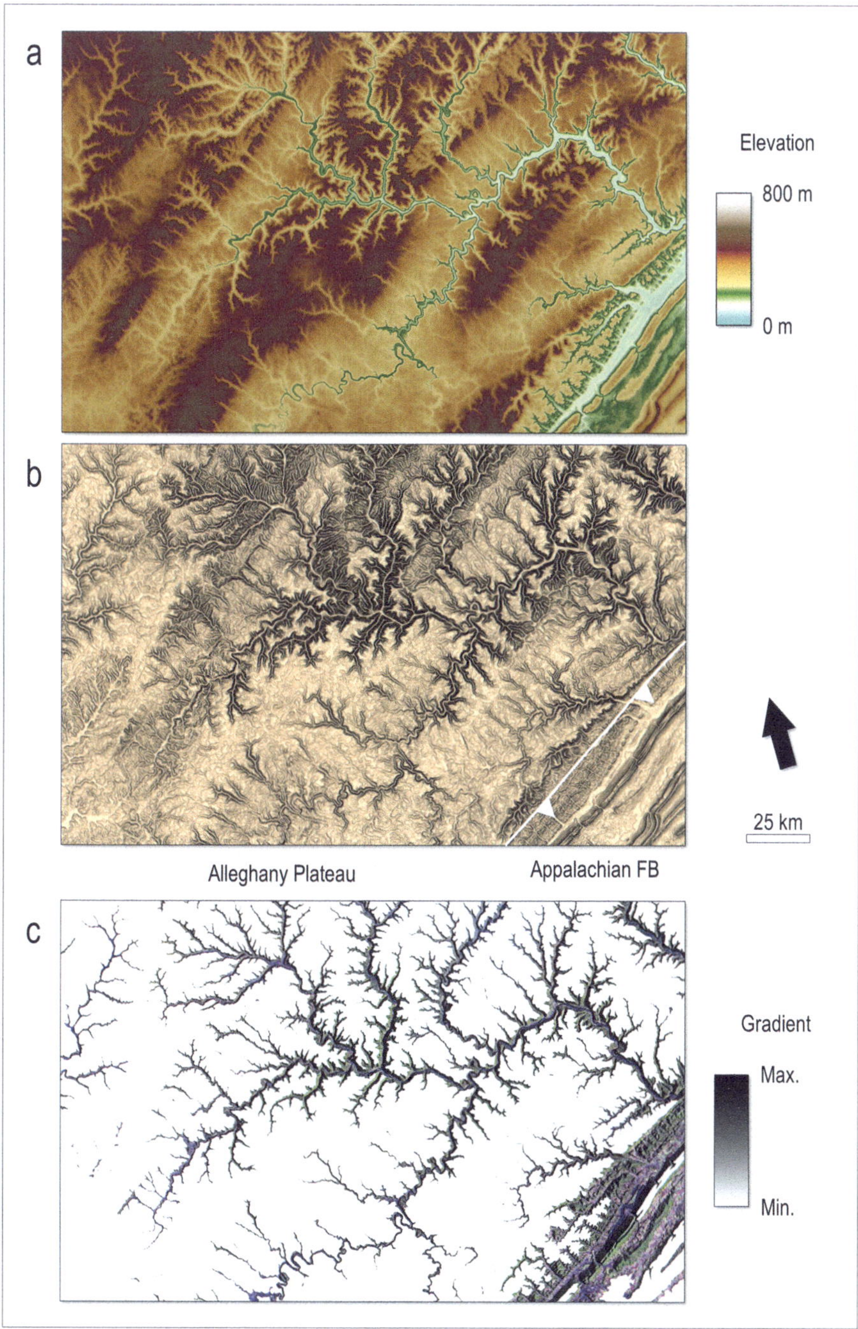

Fig. 7.7 Regional structural framework of the eastern Alleghany Plateau. (**a**) SRTM DEM, (**b**) SRTM DEM gradient, (**c**) extraction of incised river valleys from thresholding the DEM gradient, valleys are enhanced by Landsat 8 753 RGB composite. (DEM and satellite data courtesy of U.S. Geological Survey)

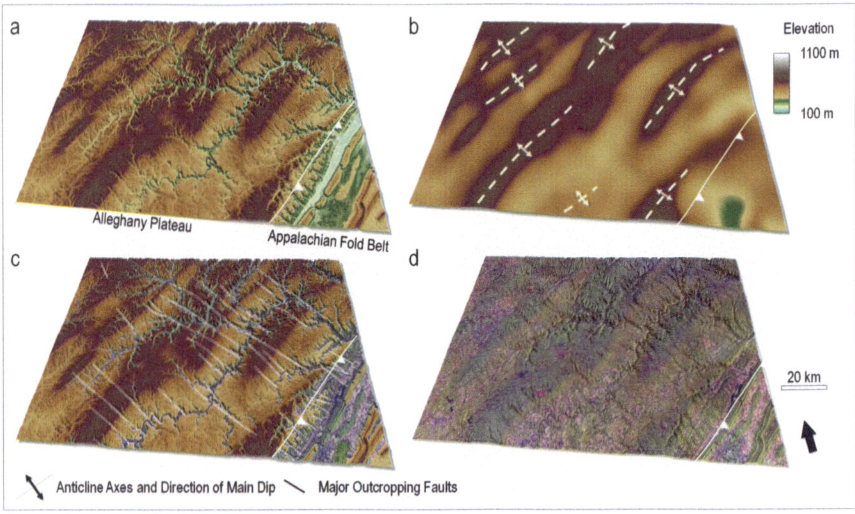

Fig. 7.8 Mapping of fold and fault contributions to the regional structural framework of the eastern Alleghany Plateau. (**a**) SRTM DEM, (**b**) 5 km low pass spatial filtering of SRTM DEM showing low anticlines, (**c**) regional fault framework from incised river valleys that are enhanced by Landsat 8 753 RGB composite, (**d**) Landsat 8 753 RGB composite merged with DEM and draped on SRTM DEM. All figures rendered at 5× vertical exaggeration with slant illumination. (DEM and satellite data courtesy of U.S. Geological Survey)

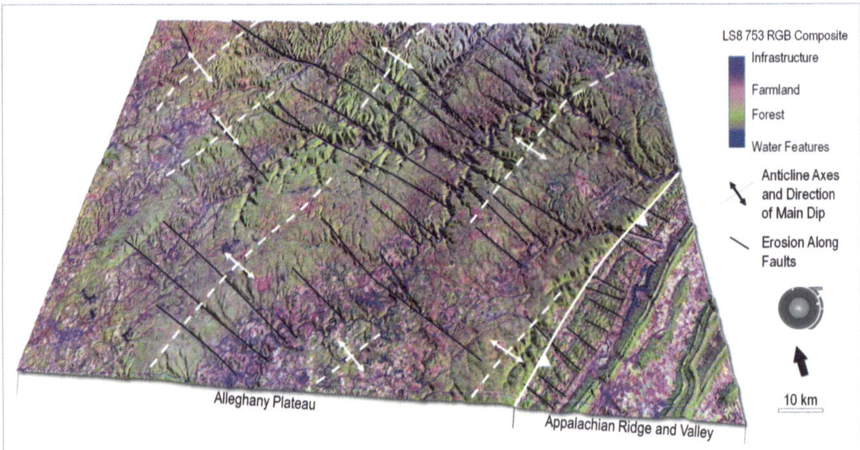

Fig. 7.9 Correlation of regional structural framework with vegetation, Landsat 8 753 RGB composite draped on SRTM DEM at 5× vertical exaggeration and slant illumination. (DEM and satellite data courtesy of U.S. Geological Survey)

7.3 Arctic Features in Northwest Siberia

The application of remote sensing data interpretation for depositional and erosional features is subdivided into two climatic zones. We start with the delineation of features in the arctic climate zone in northwest Siberia. The study area (Fig. 7.10) is dominated by historic and present glacial features. Except in the Ural Mountains hard rock outcrops are rare. Therefore, we will focus the remote sensing mapping on the analysis of topography and landuse, i.e. vegetation and its correlation with glacial geology.

The digital elevation model at high latitudes poses a challenge since the most frequently used satellites do not cover these latitudes. The most comprehensive digital elevation model is the Global Multi-Resolution Terrain Elevation Model Data 2010 GMTED (Danielson & Gesch, 2011). Where this DEM shows errors, we use the Arctic DEM (Porter et al., 2018) to replace erroneous data and fill no-data zone.

The history of the Quaternary glaciations in northwest Siberia comprises several stages that are described by Ehlers and Gibbard (2007, 2008), Ehlers et al. (2013a, b), and Jakobsson et al. (2014). Arctic features included in this study comprise moraines and permafrost. Water features are mapped to obtain information about the depth of the water table. An arctic delta is the focus of a study of the interaction

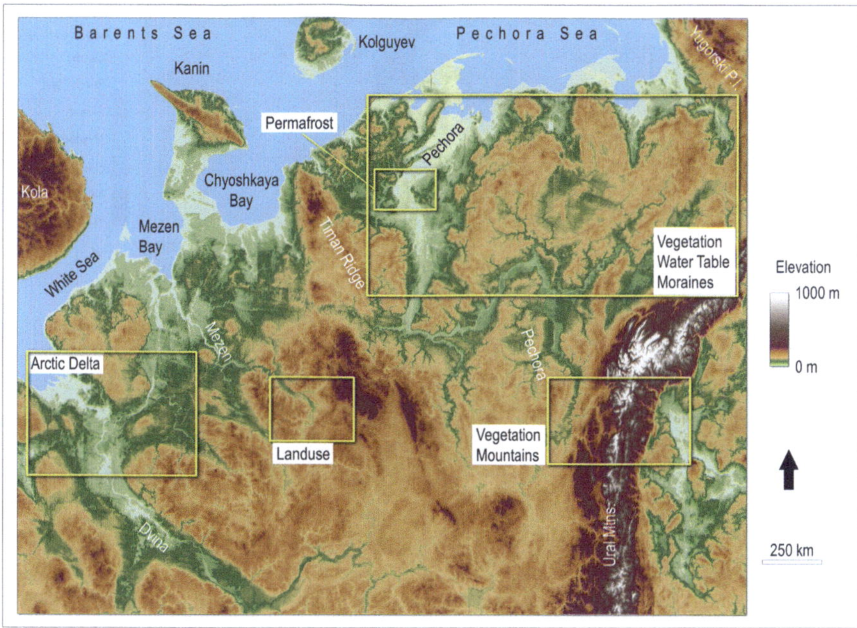

Fig. 7.10 Overview of northwest Siberia areas for the remote sensing studies of arctic features covered in this section. DEM merged from GMTED and Artic DEM. (Arctic DEM courtesy Polar Geospatial Center, GMTED DEM courtesy U.S. Geological Survey)

between river courses and glacial deposits. We conclude the mapping of arctic features through the mapping of vegetation in mountainous terrain and the interpretation of satellite imagery for human impact in arctic terrain.

The digital elevation model of the study area in northwestern Siberia (Fig. 7.11a) comprises the Ural Mountains in the east and the Arctic Ocean represented by the Barents Sea and the Pechora Sea in the north. The coastal lowlands are dominated by glacial deposits from the Quaternary ice sheets that deposited moraines. The main rivers are the Pechora in the northeast and Mezen and Dvina in the northwest. They can only be mapped from elevation through their respective valleys. Since water features play an important role in arctic remote sensing mapping, we use the band combination of Landsat 7 742 RGB (b). This RGB composite allows the mapping of water features (blue), forest (dark green), meadows (light green) and swamps (purple). At this scale, farmland and beach sands cover such small areas that they cannot be resolved. From the blue band of the Landsat 7 image we can extract water features through amplitude thresholding (c). To this map we have added the interpretation of ice margins after Svendsen et al. (2004) to correlate the distribution of lakes with the location of moraines. When we overlay the water features on the digital elevation model (d), we obtain a good basemap for the mapping of arctic features.

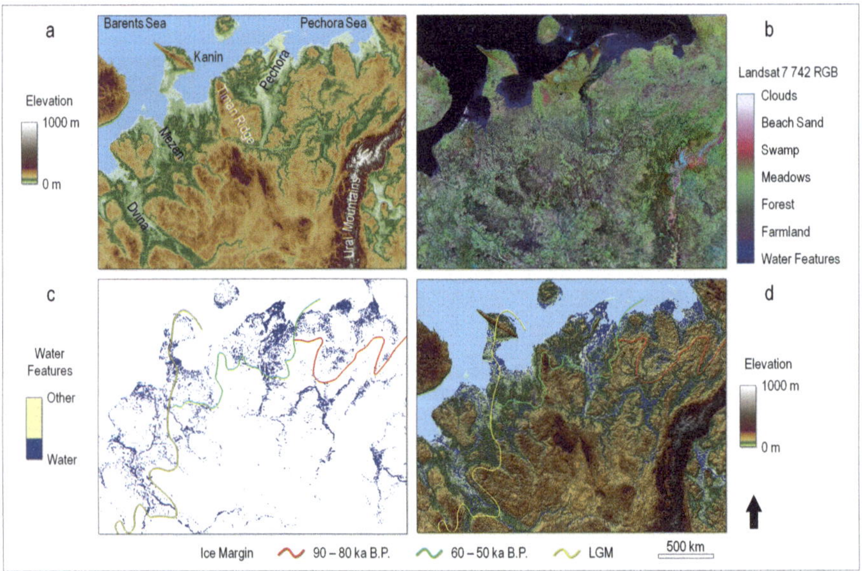

Fig. 7.11 Representation of arctic geologic features in remote sensing data. (**a**) DEM merged from GMTED and Artic DEM, (**b**) Landsat 7 742 RGB composite image, (**c**) water features extracted from Landsat 7 image, (**d**) merged DEM overlaid with water features. Outline of ice margins after Svendsen et al. (2004). (Arctic DEM courtesy Polar Geospatial Center, GMTED DEM and satellite imagery courtesy U.S. Geological Survey)

Before we continue with the geomorphologic analysis of the arctic study area, we need to have a closer look at the issues concerning digital elevation models at high latitude. The most frequently used global digital elevation model that is based on the Shuttle Tomographic Radar Model (SRTM) only covers latitudes from 56° south to 60° north, whereas the study area is located at 64°–69° north. The global ETOPO elevation model provides at best only 1 arcsec resolution, which corresponds to a spatial resolution of approximately 2 km. The GMTED elevation model offers at best a resolution of 250 m for the entire area. However, when zooming into the area of the Pechora River Delta (Fig. 7.12a) we notice in the area of the delta's coastline and the peninsula north an anomalous area of about 200 × 400 km which shows noisy values at too high elevations (yellow arrow).

To correct for such anomalies, the ArcticDEM (Porter et al., 2018, Fig. 7.12b) has been developed for all land area north of 60° north through a collaboration between the National Geospatial-Intelligence Agency (NGA) and the National Science Foundation (NSF) in the USA.

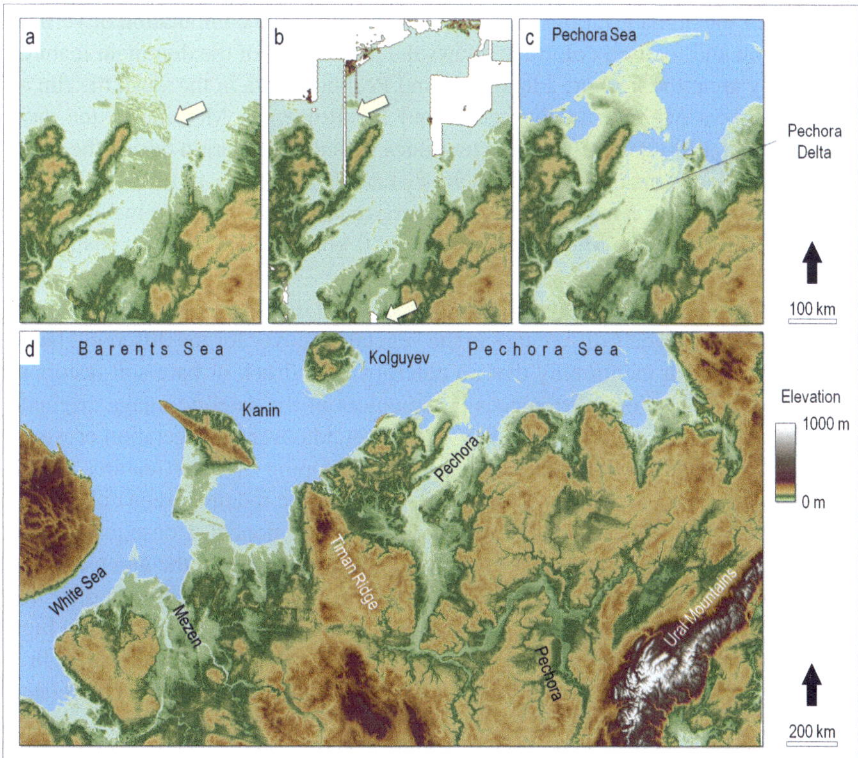

Fig. 7.12 Creation of clean arctic DEM for the study area in northwest Siberia. (**a**) GMTED, (**b**) Arctic DEM, (**c**) merged DEM from GMTED and Arctic DEM, (**d**) clean merged DEM for the entire study area. (Arctic DEM courtesy Polar Geospatial Center, GMTED DEM and satellite imagery courtesy U.S. Geological Survey)

The ArcticDEM project is hosted by the University of Minnesota, St. Paul. The basis of this DEM is stereo imagery obtained by Digital Globe's Worldview satellites which is processed into a 2 × 2 m posting elevation model using a petascale supercomputer. The project is set up as a collection of DEM strips that are updated as data become available. As a result, the DEM showed no-data areas at the time of access end of 2020. Arrows point to some of these areas.

Consequently, we merged the two DEM data sets and created a seamless digital elevation model for the entire study area (Fig. 7.12c). This DEM shows the lower Pechora River floodplain and delta as well as the peninsula north of the delta at correct elevations. We are using this merged DEM for the analysis in this section (Fig. 7.12d).

We start with the challenge of delineating glacial features in the Pechora Lowlands between the Pechora Delta and the Ural Mountains (Fig. 7.13). The study area has relatively low topography reaching only about 150 m elevation at the highest points of the glacial moraine ridges (a). Therefore, we use 50× vertical exaggeration and slant illumination in all rendering in this figure. In addition to this we scale the elevation colormap with a focus on the interval of 0–150 m thus reducing the rendering of the high elevations in the Ural Mountains. The combination of vertical exaggeration and focused colormap allows the delineation of the dominant features in the study area: the Pechora Floodplain and Pechora Delta in the west, the rim of the Laya-Adzva Moraine in the center and the Rogovaya Moraine in the east. Svendsen et al. (2004) correlate these moraines with the ice margin during the Early Weichselian glacial maximum at about 90–80 ka B.P. The interior of the 40 km wide Laya-Adzva Moraine is markedly flat compared to the undulating terrain outside the moraine. The Rogovaya Moraine is less than 10 km wide and has the shape of a narrow glacier terminus.

The landuse interpretation of the Landsat 7 742 RGB composite draped over the merged DEM (b) reveals differences in the vegetation cover for the different glacial features. It is worth mentioning that no outcrops of bedrock or bare soil occurs in the study area. Therefore, differences in the colors of the satellite image originate from changes in the spectral reflectivity of the ground cover by vegetation or water. We start the ground cover classification with the mapping of water features. These are characterized by overall low reflectivity values in the red and green color components and low to intermediate values in the blue component. Consequently, the RGB image shows them as blue. Forest generates high values in the green component and thus correspond to dark green for pine forest and light green colors for Boreal forest of sparse tree cover and meadows. Swamps show a moderately high reflectivity in the red band and are, therefore, represented by reddish tones. In a small coastal area in the far northwest we notice a cyan colored area that is related to high intensity reflections from clouds.

After extracting the landuse classes we can render them separately as solid colors (c). We note that pine forest grows particularly on the terminal moraine ridges of the Laya-Adzva Moraine. The reason may be the high sand contents of these moraine ridges, a favorable soil for pine trees. The area enclosed by the moraine ridge is

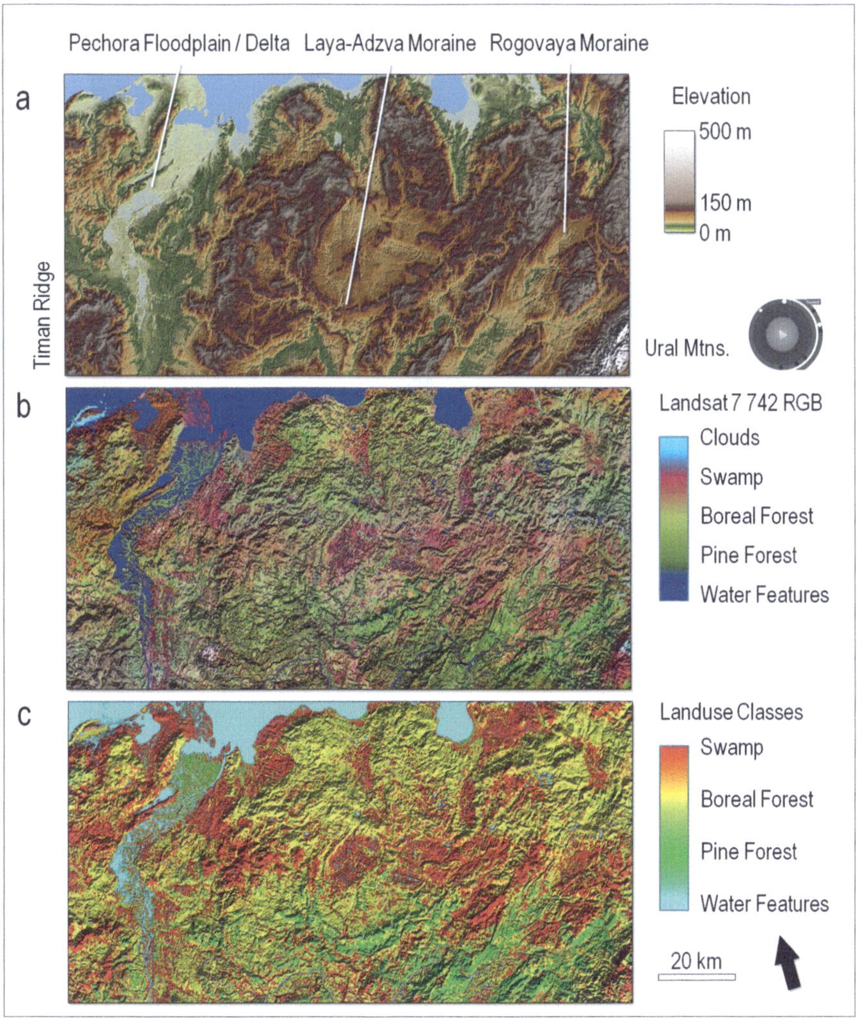

Fig. 7.13 Supervised landuse classification for northwest Siberia. (**a**) Merged DEM, (**b**) Landsat 7 742 RGB draped on merged DEM, (**c**) supervised landuse classification from Landsat 7 imagery. Outline of ice margins after Svendsen et al. (2004), 50× vertical exaggeration and slant illumination. (Arctic DEM courtesy Polar Geospatial Center, GMTED DEM and satellite imagery courtesy U.S. Geological Survey)

dominated by swamps, which is the result of the flat topography that prevents the drainage of the water.

To study this correlation further we attempt to generate a map showing the estimated depth of the water table below the surface (Fig. 7.14). For the generation of this map, we use the elevation levels of the water features and create a surface through interpolation (Laake, 2009). From the Landsat 7 742 RGB image (a) we

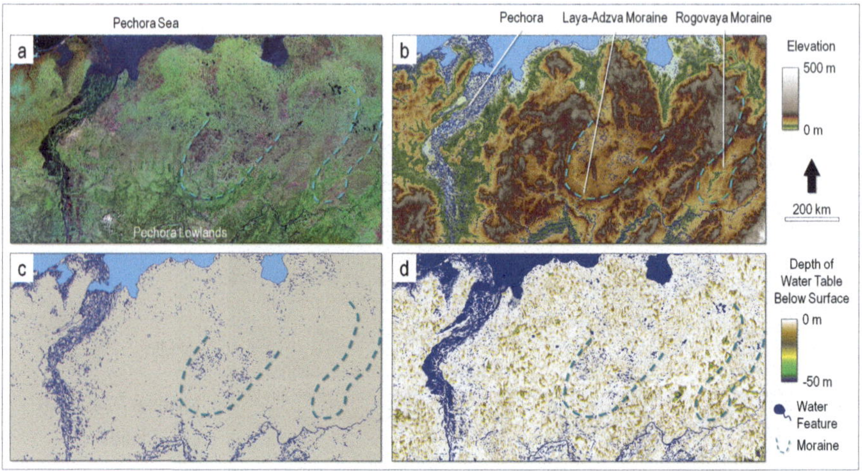

Fig. 7.14 Extraction of water features from Landsat 7 imagery and estimate of depth of water table. Glacial interpretation after Svendsen et al. (2004). (**a**) Landsat 7 742 RGB composite, (**b**) merged DEM, (**c**) water features, (**d**) estimated depth of water table below surface. Outline of ice margins after Svendsen et al. (2004). (Arctic DEM courtesy Polar Geospatial Center, GMTED DEM and satellite imagery courtesy U.S. Geological Survey)

extract the water features, i.e. the rivers and lakes (c) and generate the estimate water table level through interpolation. Then we subtract the water table level from the merged surface DEM (b) and obtain the depth of the water table below the surface. We render this map with a color scale for the depth below surface with water features showing in blue.

When we drape this map on the digital elevation model and render it at vertical exaggeration, we observe that the interior of the Laya-Adzva Moraine shows a water table at or immediately below the surface, which matches the landuse interpretation of the Landsat 7 742 RGB as swamp or water features. South of the moraines, only relatively few lakes indicate a deeper water table. The drainage pattern is oriented towards the Pechora River, the course of which follows the Early Weichselian ice margin. It broadens to form an up to 100 km wide floodplain where it enters the Middle Weichselian ice margin which Svendsen et al. (2004) date to approximately 60–50 ka B.P. The present topography of the area suggests that the Pechora Floodplain is located at the base of a glacial terminus.

The area north of the Early Weichselian ice margin has low relief and is covered densely with lakes. Glacial ground moraines left large areas of glacial till that contains a high percentage of fine silts that form sealing lake beds. Consequently, the water table in the areas of ground moraines is often at the surface or at very shallow depth.

We will focus now on the mapping of the moraine ridges (Fig. 7.15). We use the merged DEM (a) that we had already studied in Fig. 7.14a and render the surface gradient (b). Using a colormap that enhances steep edges we can highlight the slopes of the moraines. We discover that steep slopes often coincide with the 100 m

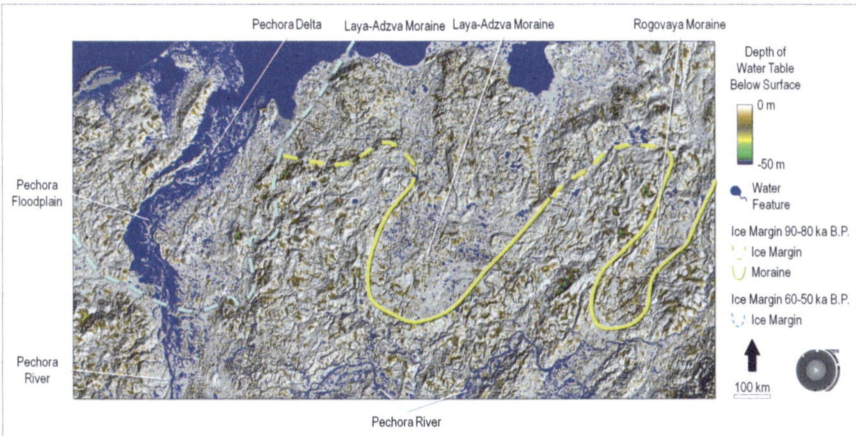

Fig. 7.15 Mapping of moraines from interpretation of glacial geomorphology, 50× vertical exaggeration and slant illumination. (Arctic DEM courtesy Polar Geospatial Center, GMTED DEM and satellite imagery courtesy U.S. Geological Survey)

elevation contour, which we consider in the design of the colormap for the elevation rendering in Fig. 7.16c. Using this approach, we can isolate the Weichselian ground moraines from Weichselian terminal and lateral moraines as well as Pre-Weichselian moraines. Adding the water features that had been extracted in the previous step supports the interpretation that most of the water features are located in terrain covered in Weichselian ground moraines.

Finally, we render the interpretation in 3D (Fig. 7.17). We drape the moraine map on the digital elevation model using vertical exaggeration. With this approach we show both, the texture of the elevated moraines in shades of brown and the texture of the ground moraines through the shading of the merged DEM with the water features nicely contrasting with the light shades of the ground moraines. To support the interpretation, we plot the Early and Middle Weichselian ice margin outlines after Svendsen et al. (2004).

The merged digital elevation model (a) shows the Pechora Floodplain bordered by the higher glacial moraine deposits of the banks. On the eastern bank of the floodplain we observe two linear areas of anomalies (arrows). Landsat 7 742 RGB. composite image (b) reveals that these features are concatenated lakes that can be attributed to degrading permafrost (Westermann et al., 2015). The floodplain itself shows the braided channel system of the Pechora River.

The mapping of permafrost features poses a challenge because permafrost has no signature in the digital elevation model or in the surface reflectance since it is a subsurface phenomenon. Indirect methods such as remote measurement of surface subsidence and surface temperature have been successfully used by Jorgenson and Grosse (2016).

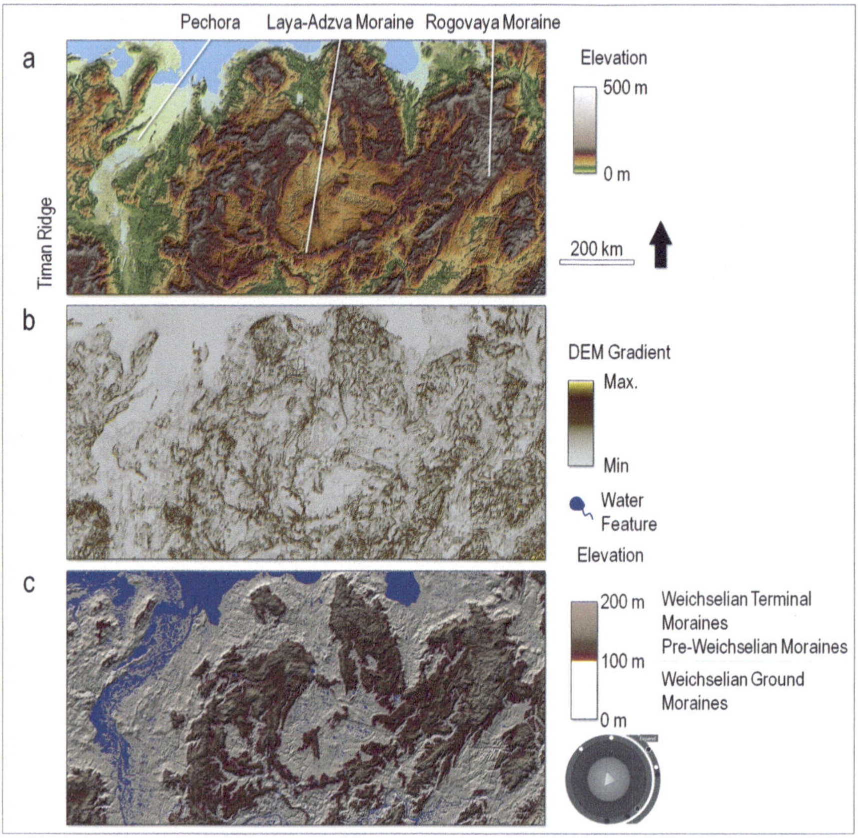

Fig. 7.16 Water features and estimated depth of water table below surface extracted from Landsat 7 imagery. (**a**) digital elevation model, (**b**) gradient of digital elevation model, (**c**) interpretation draped on merged DEM, 50× vertical exaggeration and slant illumination. Outline of ice margins after Svendsen et al. (2004). (Arctic DEM courtesy Polar Geospatial Center, GMTED DEM and satellite imagery courtesy U.S. Geological Survey)

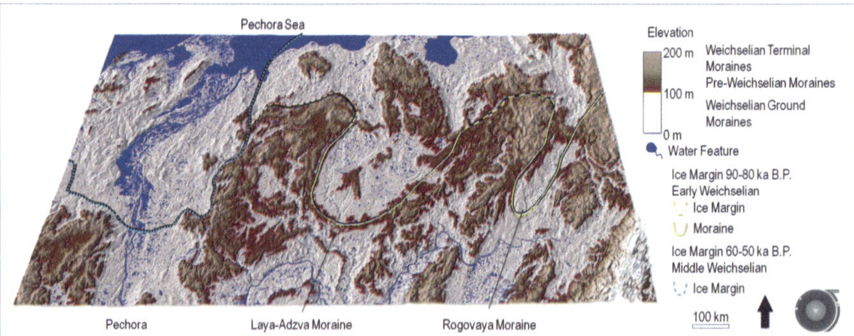

Fig. 7.17 Interpretation of glacial features from DEM and multispectral satellite imagery. Outline of ice margins after Svendsen et al. (2004). (Arctic DEM courtesy Polar Geospatial Center, GMTED DEM and satellite imagery courtesy U.S. Geological Survey)

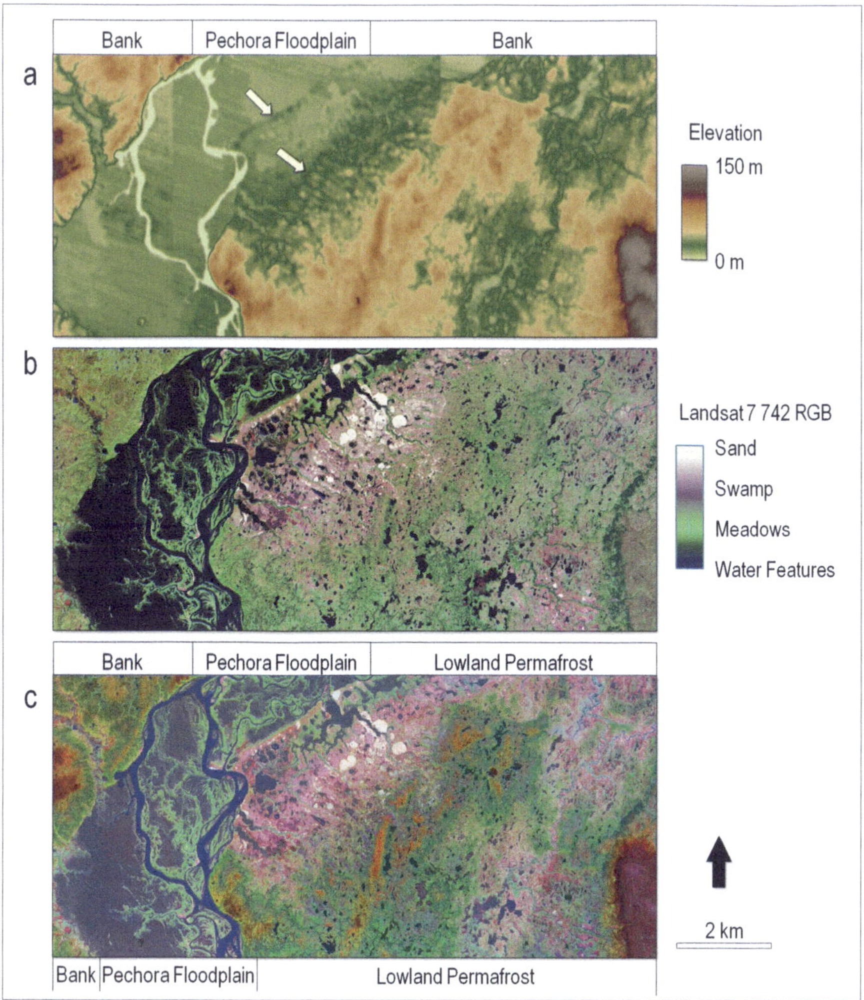

Fig. 7.18 Delineation of permafrost features in the lower Pechora River floodplain. (**a**) Merged DEM, (**b**) Landsat 7 742 RGB composite, (**c**) overlay of satellite image on DEM. (Arctic DEM courtesy Polar Geospatial Center, GMTED DEM and satellite imagery courtesy U.S. Geological Survey)

An indirect approach uses landforms that are indicative of permafrost and permafrost change (Westermann et al., 2015), an approach we will apply to the delineation of lowland permafrost near the Pechora Floodplain (Fig. 7.18). In the floodplain, these landforms are directly correlated with abandoned river channels and oxbows, whereas the lakes on the banks are rather rounded and do not show any correlation with flowing water. The landuse interpretation of the satellite image shows that the western bank of the floodplain is covered in meadows and few lakes, whereas the eastern bank shows swamps and sand patches. Most of these sand patches have the

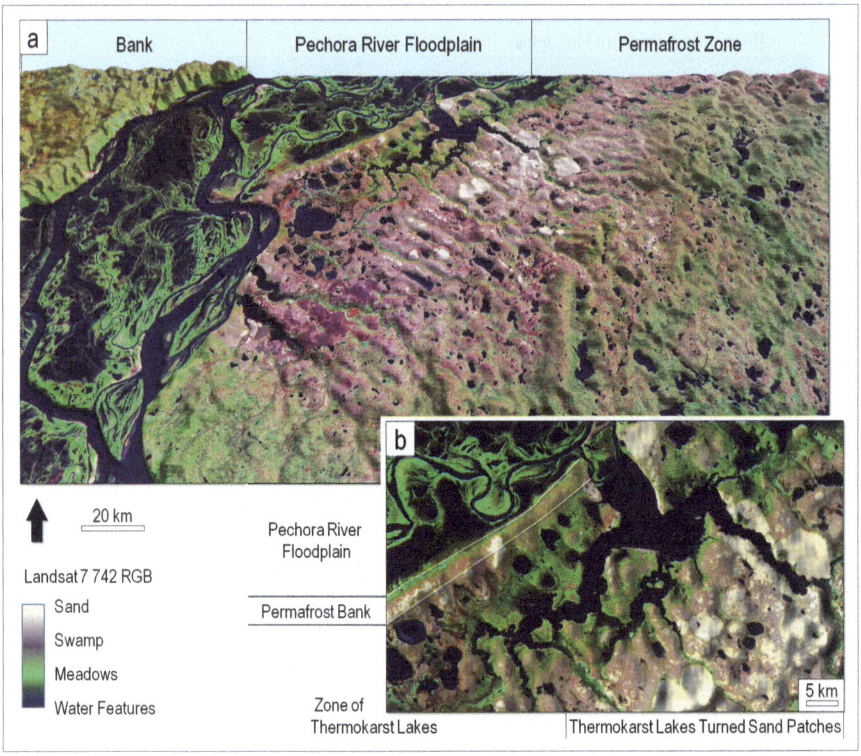

Fig. 7.19 Mapping of permafrost features in the lower Pechora River floodplain. Landsat 7 742 RGB image rendered in 3D on merged DEM at 100× vertical exaggeration. (**a**) Overview of Pechora floodplain, (**b**) zoom into permafrost features on the banks of the Pechora River floodplain. (Arctic DEM courtesy Polar Geospatial Center, GMTED DEM and satellite imagery courtesy U.S. Geological Survey)

same shape as the lakes. One possible interpretation is that lakes in the permafrost area are very shallow and their banks are formed by permafrost. Rising temperatures destabilize the banks and make them susceptible to mechanical erosion, which if it occurs leads to the drainage of the lake, a process that is called thermokarsting (Westermann et al., 2015). The formerly frozen sandy lake bed remains exposed at the surface.

When we overlay the satellite image on the DEM (c) we observe that the degraded permafrost area is located at the lower portion of the bank at only a few meters above the floodplain level. The 40–50 m high ridge to the east of this area as well as the lowland to the east of the ridge do not show any sign of thermokarsting.

Draping the overlay of DEM and satellite image on the merged DEM and rendering it in 3D at 100× vertical exaggeration eases the interpretation of permafrost features (Fig. 7.19a). A shallow ridge separates the lakes from the Pechora River Floodplain. Most likely this ridge consists of frozen ground that provides the mechanical protection against erosion. However, permafrost degradation as a consequence of rising temperatures results in thawing of the permafrost thus removing the mechanical protection against erosion. This inset (b) shows a location where the permafrost bank has thawed, and the river has breached the bank thus creating a lake in its place. Several of the small round thermokarst lakes have already started to merge as a result of deteriorating permafrost thus forming a river-like lake behind the permafrost bank. We will revisit thermokarsting in the context of logistics and infrastructure planning where thermokarst presents a severe hazard.

In the next example we study the Dvina Delta into the White Sea and aim at understanding the interaction of the delta formation and glacial deposits (Fig. 7.20). The Landsat 7 742 RGB composite shows the Dvina River in dark blue forming a broad floodplain, in which the city of Archangelsk is located. Forested ridges indicate the presence of glacial moraines following the interpretation in Fig. 7.13. The merged DEM (b) confirms this observation. The Kuloi Plateau to the north and a corresponding plateau in the south indicate the margins of a glacial terminus. This is supported further by a moraine lobe in the southeast.

When we overlay the satellite image on the DEM and create the geomorphological map (c) we can interpret the forest and water features and their correlation with glacial processes. A glacier coming from the White Sea carved out the lower Dvina valley and deposited the terminal moraine that is indicated by the yellow arrows. The ground moraine left a shallow depression which the Dvina River filled with its floodplain and delta. The flanks of the two plateaus can be expected to show lateral moraines which may contain a significant portion of glacial till, that may contribute to landslide hazards.

Human impact can be studied best using the Landsat 7 742 RGB composite (Fig. 7.21b, d). Human activity is typically related to some form of clearing vegetation thus exposing bare soil, that is detected as purple areas in the satellite image. To understand the type of human activity, we study the merged DEM in the upper Mezen River valley (a). The river has carved its valley into the ground moraine of the glacial terminus. The terminal moraine forms a barrier across the valley forcing the river to flow around the moraine. We zoom into the area where human activity has left a rectangular pattern of forest clearings. The DEM (a) shows that the area is located on the terminal moraine. The satellite image (d) shows rectangular patterns that represent forest clearings and irregular patches of darker purple color that represent farmland. When we create the geomorphological image from the overlay of the satellite image on the DEM we notice that the forest clearings are located on the moraine ridge. This suggests that the rectangular patterns are used for managed pine forest, whereas the farmland is located in the glacial tills of the ground moraine.

Using this information. we can now generate a landuse classification into water features, forest, meadows, farmland and managed forest, i.e. forest clearings (Fig. 7.22a). To obtain a clearer interpretation of the human activity in relation to the glacial features, we drape the satellite image on the DEM and overlay only the landuse classes of farmland and managed forest (orange to purple tones). Thereby we retain as much of the visual impression of the satellite image whilst highlighting the areas of human activity. Slant illumination enhances the 3D impression of the rendering.

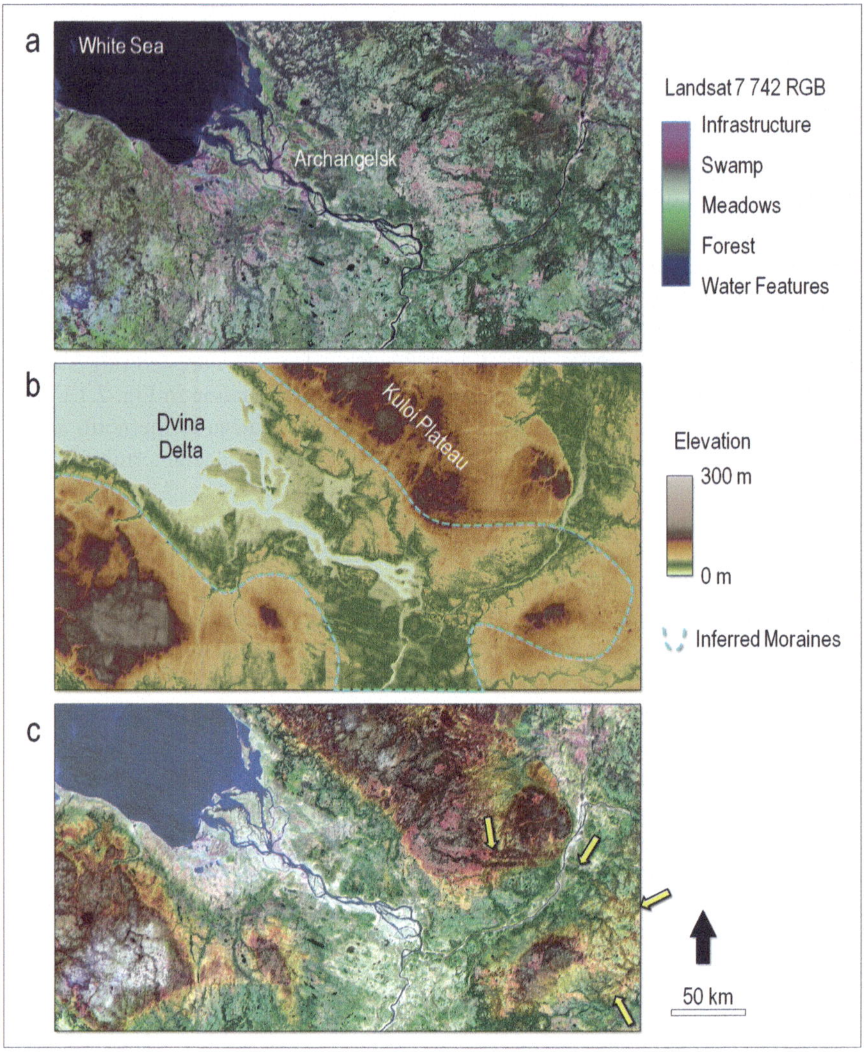

Fig. 7.20 Remote sensing analysis of the Dvina Delta considering glacial processes. (**a**) Landsat 7 742 RGB composite, (**b**) glacial interpretation of merged DEM, (**c**) overlay of satellite image on merged DEM at 20× vertical exaggeration. (Arctic DEM courtesy Polar Geospatial Center, GMTED DEM and satellite imagery courtesy U.S. Geological Survey)

Fig. 7.22 Correlation of managed forest with geomorphology. (**a**) Supervised landuse classification from Landsat 7 742 RGB composite image, (**b**) draping of Landsat 7 742 RGB image on merged DEM at 10× vertical exaggeration and slant illumination. (Arctic DEM courtesy Polar Geospatial Center, GMTED DEM and satellite imagery courtesy U.S. Geological Survey)

Fig. 7.21 Study of human impact on arctic landscape. (**a**) Merged DEM at 50× vertical exaggeration, (**b**) Landsat 7 742 RGB composite, (**c**) merged DEM for study area, (**d**) Landsat 7 742 RGB composite highlighting managed forest areas, (**e**) overlay of satellite image on DEM. (Arctic DEM courtesy Polar Geospatial Center, GMTED DEM and satellite imagery courtesy U.S. Geological Survey)

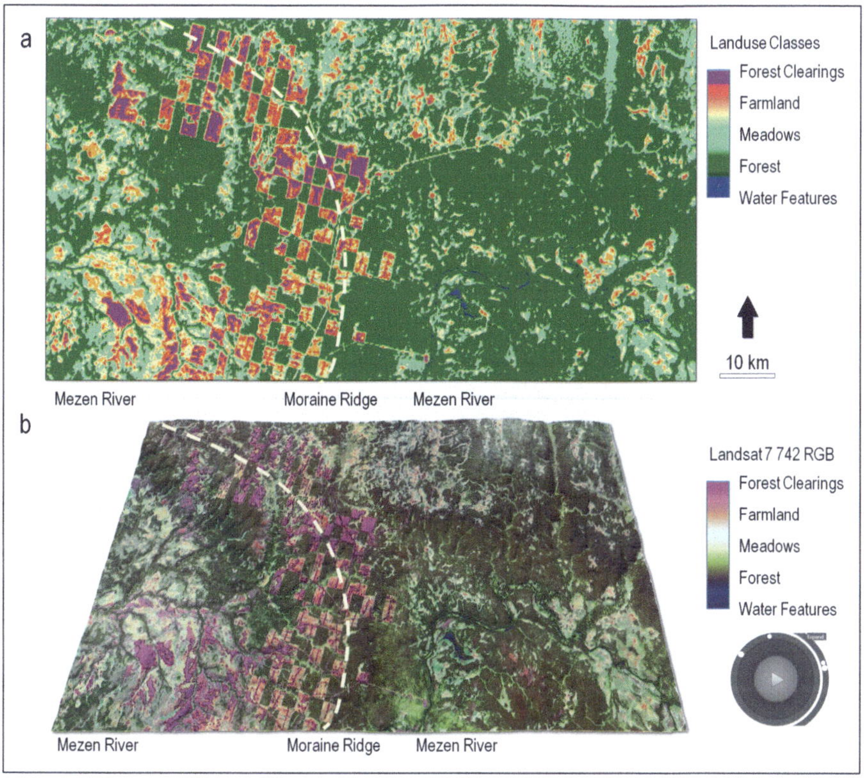

Fig. 7.22 (Refer caption on p. 106)

We conclude this section with a brief landuse study in the Ural Mountains (Fig. 7.23). The digital elevation model (a) shows the mountain range in the center with the Pechora River valley in the west and the Ob Floodplain in the east.

The folded ridges of the Ural Mountains exceed elevations of 1000 m, which is significantly higher than elevation in the glacially dominated lowlands to which the

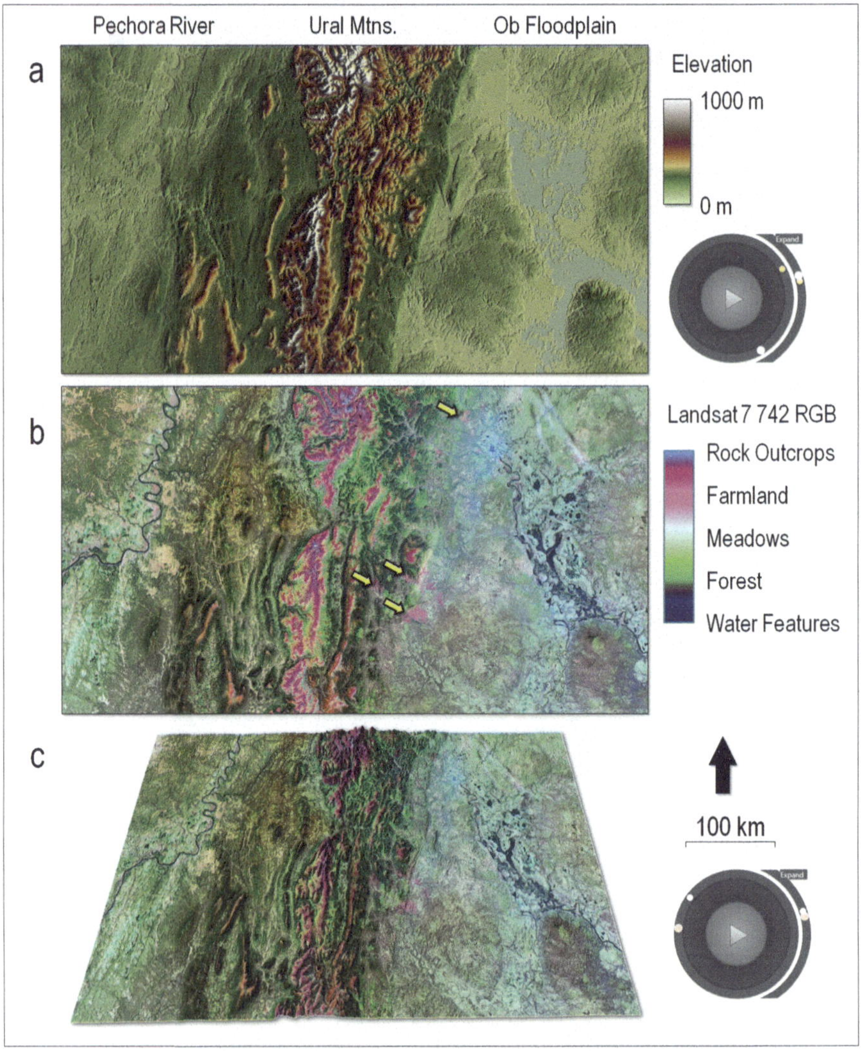

Fig. 7.23 Discrimination of rock outcrops from farmland in the northern Ural Mountains, Siberia. (**a**) Merged DEM at 25× vertical exaggeration, (b) Landsat 7 742 RGB composite, (**c**) 3D rendering of satellite imagery on merged DEMat 10× vertical exaggeration, slant illumination used for (**a**, **c**). (Arctic DEM courtesy Polar Geospatial Center, GMTED DEM and satellite imagery courtesy U.S. Geological Survey)

Pechora River valley and the Ob Floodplain belong. As said before, the Ural Mountains contain the only bed rock outcrops. On the Landsat 7 742 RGB composite (b), rock outcrops show as dark purple to gray colors. The difference in color-shade between farmland (yellow arrows) and rock outcrops is minimal. Draping the satellite image on to the DEM and rendering it in 3D helps to distinguish rock outcrops from farmland (c). The lowlands without rock outcrops are shown in greenish and brownish shades. The mountain ridges are detectable through the topography and the cover in forest. Only purple shades above the tree line should be interpreted as rock outcrops, whereas purple patches in the lowlands and in areas with minor ridge texture are most likely farmland.

7.4 Arid Desert Features on the Colorado Plateau, U.S.A.

The Colorado Plateau in the southwest USA is presently located in a hot arid climate. Consequently, few of the geologic features are obscured by vegetation and are hence exposed to remote sensing mapping. Incised into the plateau, the Grand Canyon provides a variety of geologic features which we will study for the representation in remote sensing data.

The first geological exploration of the Grand Canyon was conducted by J.W. Powell in 1869 (Rabbitt et al., 1969). A short summary of the geology has been prepared by Karlstrom et al. (2012), the structural geology has been summarized by Gettings and Bultman (2005). Physiographic (Billingsley & Hampton, 1999) and geological maps (Billingsley, 2000) and a geologic field photograph map (Billingsley et al., 2019) are available through the U.S. Geological Survey.

In Fig. 7.24 we provide satellite overview map of study area in the eastern Grand Canyon. The Landsat 8 753 RGB composite shows the Colorado Plateau consisting of the arid Coconino Plateau in mauve tones, the forested Kaibab Plateau (green) and the dune covered Navajo Plateau (yellow). Into the plateau, the Colorado has incised the Grand Canyon, a wide system of erosional canyons that cut deep into the rock layers in the subsurface. The purple tones inside the canyon indicate that the composition of the rock within the canyon differs from the rocks exposed on the plateau. In the east, the Little Colorado River Canyon shows very distinct linear erosional features, whereas Black Knob in the far southeast exposes volcanics.

The Landsat 8 753 RGB composite, a combination of a short-wave infrared, very near infrared and visible red spectral bands provides a good contrast of vegetation, sand, carbonates and sandstone. However, the different formations in the Grand Canyon appear quite undifferentiated. Next we include a thermal infrared band and generate the 10-7-4 RGB composite of a thermal, short-wave and near infrared spectral bands (Fig. 7.25).

With this band combination we obtain more contrast of the different formations in the Grand Canyon as well as details of the texture of the old alluvial fan deposits and on the Pennsylvanian rocks on the eastern Coconino Plateau. We trade this

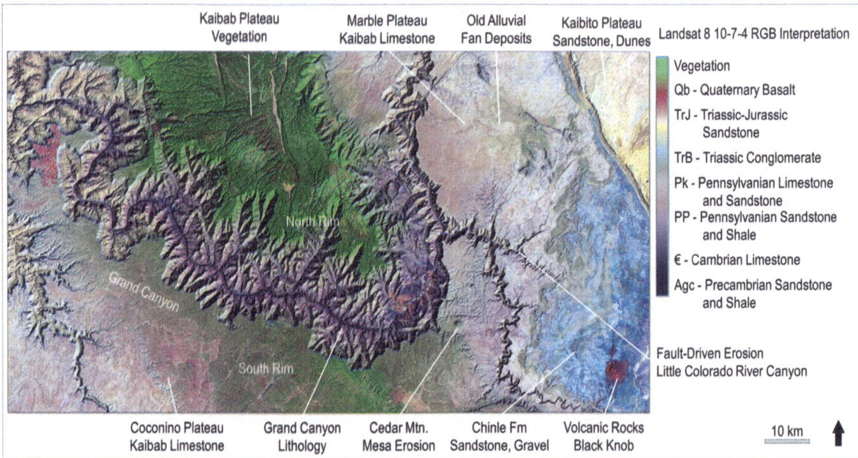

Fig. 7.24 Overview of eastern Grand Canyon and adjacent areas, Landsat 8 753 RGB composite. (Satellite data courtesy of U.S. Geological Survey)

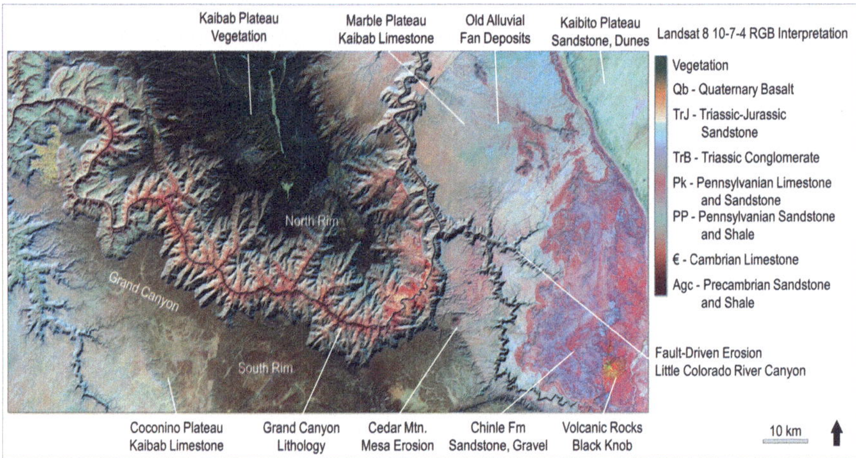

Fig. 7.25 Lithology mapping of Grand Canyon and adjacent areas, Landsat 8 10-7-4 RGB composite. (Satellite data courtesy of U.S. Geological Survey)

improvement with some degradation related to the mapping of vegetation: by replacing band 5 with band 4, we loose the strong contrast of the vegetation on the Kaibab Plateau. Including the thermal infrared band also masks some lithology mapping capability by thermal rock properties. This is particularly true for the Triassic conglomerate on the southeastern Coconino Plateau. The 753 RGB shows a strong contrast of the conglomerate signature versus the volcanic rocks of Black

Knob. The 10-7-4 RGB image reduces this contrast because the heat weathering of the conglomerates has created a surficial darkening called desert varnish that has a strong thermal infrared signature. We suggest that for a comprehensive interpretation of the satellite imagery for landuse and lithology both band combinations should be used.

We return to the 753 RGB composite and focus on the lithology mapping in the eastern portion of the study area (Fig. 7.26). The small inset map provides the overview of the three local case studies. For the first case we go close to the edge of the Kaibito Plateau (a). Sand dunes and sand sheets (yellow), the erosion products of the Navajo Sandstone, cover the plateau. A regional normal fault causes the drop to the Marble Plateau thus forming the Echo Cliffs that expose the Navajo Sandstone.

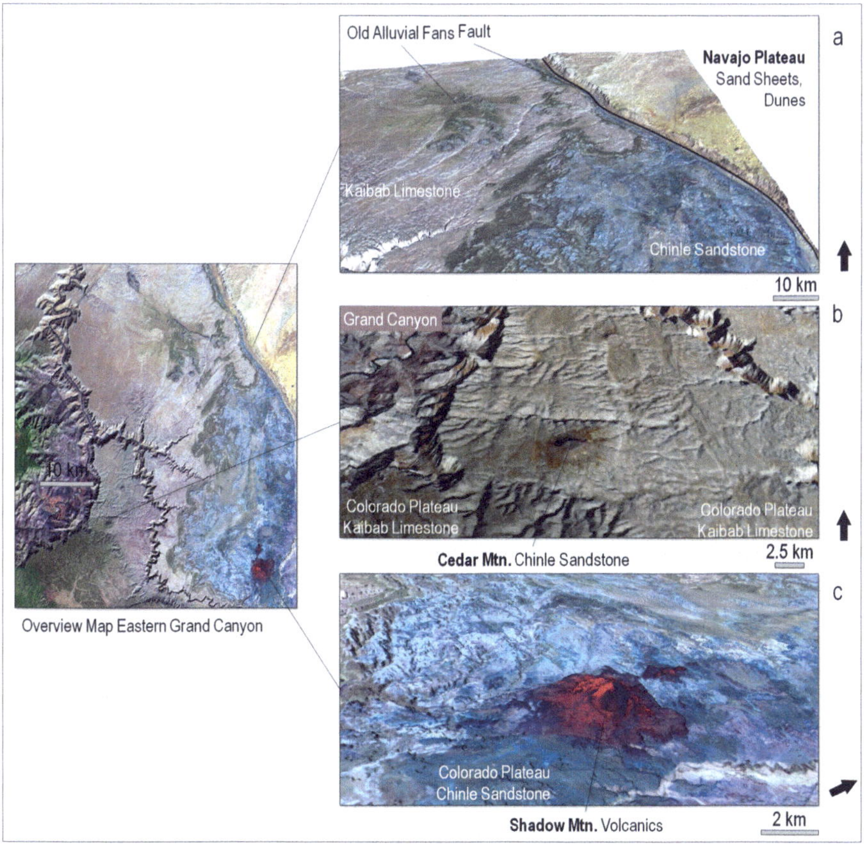

Fig. 7.26 Lithology mapping on plateaus adjacent to Grand Canyon. (**a**) Navajo sandstone plateau with sand sheets LS8 753 RGB composite, (**b**) Cedar Mountain mesa on Colorado Plateau LS8 432 RGB composite, (**c**) Shadow Mountain volcano on Colorado Plateau LS8 753 RGB composite. All satellite images are draped on the digital elevation model, 2× vertically exaggerated. Slant illumination applied. (SRTM DEM V3 and Landsat 8 imagery courtesy of U.S. Geological Survey)

The Marble Plateau shows two different surface rock types, the Kaibab Limestone in the north (mauve tones) and the overlying Chinle Sandstone in the south (bluish tones). We can also distinguish some old alluvial fan deposits through their green-brownish color.

The second case focuses on Cedar Mountain, an inselberg in the form of a flat mesa (b). The Chinle Sandstone (dark brown) that covers the top of the mesa is the remnant of a large flat plateau that has been eroded most likely through combined fluvial and eolian erosion, the products of which can be seen surrounding the mesa.

The third case shows a very different lithology. The volcano that created Black Knob (c) and the lava flows surrounding it is evidence of significant Tertiary volcanic activity that had its center in the San Francisco Mountains near Flagstaff. The dark basaltic rocks show their main radiation in the thermal and short-wave infrared spectral bands which makes Black Knob appear red in the 742 RGB composite.

In addition to the lithology and landuse analysis we can also study how remote sensing data can assist with the delineation of the outcropping structural framework (Fig. 7.27). We use the digital elevation model (a) and its gradient (b) to assess the correlation of the morphology and will try to infer faults that are affecting the direction of the erosion.

The Grand Canyon erosional canyon network appears chaotic at first glance but setting the elevation colormap for the DEM such that the elevation of the smooth plateaus shows as a brown to white surface, then the dendritic configuration of the canyons becomes visible. The gradient even enhances the dendritic character of the canyon network. On the DEM we observe for some canyons an almost straight main

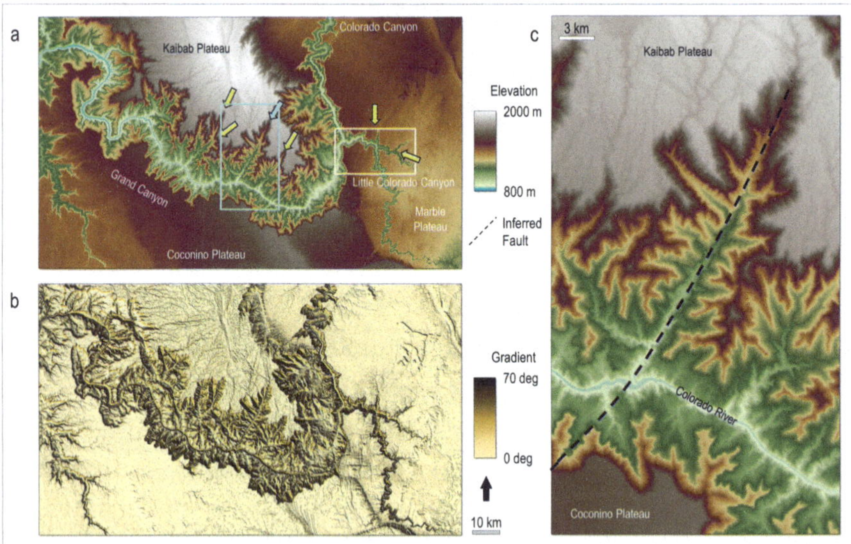

Fig. 7.27 Digital elevation model and structural framework of eastern Grand Canyon area. (**a**) Digital elevation model, (**b**) gradient of digital elevation model, (**c**) zoom into structural mapping from Kaiba to Coconino Plateau. (SRTM DEM V3 data courtesy of U.S. Geological Survey)

axis (cyan box in a and c) that we interpret as a weak zone along a fault (dashed line) that had weakened the rock and eased erosion.

We can study this approach at the canyon of the Little Colorado River (Fig. 7.27a yellow box and Fig. 7.28). The Little Colorado River approaches the Grand Canyon from the east and eroded a separate canyon into the limestone of the Marble Plateau.

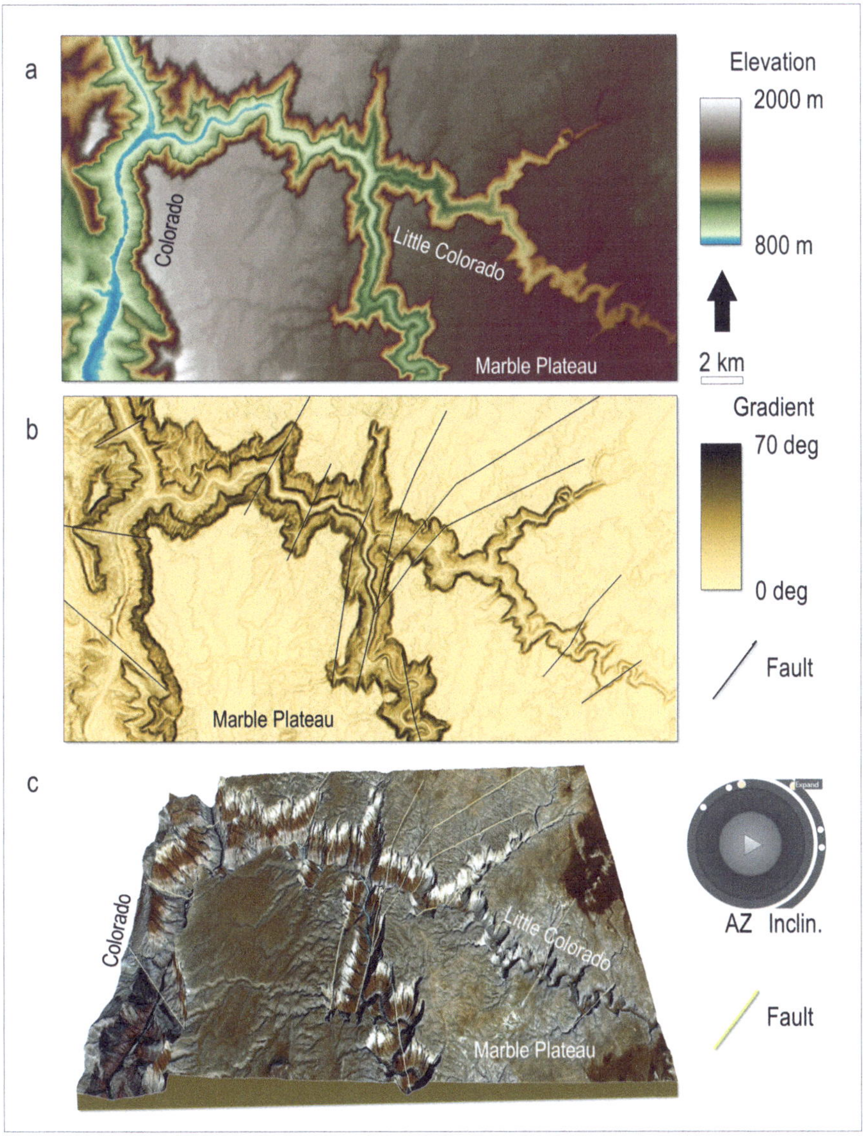

Fig. 7.28 Fault-driven erosion in the Little Colorado River Canyon. (**a**) Digital elevation model, (**b**) gradient of digital elevation model with fault interpretation, (**c**) 3D rendering of Landsat 8 432 RGB pansharpened with band 8 and draped on digital elevation model with fault interpretation. 3× vertical exaggeration and slant illumination

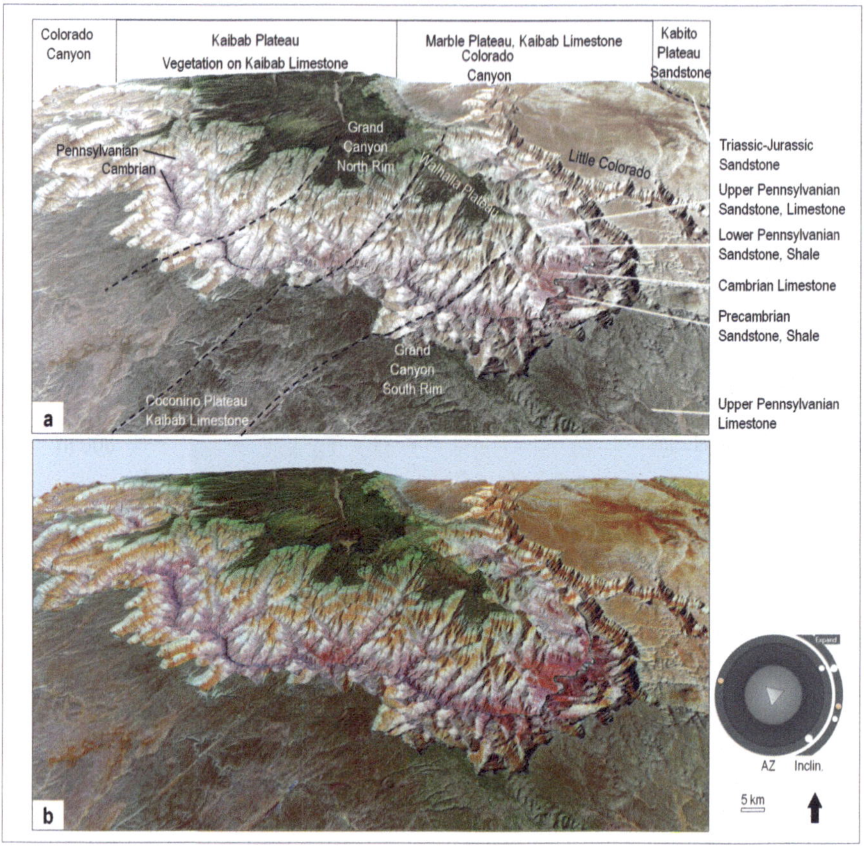

Fig. 7.29 Elevation restriction of vegetation on Kaibab Plateau in arid desert. Merge of Landsat 8 band combinations 753 RGB (**a**) and 10-5-8 RGB (**b**) draped on digital elevation model at 3× vertical exaggeration and slant illumination. (SRTM DEM and satellite imagery courtesy of U.S. Geological Survey)

The digital elevation model (a) shows straight sections of the canyon. The gradient of the DEM (b) enhances the edges of the plateau as well as the river course at the bottom of the canyon. The best approach to identify canyon sections that had been eroded most likely along faults is from the rendering in 3D (c). We can rotate and dip the 3D view such that the view is aligned with a straight section of the canyon. When the texture of the plateau shows a continuation of the orientation of the incised canyon, then we can consider it very likely that we have marked the location of a fault (light lines).

For the conclusion of this section we return to the main Grand Canyon in the area of the Grand Canyon National Park (Fig. 7.29). We overlay the two Landsat 8 band composites 753 RGB for vegetation and lithology of the plateau with 10-7-4 RGB for the canyon lithology, drape it over the DEM and render it at 3× vertical exaggeration. The combination of the satellite band composites highlights both the plateau and its cover as well as the canyon with the varying rock layers at depth. The vertical exaggeration enhances the structural expression showing the straight sections of canyons aligned with straight lineaments on the plateau (dashed lines).

References

Structural Geology of Appalachians and Alleghany Plateau

Danielson, J.J. & Gesch, D.B., (2011). Global multi-resolution terrain elevation data 2010 (GMTED2010): U.S. *Geological Survey Open-File Report*, 2011–1073, 26 p.

Davis, D. M., & Engelder, T. (1985). The role of salt in fold-and-thrust belts. *Tectonophysics, 119*, 67–88.

Freedman, J., Wise, D. U., & Bentley, R. D. (1964). Patterns of folded folds in the Appalachian Piedmont along Susquehanna River. *Geological Society of America Bulletin, 75*(7), 621–638. https://doi.org/10.1130/0016-7606(1964)75[621:POFFIT]2.0.CO;2

Gillespie, P., & Kampfer, G. (2017). Mechanical constraints on kink band and thrust development in the Appalachian Plateau, USA. In J. P. Turner, D. Healy, R. R. Hillis, & M. J. Welch (Eds.), *Geomechanics and geology* (Vol. 458, pp. 245–256). Geological Society, London, Special Publications. https://doi.org/10.1144/SP458.12

Porter, C., Morin, P., Howat, I., Noh, M.-J., Bates, B., Peterman, K., Keesey, S., Schlenk, M., Gardiner, J., Tomko, K., Willis, M., Kelleher, C., Clouties, M., Husby, E., Foga, S., Nakamura, H., Platson, M., Wethington, M. Jr., Williamson, C., Bauer, G., Enos, J., Galen, K., Kramer, W., Becker, P., Doshi, A., D'Souza, C., Cummens, P., Laurier, F. & Bojesen, M. (2018). ArcticDEM, Harvard Dataverse, V1, https://doi.org/10.7910/DVN/OHHUKH, [accessed 15 Feb 2021].

Valentino, D. W., Peavy, S. T., & Valentino, R. W. (2004). Alleghanian orogenic-float on the Martic thrust during dextral transpression, central Appalachian Piedmont. *Journal of Geodynamics, 37*, 613–631. https://doi.org/10.1016/j.jog.2004.02.007

Yeh, M. W., & Bell, T. H. (2004). Significance of dextral reactivation of an E-W transfer fault in the formation of the Pennsylvania orocline, central Appalachians. *Tectonics, 23*, TC5009. https://doi.org/10.1029/2003TC001593

Remote Sensing Case Study Northwest Siberia, Russia

Ehlers, J., & Gibbard, P. L. (2007). The extent and choronology of Cenozoic global glaciation. *Quaternary International, 164–165*, 6–20. https://doi.org/10.1016/j.quaint.2006.10.008

Ehlers, J., & Gibbard, P. L. (2008). The extent and choronology of quaternary glaciation. *Episodes, 31*(2), 211–218.

Ehlers, J., Astakhov, V., Gibbard, P. L., Mangerud, J., & Svendsen, J. I. (2013a). Late Pleistocene glaciations in Europe. In S. A. Elias (Ed.), *Encyclopedia of quaternary science*. Elsevier. 2007, 3576 p. ISBN 978-0-444-52747-9.

Ehlers, J., Astakhov, V., Gibbard, P. L., Mangerud, J., & Svendsen, J. I. (2013b). Middle Pleistocene in Eurasia. In S. A. Elias (Ed.), *Encyclopedia of quaternary science*. Elsevier. 2007, 3576 p. ISBN 978-0-444-52747-9.

Jakobsson, M., Andreassen, K., Bjarnadottir, L. R., Dove, D., Dowdeswell, J. A., England, J. H., Funder, S., Hogan, K., Ingolfsson, O., Jennings, A., Larsen, N. K., Kirchner, N., Nilsson, J., O'Regan, M., Polyak, L., Nørgaard-Pedersen, N., & Stein, R. (2014). Arctic Ocean glacial history. *Quaternary Science Reviews, 92*, 40–67. https://doi.org/10.1016/j.quascirev.2013.07.033

Laake, A. (2009). Hybrid near-surface modeling for seismic property estimation in arctic areas. In *71st EAGE Conference and Exhibition Incorporating SPE EUROPEC 2009, CP127-00312*. https://doi.org/10.3997/2214-4609.201400319

Svendsen, J. I., Alexanderson, H., Astakhov, V. I., Demidov, I., Dowdeswell, J. A., Funder, S., Gataullin, V., Hernriksen, M., Hjort, C., Houmark-Nielsen, M., Hubberten, H. W., Ingolfsson, O., Jakobsson, M., Kjær, K. H., Larsen, E., Lokrantz, H., Lunkka, J. P., Lysâ, A., Mangerud, J., … Stein, R. (2004). Late Quaternary ice sheet history of northern Eurasia. *Quaternary Science Reviews, 23*, 1229–1271. https://doi.org/10.1016/j.quascirev.2003.12.008

Arctic Remote Sensing and Permafrost Mapping

Jorgenson, M. T., & Grosse, G. (2016). Remote sensing of landscape change in permafrost regions. *Permafrost and Periglacial Processes, 27*, 324–338. https://doi.org/10.1002/ppp.1914

Westermann, S., Duguay, C. R., Grosse, G., & Kääb, A. (2015). Chapter 13: Remote sensing of permafrost and frozen ground. In M. Tedesco (Ed.), *Remote sensing of the cryosphere* (pp. 307–344). Wiley. ISBN 978-1-118-36885-5.

Colorado Plateau

Rabbitt, M. C., McKee, E. D., Hunt, C. B., & Leopold, L. B. (1969). *The Colorado River region and John Wesley Powell*, U.S. Geological Survey Professional Paper 669, 164 p. https://doi.org/10.3133/pp669

Geology

Billingsley, G. H., Goodwin, G., Nagorsen, S. E., Erdman, M. E., & Sherba, J. T. (2019). *Geologic field photograph map of the Grand Canyon region, 1967–2010: U.S. Geological Survey General Information Product 189*, 11 p. https://doi.org/10.3133/gip189.

Gettings, M. E., & Bultman, M. W. (2005). A predictive penetrative fracture mapping method from regional potential field and geologic datasets, southwest Colorado Plateau, U.S.A. *Earth, Planets and Space, 57*, 701–715. https://doi.org/10.1186/BF03351850

Karlstrom, K. E., Timmons, J. M., & Crossey, L. J. (2012). Introduction to Grand Canyon geology. In J. M. Timmons & K. E. Karlstrom (Eds.), *Grand Canyon Geology: Two billion years of earth's history: Geological Society of America special paper 489* (pp. 1–6). https://doi.org/10.1130/2012.2489(00)

Geologic Maps

Billingsley, G. H. (2000). *Geologic map of the Grand Canyon 30'x60' quadrangle, Coconino and Mohave Counties, Northwestern Arizona*, U.S. Geological Survey Geologic Investigation Series I-2688.

Billingsley, G. H., & Hampton, H. M. (1999). *Physiographic rim of the Grand Canyon, Arizona*, U.S. Geological Survey Open File Report 99-30. https://doi.org/10.3133/ofr9930

Part III
Remote Sensing for Hydrocarbon Exploration

Chapter 8
Remote Sensing for Hydrocarbon Exploration

Abstract In this section we combine the knowledge of the information remote sensing data provide and integrate different remote sensing products into geologic models of the surface and subsurface to assist all phases of the hydrocarbon exploration. We follow the exploration and production workflow and study solutions that are tailored for individual steps. We start with the aspects of frontier exploration generating regional and basin models and complement these with petroleum system models. The integration of surface and subsurface data shows the impact remote sensing has on subsurface interpretation and modeling even when the raw remote sensing data do not penetrate the earth surface to considerable depth.

In the next stage when the exploration has determined an area of interest within the basin and subsurface geophysical data are to be acquired, we show how remote sensing can contribute to the planning, execution and data quality assessment of seismic data. We conclude that section with a method of creating velocity models for seismic data processing from remote sensing data.

Upon completion of the subsurface 3D modeling based on geophysical data, the decision will be made where to drill and how to develop the field. In this context we study the application of remote sensing data for the assessment of various environmental aspects of infrastructure planning.

We conclude the section with remote sensing contributions to the method of geologic analogs that provide geometrical, structural and depositional concepts for subsurface reservoir modeling. Remote sensing is particularly suited to inspire the geomodeler with realistic concepts for the interpretation of subsurface geologic settings under the assumption that geologic processes in the past have been driven by the same parameters as we observe them in remote sensing data today.

Keywords Frontier exploration, basin modeling, petroleum system modeling, integration of surface and subsurface data, remote sensing for gravity · Logistic planning, environmental hazards, multi-temporal remote sensing · Climatic effects, permafrost, flooding, sand dune encroachment, monsoon mapping · Geologic analogs, structural analogs, rifting, pull-apart basins, relay ramps, depositional analogs, carbonate environment, evaporite deposition, karst, glacial features

A. Laake, *Remote Sensing for Hydrocarbon Exploration*, Springer Remote Sensing/Photogrammetry, https://doi.org/10.1007/978-3-030-73319-3_8

8.1 Introduction

We will study concepts using remote sensing data, their interpretation and integration with geophysical and geologic data for all stages of exploration and production.

For frontier exploration we focus on large areas to build regional models. Using the concept of conjugate margins in the central and south Atlantic, we show how remote sensing can assist finding corresponding locations on the African and on the South American margins using remote gravity and global digital relief models to interpret the regional structural framework. In Libya, we delineate the basin outlines, derive key basin elements such as surface and top basement as well as regional fault zones from the remote sensing data. Based on this outline model we complete a 2D petroleum system modeling study for the same area.

An important aspect for the use of remote sensing data in exploration is the integrated interpretation of satellite and subsurface geophysical data. This technique uses the concept of geomorphology and amends it with the subsurface data that are available from geophysical measurements which is demonstrated on the coastal plains of the Gulf of Suez in Egypt. We also apply this integrated technique to estimate the impact of volcanism on the prospectivity in the western part of the Sirt Basin in Libya.

For seismic logistics and planning we use remote sensing data to remotely characterize the terrain as an input to the planning of seismic surveys on the Payun volcanic plateau in the Andean front ranges in Argentina. We estimate the risk of vehicles getting stuck in coastal sabkha in the UAE and assist challenging seismic operations in the sand dunes of east Algeria. Geomorphologic interpretation of remote sensing data from the edge of the Tademait Plateau in southern Algeria yields insights into the impact of the surface geology on the seismic data quality. We conclude the section with an estimate for the surface velocity on the Ghazalat plateau west of the Quattara Depression in the Western Desert of Egypt, where we show how the interpretation of remote sensing data can assist the building of near-surface velocity models and the estimation of static corrections in complex terrain.

To support the planning of well locations and oilfield infrastructure, we study various climatic and environmental parameters using multi-temporal satellite imagery. We start in arctic Alaska where we study seasonal changes of ice coverage and their impact on logistics as well as long-term changes related to permafrost thawing and its impact on infrastructure planning. In the moderate temperature humid climate we estimate the extent flooding in the Arkansas White River and Black River floodplains can have on infrastructure planning and repeat the study in the western Mississippi Delta to map and predict the extent of flooding. Long-term climatic change in the arid desert and monsoon climatic zone of the Chad can assist with logistics through predictions of the areal extent of the monsoon time window. We conclude the section with the multi-temporal mapping of sand dune encroachment in the Grand Erg of southern Tunisia using an RGB-rendering method for the simultaneous study of multi-temporal satellite images.

Finally, we identify modern surface geologic analogs and their expression in remote sensing data to assist the understanding of subsurface geologic setting. This is particularly helpful since the subsurface geological setting, i.e. the study of fossil structural and depositional settings in the subsurface can only be accessed through seismic data, the interpretation of which is inherently ambiguous. Reservoir geologists who assess an exploration opportunity will, therefore, turn to modern representations of the subsurface geological setting to constrain the subsurface models from wells and seismic and thereby reduce the risk of uncertainty of the opportunity assessment.

We study geologic analogs in three different groups differentiated for the geological processes involved. For structural analogs we cover relay ramps with the Arches National Park ramp in UT, U.S.A., karsting processes along faults in Hassi R'Mel in Algeria, salt tectonics in the Zagros Mountains in south Iran and in the Mississippi Delta in LA, U.S.A. Rift tectonics in the Afar Triangle in east Africa, pull-apart basins in the Sinai in Egypt and fold and fault belts of the Appalachian Mountains U.S.A. complete the study of structural analogs.

Depositional analogs are covered through examples from shallow water and evaporite settings of the Abu Dhabi coast UAE, inland sabkha of Umm As Samim, Oman, and Lake MacLeod in west Australia. Carbonate depositional settings are represented by the reefal processes on the Caicos Islands in the Atlantic in the eastern Caribbean. Delta and channel depositional processes are studied through onshore and offshore deltas of the Irrawaddi in Myanmar, floodplain channels in Arkansas and intra-cratonic deltas with the modern example of the Okavango Delta in Botswana and the Tertiary example of the Kufrah River in Libya.

Erosional settings start with limestone karsting in the Karst Mountain range in Slovenia and Hassi R'Mel in Algeria. Glacial erosion is studied in Jutland, Denmark and the Great Lakes in north America with focus on special glacial features that can indicate the direction of the ice advance, which is important to reconstruct fossil glacial features where only spatially sparse information can be obtained from subsurface data.

Chapter 9
Frontier Exploration

Abstract Frontier exploration means assessing an area for hydrocarbon prospectivity that lacks information and data. Satellite remote sensing provides the first set of potential fields data to enable the delineation of basins and build large regional geological models in 3D. These models can be used to execute exploration strategies such as the conjugate margin concept which transfers knowledge from known basins on margin to their unknown counterpart on the conjugate margin along fracture zones.

Remote sensing also plays a major role in providing the boundary conditions for basin and petroleum system models. These form the framework for more detailed interpretation once seismic data and well logs become available.

At field scale, interpretation carried out on remote sensing data at the surface can be used as in-situ geologic analogues to guide the structural and depositional interpretation of 3D seismic data in the subsurface.

The integration of satellite remote sensing and sparse regional 2D seismic sections give insights into the shallow structural framework through the mapping of shallow buried channels, which in turn often follow regional faults. In volcanic areas, the interpretation and integration of multiple satellite remote sensing data allows delineating the subsurface extent of volcanic rocks, thus giving clues for estimating the potential impact volcanism may have had on the maturity of hydrocarbons in the subsurface.

Keywords Regional geologic models · Basin depth from gravity · Basin structure from gravity and magnetics · Conjugate margin exploration · Geologic analogues · Shallow geology from surface-subsurface integration · Basin modeling · Petroleum system modeling

A. Laake, *Remote Sensing for Hydrocarbon Exploration*, Springer Remote Sensing/Photogrammetry, https://doi.org/10.1007/978-3-030-73319-3_9

9.1 Introduction

Frontier exploration aims at unlocking information about the subsurface hydrocarbon prospectivity in areas with little or no prior geological knowledge. A complete frontier exploration workflow starts at the regional scale to identify basins and then zooms into a selected basin to identify structures that can be studied further to map leads and mature them to prospects with the ultimate goal to determine locations for the drilling of wells. At each step along this workflow, the information obtained from the underlying data becomes more accurate, which in turn allows to determine the uncertainty and risk associated with certain locations that are considered candidates for drilling. This workflow resembles the maturation of opportunities from a first idea to a location for which uncertainty and risk are supported well from the integration of all data that are becoming available.

After assessing the available data, a typical approach starts with outlining potential basins and studying their internal structure. Due to the global coverage, satellite remote sensing data are the first choice for this task. Global relief data deliver the topography of the land surface onshore and of the seafloor offshore (Amante & Eakins, 2009). Gravity data (Sandwell & Smith, 2009, Sandwell et al., 2014) provide clues for the depth of basement and the combination of gravity and magnetic data (Dyment et al., 2015) allow regional structural delineation that can be represented in a regional geologic model. For calibration, data from deep sea drilling projects (IODP, n.d.) have been interpreted for marine sediment thickness (Straume et al., 2019) and crustal age (Muller et al., 2008). We illustrate this process at the case of the equatorial Atlantic Ocean (Sect. 9.2).

Once a specific basin has been selected for exploration, regional 2D seismic data are acquired, often supported by airborne or shipborne gravity and magnetic data. With these data, specifically the seismic data that provide sections through the basin from surface to basement, the coarse regional model can be refined to a basin model. Remote sensing data contribute significantly to the basin model, particularly for onshore basins where the concept of geomorphology can be used to link the surface texture with the subsurface structure. We show the process of generating a model for three Libyan basins from remote sensing data (Sect. 9.3).

At this stage, the first potential structures catch the eye of the geoscientist. From their experience, they will associate some of these structures with known structures of similar geologic setting and use the method of geologic analogues to introduce additional information into the assessment. This approach is explained through a case history from the Gulf of Suez (Sect. 9.5). Based on the information, often the first exploration well is drilled to validate the model. The logs from this well or from wells in a neighboring more explored basin provide the full calibration of the geophysical data for petrophysical, sedimentological and fluid aspects. From the integration of non-seismic, seismic and well data a refined basin structural model can be built. When complemented with information about the depositional history of the subsurface formations and models about heat flow and tectonic evolution, the structural model can be converted into a petroleum system model. We show the approach of extending a petroleum system model from a mature basin into a neighboring frontier basin at the case study from the western Sirt basin in Libya (Sect. 9.4).

The completion of a basin petroleum system model typically warrants the acquisition of 3D seismic data, that image subsurface geologic structures at high resolution. The mapping of leads and their maturation to prospects and locations suggested for drilling requires the detailed processing and interpretation of the seismic data and their calibration and integration with well data to achieve a reliable input for computation of uncertainty and risk, which is necessary to support the decision to invest into the drilling of wells and the potential development of the target area into a producing field. We show two cases of integration. The first case integrates remote sensing data, shallow seismic data and deep seismic data using geologic analogues from the surface to guide the exploration downwards. The second case integrates satellite radar data with shallow seismic data to map buried channels that follow regional faults in an area where sand dunes mask the fault expression at the surface.

The final example covers the special risk that is associated with volcanism. On one hand, volcanism indicates an increased heat flow in the subsurface, which may accelerate the maturation of kerogen to hydrocarbons leading to younger source rock maturing faster. On the other hand, volcanism also bears the risk of overheating already mature older source rock thus transforming the oil phase into a gas phase or destroying the hydrocarbons altogether. For the estimation of the impact of volcanism it is important to understand the extent of the volcanic rocks in the subsurface as well as the age of the volcanic activity. We show at the example of the extensive Haruj volcanic system that large portions of the volcanic rocks are just thin surface flows without severe impact on the subsurface heat flow (Sect. 9.6).

For the case studies used in this section, Table 9.1 shows an overview of the locations of which are shown in the map in Fig. 9.1.

Table 9.1 Overview of case studies demonstrating remote sensing applications for frontier exploration

Section	Geologic setting	Area of interest	Remote sensing input for
10.2	Rifting, conjugate margins	Equatorial Atlantic	Conjugate margin modeling
10.3	Basins	Libya	Basin modeling
10.4	Basins	Libya	Petroleum system modeling
9.5.1	Rifting	Gulf of Suez, Egypt	Surface-subsurface integration
9.5.2	Radar rivers	Gilf Kebir, Egypt	Ground penetration of satellite radar
9.6	Shield volcano	Haruj, Libya	Impact of volcanism on HC prospectivity

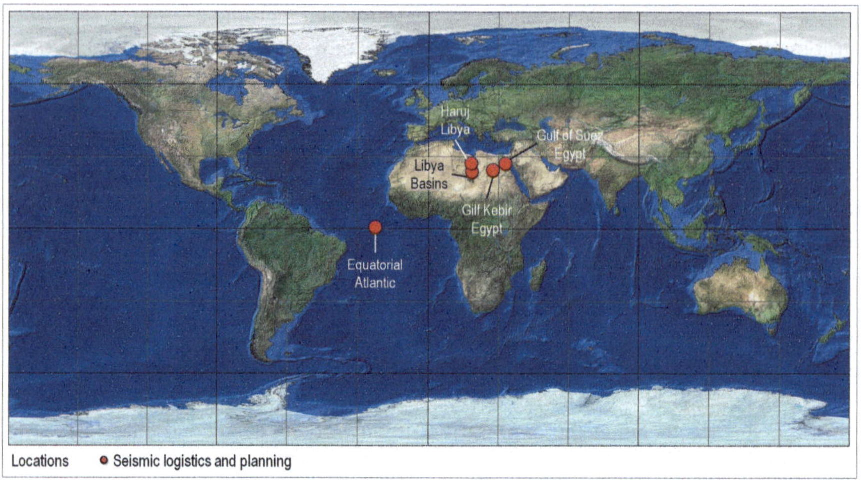

Fig. 9.1 Map of the frontier exploration case studies. (Satellite imagery courtesy of U.S. Geological Survey)

9.2 Regional Modeling of the Equatorial Atlantic Conjugate Margin

Frontier exploration into unknown basins benefits from the opportunity to transfer the experience gathered in known basins. One example for this approach is the conjugate margin exploration concept (Mohriak et al., 2013). Conjugate margins develop along the lineaments of continental breakup by rifting. Subsequent seafloor spreading widens the initial rift valley to become an ocean. The continental margins either side of the spreading center often show a common environment for sediment deposition. However, the further the seafloor spreading progresses and the wider the ocean grows, this commonality decreases as each continental margin develops independently with respect to sedimentary deposition. Satellite radar altimetry allowed the interpretation for free air gravity on a global scale (Smith & Sandwell, 1994, 1997; Cande et al., 2000; Sandwell & Smith, 2009; Sandwell et al., 2014) and has since been applied to marine exploration (Fairhead et al., 2001). The first global digital age grid based on ship tracks has been published by (Muller et al. 1997).

Critical aspects for the conjugate margin concept are the detection of the plate boundaries, both present (Bird, 2003) and historic (MacLeod et al., 2017). For the delineation of the plate boundaries and the ocean basins, Crosby et al. (2006) documented the correlation between ocean depth, age and gravity data, which was extended to the mapping of basement structures from satellite gravity data by Braitenberg et al. (2006).

In our regional modeling study we focus on the equatorial Atlantic with its African and South American passive margins. Gibbs et al. (2003) presented models for the syn-rift evolution comprising also paleo-topographic models. The geodynamics has since further been refined (Lentini et al., 2010). Heine et al. (2013) give a detailed overview of the plate tectonic evolution of the south and equatorial Atlantic. Finally, the conjugate margin exploration concept has been studied for its limitations as analogues particularly for trap and production aspects (Zanella et al., 2018).

Deep sea drilling programs Mohole (NAS, 2020), Offshore Drilling Program ODP (Consortium for Ocean Leadership 2007), Deep Sea Drilling Project DSDP (DSDP, n.d.), and IODP (IODP, n.d.) delivered cores and logs from deep sea wells as hard data for calibration.

The goal for using remote sensing data in large area geologic assessments is the creation of regional geologic model as a background for further, more detailed investigations. Such a model comprises the top and base of the model, i.e. in the offshore case, the seafloor and the top of basement, as well as internal structure such as regional faults. If possible, information about the geologic age of regional strata may also be included. Typically, we use globally available remote sensing data for this approach (Fig. 9.2). The resolution of these data is typically in the range of a few 100 m to a few kilometers.

The digital relief model provides both, the digital elevation onshore as well as the bathymetry offshore. The satellite gravity anomaly gives an estimate of the top of basement, i.e. the thickness of the sedimentary fill of a basin. Regional faults are represented in the DRM if these outcrop at the surface and in the gravity map if these are affecting a large segment of the subsurface. The lateral gradient extracts these lineaments in both cases. Regional structures often also leave a trace in magnetic anomaly data, but their expression is generally weaker than on DRM and gravity data. Therefore, we have not considered the lateral gradient of the magnetic anomaly in the data sets used for this analysis. Where magnetic field reversals occur,

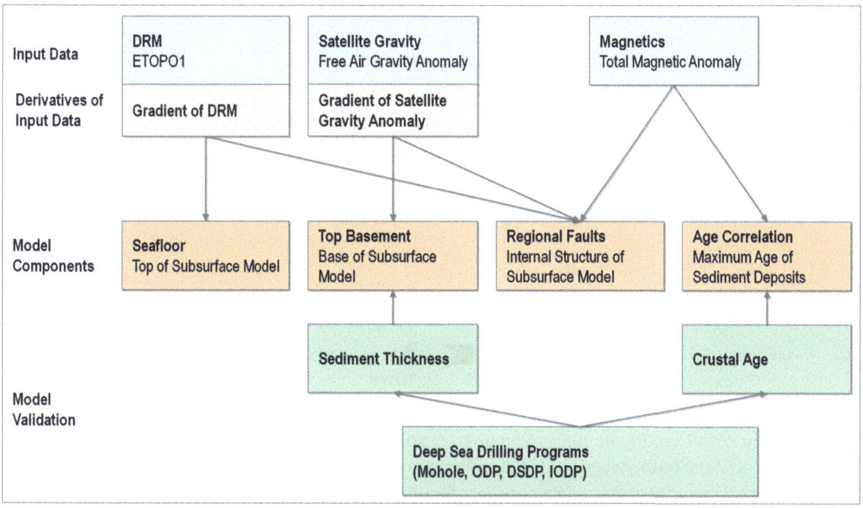

Fig. 9.2 Workflow for remote sensing based conjugate margin exploration

i.e. in areas of seafloor spreading, anomalies in the magnetic field data also reveal a chronology that can assist dating the age of basin forming.

We illustrate the concept of conjugate margin exploration at the southern and equatorial Atlantic Ocean (Fig. 9.3) overlaying satellite derived free air gravity (Sandwell et al., 2014) on the continents and sediment thickness (Straume et al., 2019) offshore. For the present day setting the seafloor topography (Amante & Eakins, 2009) has been amended to show the location and structure of the Mid Atlantic Ridge (MAR). Western Gondwana broke up as a result of intraplate rifting during the Jurassic. We show reconstruction at the boundary of Jurassic to Cretaceous as 120 my with the locations of present-day sediments accumulations for reference. During this time the depositional environments of either conjugate margin were rather similar. The use of one margin as a geologic analogue for the other margin appears appropriate.

80 my later at the period of Mid Paleogene the Atlantic Ocean had fully opened and the two margins were separated by about 1000 km. This means, that the sedimentation on either margin became increasingly decoupled from the corresponding location on the opposite margin. The value as an analogue decreases, particularly with respect to reservoir and trap formation.

Today (0 my Holocene), the Atlantic Ocean completely decoupled the two conjugate margins. Hence, present data deposition is completely independent and can no longer be used as an analogue.

The reconstruction of the opening of the South Atlantic became possible when satellite gravity provided the detailed texture of the seafloor through large continuous fracture zones connecting locations on the African margin across the Atlantic to the corresponding location on the South American margin (Smith & Sandwell,

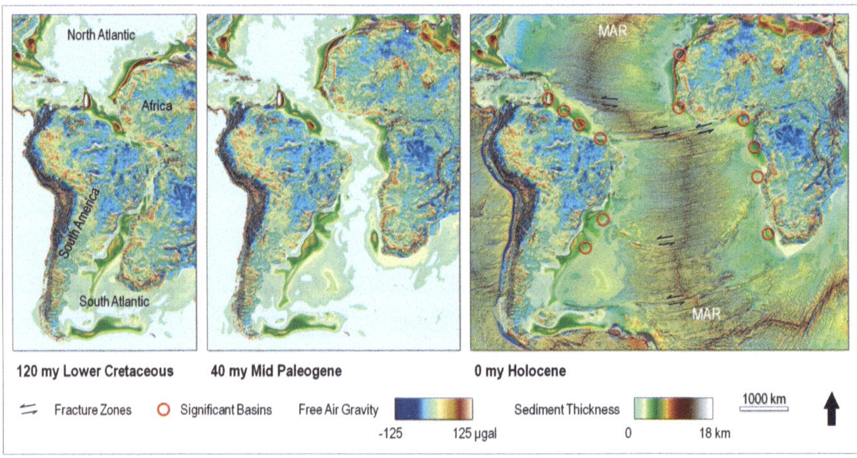

Fig. 9.3 Concept of conjugate margin analogues applied to the South Atlantic for three stages of the opening of the South Atlantic Ocean using gravity and sediment thickness data. *MAR* Mid Atlantic Ridge. (Satellite gravity data courtesy Sandwell et al., University of California San Diego, sediment thickness data courtesy of NOAA)

1994, 1997). The basins on the African margin (red circles) could now be connected to their counterparts on the South American margin.

Historically, the exploration for hydrocarbons did not start and evolve simultaneously on both margins of the Atlantic. Offshore exploration began in the 1960s and 1970s in the Niger Delta and continued both towards the north to Ghana and south to Angola. The exploration offshore Brazil in the late 1990s used the conjugate margin concept to benefit from the expertise gathered offshore Angola. When significant discoveries were made offshore Ghana, the same concept has been used to search for similar petroleum system concepts in Guyana in the late 2000s.

9.2.1 Data Input and Preparation

We will follow this concept by building a regional geological model for the conjugate margins of the equatorial Atlantic Ocean. We use three remote sensing input data sets (Fig. 9.4): the DRM from ETOPO1 (Amante & Eakins, 2009), satellite gravity anomaly (Sandwell et al., 2014), and global magnetic anomaly (Dyment et al., 2015). The seafloor bathymetry (a) reveals the Mid Atlantic Ridge (MAR) as well as lineaments associated with the fracture zones. The fracture zones are highlighted by the gravity anomaly (b) where individual fracture zones (FZ) like the Ascension FZ (AFZ) and the Romanche FZ (RFZ) can clearly be delineated. These served as the key links between the African and the South American margins.

The gravity data also reveal that at certain areas near both margins deep basins can be expected. These are represented by the blue areas showing a strong negative gravity anomaly.

The magnetic anomaly data (c) reveal strong banded texture, which is related to the reversal of the earth magnetic field through history. We have overlaid the isochrons for the base Cretaceous (120 my), the top Paleogene (40 my) and present day, i.e. the Mid Atlantic Ridge. Whilst these bands are useful to date the spreading of the Atlantic seafloor, they overlay a banded noise on the structural lineaments that show as offsets in the banded signal.

Since we observed lineaments related to the regional fracture zones on the DRM and gravity anomaly data, we can compute the lateral derivatives for both data sets (Fig. 9.5a, b). For orientation we have plotted the Mid Atlantic Ridge as a red line. The gradient of the seafloor bathymetry (a) reveals the structural framework as it outcrops at the seafloor. Towards the MAR, the absence of sediments creates an increasingly rough texture, which blurs the image of the fracture zones. The gradient of the gravity anomaly (b) is clearer since it averages over a thick package of rocks thus delivering a smoother and better defined image of the fracture zones. The reason why large fault zones show on gravity and consequently on gravity gradient data is that fracture zones result in a destruction of the rock matrix leaving a lower density rocks within the fracture zone.

The other two images in Fig. 9.5 show data from wells of the Deep Sea Drilling Program DSDP (IODP, n.d.) that are used for calibration of the remote sensing data.

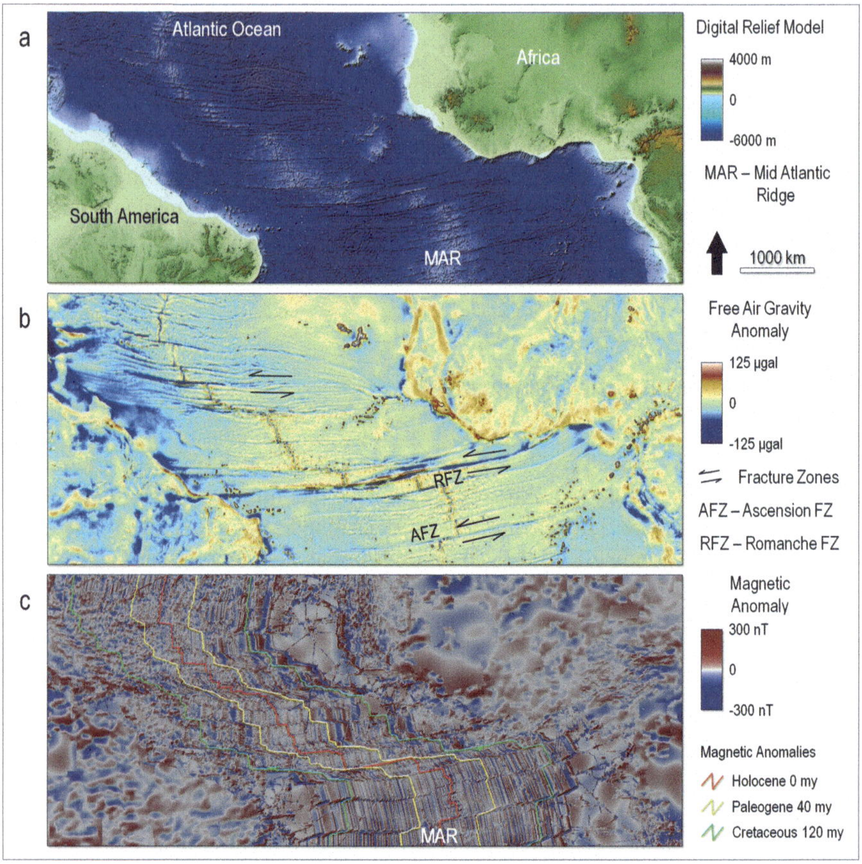

Fig. 9.4 Remote sensing input data used for conjugate margin study of the equatorial Atlantic Ocean. (**a**) Digital relief model, (**b**) satellite gravity anomaly, (**c**) magnetic anomaly. Arrows indicate major fracture zones, colored lines show geologic age of major magnetic anomalies. (ETOPO1 data Amante and Eakins (2009), satellite gravity data courtesy Sandwell et al., University of California San Diego, magnetic data courtesy WDMAM (Dyment et al., 2015))

The crustal age (c) is taken at deep well locations and gridded using magnetic anomalies. The result is a map of crustal age which also shows the offsets induced by the fracture zones. The sediment thickness map (d) takes its data input from the sedimentary column between seafloor and basement at the well locations. Shipborne data assist in the interpolation between the well locations.

9.2.2 Generating Geologic Models

The integration of the input data, the derivatives of the input data and the calibration data allows to create the boundary surfaces of the model, i.e. the seafloor from bathymetry and the topography of the top of the basement from the combination of

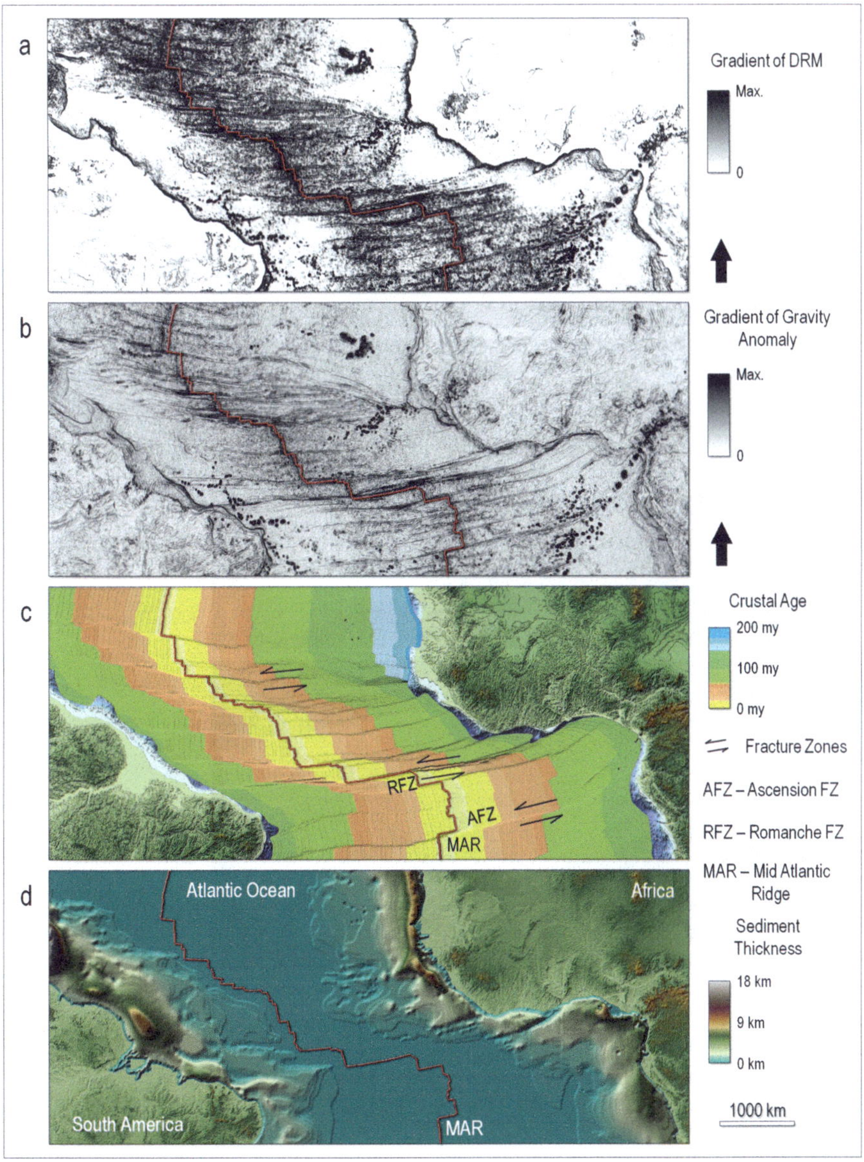

Fig. 9.5 Data derivatives and calibration data. (**a**) Lateral gradient of DRM, (**b**) lateral gradient of satellite gravity anomaly, (**c**) crustal age inserted into DRM, (**d**) sediment thickness inserted DRM. Arrows indicate major fracture zones, colored lines show geologic age of major magnetic anomalies. (ETOPO1 data Amante and Eakins (2009), satellite gravity data courtesy Sandwell et al., University of California San Diego, crustal age courtesy NOAA (Muller et al., 2008), sediment thickness courtesy NOAA (Straume et al., 2019))

gravity and sedimentary thickness data (Fig. 9.6). We can derive the regional structure of the basin fill from the lateral gradient of the gravity anomaly with a calibration at the seafloor from the gradient of the seafloor topography.

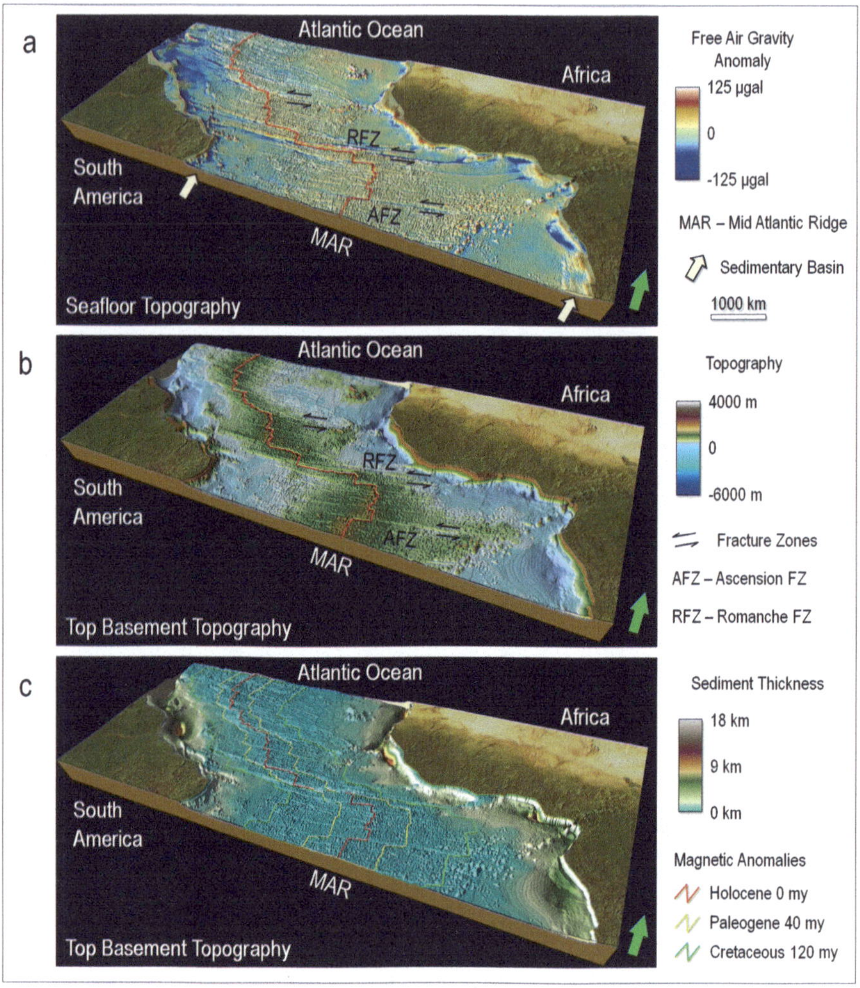

Fig. 9.6 Geologic model for the equatorial Atlantic conjugate margin. (**a**) Satellite gravity anomaly draped on DRM, multispectral satellite image onshore, (**b**) topography of top basement, (**c**) sediment thickness draped on top basement topography. Thin arrows indicate major fracture zones, colored lines show geologic age of major magnetic anomalies. (ETOPO1 data after Amante and Eakins (2009), satellite gravity data courtesy Sandwell et al., University of California San Diego, crustal age courtesy NOAA (Muller et al., 2008), sediment thickness courtesy NOAA (Straume et al., 2019), Landsat 5 imagery courtesy USGS)

When draping the gravity anomaly onto the seafloor (a), we get a clear impression of the fracture zones connecting the Africa margin to the South American margin. To focus the attention to the offshore, we have draped a satellite image on the onshore portions of the model. Sedimentary basins can be observed on either margin, but at the scale of the model, these leave only a very small expression on the regional model.

Next we map the top of basement topography for the offshore section (b). Near the center of the Atlantic Ocean on either side of the MAR, the texture of Top Basement very much resembles the seafloor texture because of limited sediment cover. Towards the continental margins, however, deep basins become visible (dark blue colors). We can highlight these sedimentary basins by draping the sediment thickness map onto the Top of Basement surface (c). The deep basins are now showing in brown to dark green colors for deep and shallow basins, respectively. The thin sediment cover of the central Atlantic basin is showing in cyan blue colors.

To obtain an age estimate for the formation of these basins we overlay the isochrons from the magnetic anomalies. The correlation of the basins with the isochrons shows that most basins have formed in the Cretaceous and filled with sediments until present time.

We return now to the conjugate margin concept to better understand how the width of the paleo Atlantic Ocean would have impacted the correlation between the two margins.

We run a simple plate tectonic reconstruction using the isochrons to align the African and South American plates in earth history (Fig. 9.7). More sophisticated reconstruction is possible using the GPlates platform (GPlates, n.d.; Müller et al., 2016, 2018). To reconstruct the stage of opening of the Atlantic, we drape the isochron maps onto the seafloor topography (series 2) to cut up the plates along the location of the paleo MAR for the Base of Cretaceous (C), the Upper Paleogene (B) and the Holocene (A). Draping the gravity anomaly on the seafloor shows the fracture zones along which to rotate the two plates relative to another to obtain the reconstruction (series 1).

At the boundary of Jurassic to Cretaceous (C), the rift was relatively narrow, the deposition from one margin had an impact on the deposition on the corresponding margin. We have to consider, that at this time the Niger and Congo rivers had not deposited their deltas because they had been connected to the Amazon basin on the South America plate flowing west into the Pacific Ocean. During the Cretaceous, a major source rock of the West African petroleum systems was deposited which forms a major component of petroleum systems developing along the conjugate Atlantic margins.

In the Upper Paleogene (B), the Atlantic had fully opened providing deposition space for passive margin sediments on both margins. On the African plate, the Niger and Congo rivers were given space to develop their deltas. On the South American plate, however, the Amazon reversed its course only about 30 my later. Therefore, the deposition of the Amazon and Orinoco fan deltas occurred significantly later than on the African margin. The delta fans at both margins form the reservoir rocks today.

At present (A), the Atlantic has separated the two margins by about 3000 km. The deposition is completely independent with the Amazon and Orinoco as main sediment sources on the South American margin and Niger and Congo as main sediment sources on the African margin. Under the sediment load, the basins deepened thus maturing the kerogen in the source rock to hydrocarbon, which subsequently migrated into the reservoir rocks and traps in the overburden. Since this process

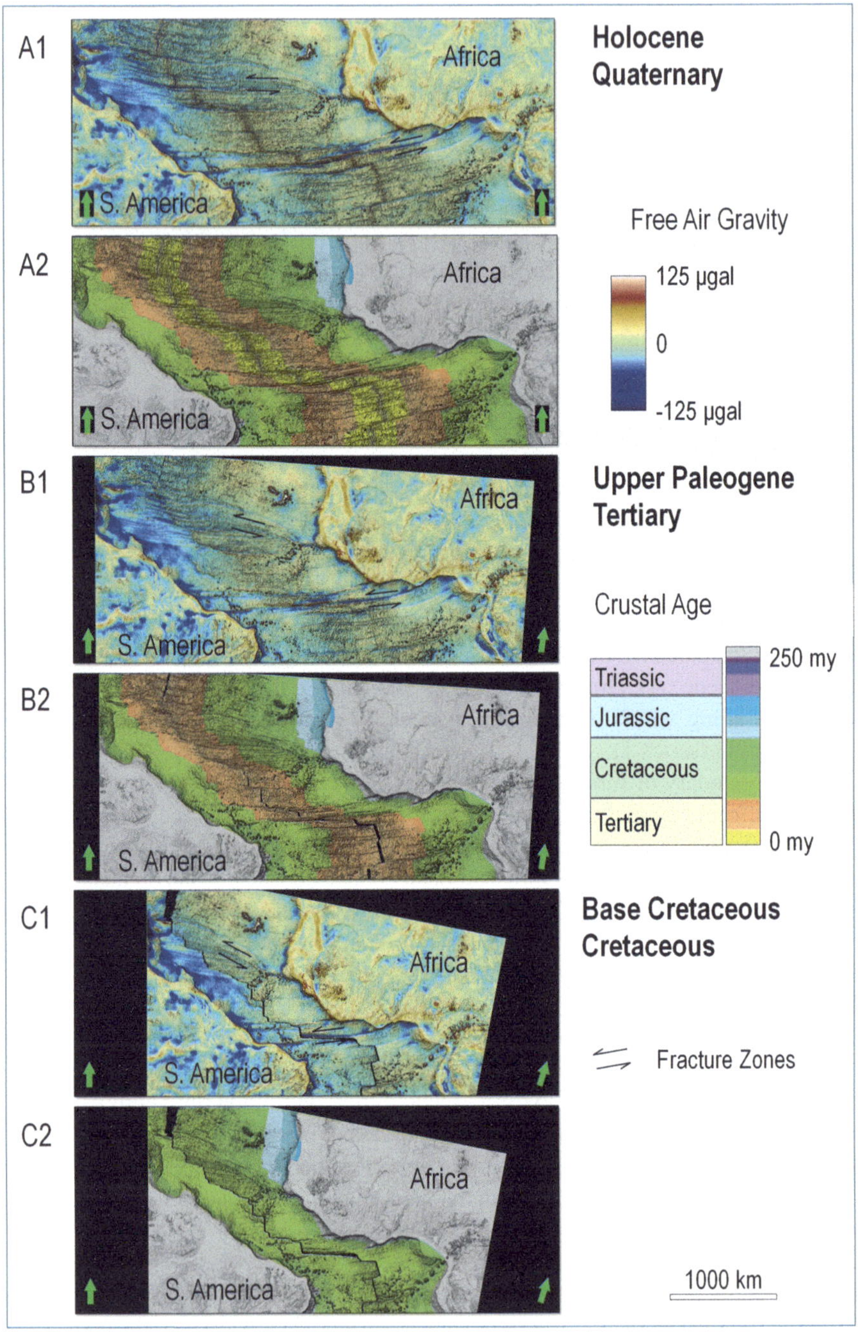

Fig. 9.7 Plate tectonic reconstruction of the equatorial Atlantic conjugate margin using satellite gravity anomaly (1) and crustal age (2). (**a**) Quaternary age, (**b**) Upper Paleogene 40 my, (**c**) base Cretaceous 120 my. (ETOPO1 data after Amante and Eakins (2009), satellite gravity data courtesy Sandwell et al., University of California San Diego, crustal age courtesy NOAA (Muller et al., 2008))

requires the presence of the Cretaceous source, finding it on one margin increases the chance of finding it on the corresponding side of the conjugate margin.

9.2.3 Executing the Conjugate Margin Exploration Concept

In the case of the equatorial Atlantic margin, the conjugate margin exploration concept started at the discoveries on the African margin, i.e. in the Niger, Congo and Guinea basins, used the fracture zones like RFZ and AFZ to find the corresponding locations on the South American conjugate margin to search for significant sediment accumulations in these target areas (Fig. 9.8). Basins with a sediment thickness above 3 km are marked in yellow (a).

An even more intuitive way of guiding the conjugate margin exploration concept is shown in (b). The sediment thickness is plotted as positive elevation starting at mean sea level (MSL) thereby highlighting potential opportunities. Using the fracture zones on the gravity anomaly map, the Niger basin can be correlated with the northeastern Brazil offshore basins and the Ghana and Guinea basins with Guyana and Trinidad. After identifying the correlations in 3D, we return to map view (c) for a more detailed correlation. The first discoveries have been made on the African margin in Congo, Nigeria and Ghana in the 1960s and 1970s. The advance of satellite gravity mapping allowed mapping the fracture zones that were used in the exploration offshore Brazil with the discovery of the Pernambuco, Potiguar and Foz do Amazon Basins. Once new plays in Ghana were discovered in the late 1990s, the same concept has been employed to explore offshore Guyana and Suriname in the 2000s. On the other hand, the fields offshore Trinidad may initiate exploration activities along the Guinea coast.

9.2.4 Conclusions

Let us consider a target area for frontier exploration in Guyana and Suriname, which was the case in the 2010s and apply the conjugate margin exploration concept. At that time, considerable discoveries had been made in the Ghana basins. Therefore, tracing the fracture zones back from the target area on the South American margin in Guyana back to the corresponding section of the African margin points to the Ghana and southern Guinea basins (Fig. 9.9 map insert). The petroleum system model, specifically the source rock and maturation aspects, are then carried across back to the Guyana basins.

Interestingly, the Guyana basins show an even thicker basin fill than the conjugate basin on the African margin (Fig. 9.9 main figure, dark red areas), which provides a good chance for successful exploration. From the 3D model and the isochrons we can obtain that the basins had formed in the late Jurassic and received continuous sediment fill until today. The success story of this exploration concept is

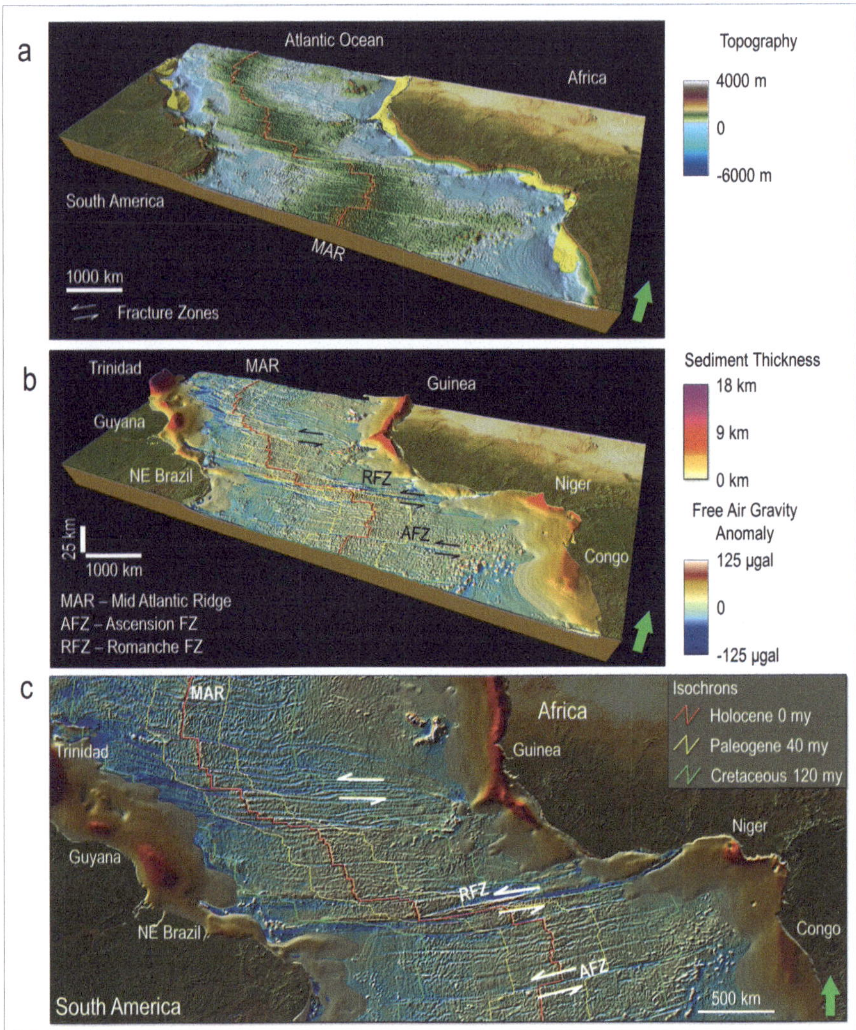

Fig. 9.8 Exploration potential on equatorial conjugate Atlantic margins. (**a**) Top of basement topography, sedimentary accumulations highlighted in yellow, (**b**) satellite gravity anomaly draped on top basement horizon. Sediment thickness shown to highlight the exploration opportunities. (ETOPO1 data after Amante and Eakins (2009), satellite gravity data courtesy Sandwell et al., University of California San Diego, sediment thickness courtesy NOAA (Straume et al., 2019), Landsat 5 imagery courtesy USGS)

well-known: some of the most important discoveries in recent years have been made in this area in the late 2010s. These are starting production 2020. The exploration offshore Suriname is ongoing.

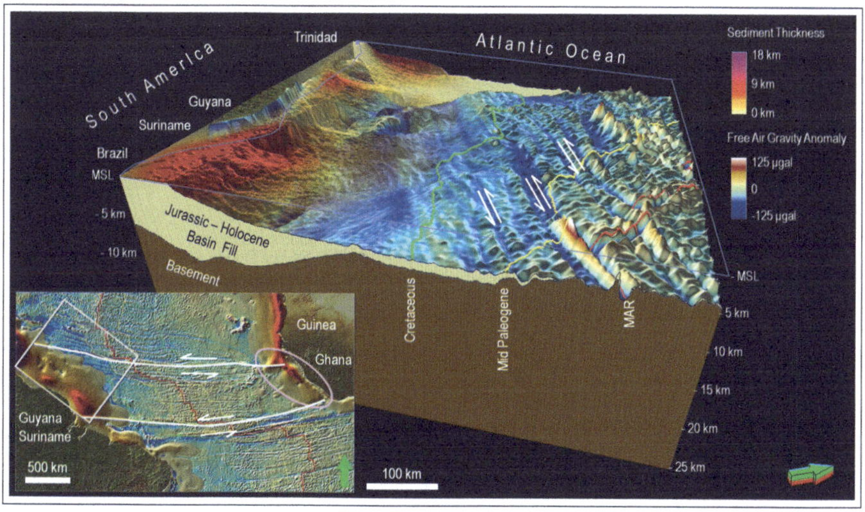

Fig. 9.9 Example for a target area of interest identified using the conjugate margin exploration concept. (ETOPO1 data after Amante and Eakins (2009), satellite gravity data courtesy Sandwell et al., University of California San Diego, sediment thickness courtesy NOAA (Straume et al., 2019))

9.3 Basin Modeling of Libyan Basins

Basin modeling aims at interpreting different measurements to delineate the outline of sedimentary basins and to infer basin structure. To achieve the derivatives of the input, data are created such as gradients and spatial filters. The results are jointly interpreted and integrated and provide information about the basin outline, basin depth and thickness of basin fill as well as shallow and deep structural information. These three aspects together deliver the basin model (Fig. 9.10). In this context the term "basement" means crystalline rock formations. The non-crystalline rock strata between the top of basement and the surface are considered basin fill.

The Libyan Basins in the North Africa are used as an example for a basin modeling workflow applied to multiple basins over an area of about 3.8 million km^2. The area is part of the Sahara Desert which is devoid of vegetation. Approximately 60% of the study area is composed of rock outcrops, which are accessible to multispectral electromagnetic remote sensing analysis. The remaining 40% are covered by drifting sands and dunes which mask the underlying rocks. However, most of the sand cover is sufficiently thin to allow microwave radar map buried structures such as paleochannels and structural lineaments. The study area is also covered by free air gravity data and a surface geologic map.

The first step is the selection of the input data. Different measurements reveal different aspects of a sedimentary basin. Figure 9.11 shows a schematic geologic cross-section through a basin with rock layers of different age and composition. The

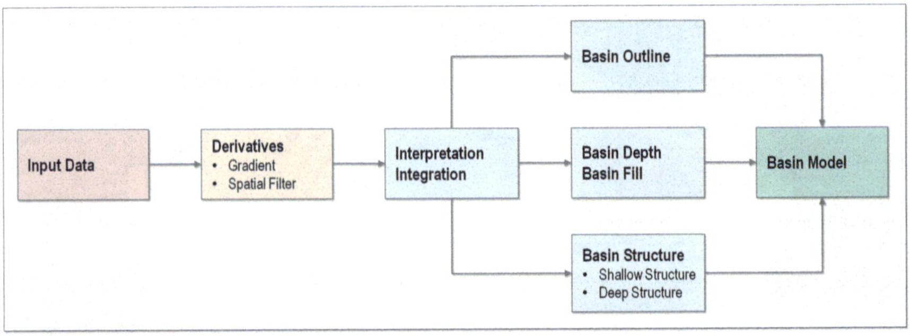

Fig. 9.10 Schematic workflow for basin modeling using remote sensing data

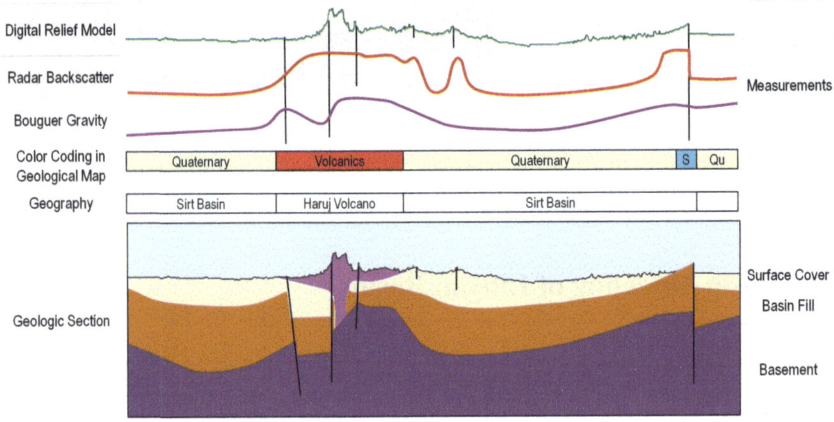

Fig. 9.11 Schema of the correlation of remote sensing measurements with surface and subsurface geology

basin is structured by tectonic faulting which is indicated by traces of normal faults in the section. Above the section schematic characteristics of different data sets are shown.

Starting at the top of the section, the digital relief model (DRM) maps the elevations, i.e. the topography as well as surface lineaments. Multi-spectral satellite imagery offers an insight into surface rocks through the corresponding electromagnetic absorption spectra (Chap. 2). The surface geologic map correlates the surface rock formations with geologic age and genetic origin such as volcanism. The backscatter intensity of microwave radar is maximum for hard rock, but in soft sand basins may also penetrate into the ground thus revealing buried channels and their link to fault lineaments that are hidden by the sand. Anomalous satellite gravity data obtained from satellite radar altimetry show larger than average values where high-density basement rocks are near the surface but are also affected by surface volcanic rocks of limited thickness.

When using satellite gravity data, one needs to consider that these are typically free air gravity measurements. This means that they include a contribution from surface elevation changes that need to be removed to obtain data that are interpretable for basin depth. Using this approach, the free air gravity data are reduced to mean sea level, i.e. the Bouguer anomaly data. An approximation of the thickness of the basin fill and thus the depth to top of basement can be computed assuming an average rock density of the basin fill, an approach that is used here. More accurate computation requires spectral filtering and inversion of the gravity data (Li & Oldenburg, 1998; Camacho et al., 2002; Sjöberg & Bagherbandi, 2017; Wu et al., 2014).

Structural lineaments at and near the surface are derived from gradients of the input data that are typically computed using edge detection algorithms. Deeper structural lineaments can be extracted from spatial filtering of the Bouguer gravity anomaly. Figure 9.12 lists the data sets that have been used for this basin study. Also shown are the derivatives and the interpretation results that have been integrated to build the basin model.

The integration of multiple data sets follows concepts used in geomorphology. The topography of a basin is the expression of the thickness and lithological composition of subsurface layers and their orientation. Tectonic processes such as folding and faulting place the subsurface rock layers at an angle towards the surface and change their exposure to weathering. As a result, hard layers tend to weather less than soft rock layers and thus become prominent features such as mesas. Folding creates for example typical bulls-eye shaped surface expressions. Faulting generates breaks in the outcropping surface layer package, which typically results in tectonic lineaments. Drainage often follows structural lineaments, which in turn makes drainage patterns a good indicator for tectonic lineaments. This principle also applies to paleo-drainage that is presently buried under dry sand (compare Sect. 2.4.6).

Surface rocks can be characterized from different aspects. For the delineation of basins, outcrops of basement rocks are hard evidence. Where basement outcrops are absent, outcrops of Paleozoic rocks give an indication that basement rocks are closer to the surface than in areas with Mesozoic and Cenozoic rock outcrops. Typically, older rocks and basement rocks are harder than younger rocks due to diagenesis and metamorphism. Radar backscattering at the surface allows mapping such rocks. Where older and basement rocks have a darker color, multispectral satellite imagery can provide indications for their mapping. However, often young volcanic rocks

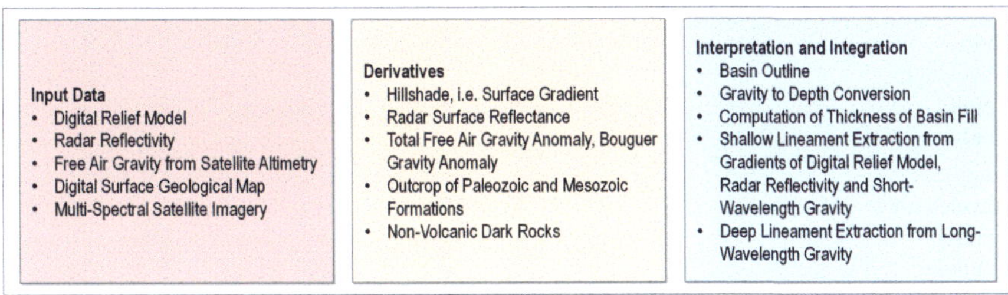

Fig. 9.12 Data sets, processing, interpretation and integration in basin modeling workflow

also show very dark color. Using the surface geologic map can help discriminating dark young volcanic rocks from older and basement rocks. Table 9.2 lists the data sets and their derivatives that can be used to derive specific characteristics of sedimentary basins.

For the integration of the interpretation of different data sets and their derivatives, the following methods for integration of input data and derivatives proved to be useful:

- **Overlay of multiple data sets**—these highlight features where different data sets are spatially correlated.
- **Draping the derivative over the DRM**—this integration is achieved by rendering the derivative, for example the gradient, as a surface attribute on the DRM. Vertical exaggeration and slant illumination highlight the DRM structure whereas the gradient emphasizes relative elevation changes.
- **Spatial masking of one data set with another**—this approach is used when one data set delineates one specific character of the studied layer. For example, the

Table 9.2 Overview of mapping capabilities of input and derivatives. X—mapping capability, (X)—indications

Data Set	Topo-graphy	Hardrock outcrop	Volcanic outcrop	Fault-related rivers	Basin outline	Basin depth	Shallow faults	Deep faults	Figure
DRM	X								9.13
DRM Gradient		(X)			X		X		
Radar Backscatter		X	(X)	X					9.14
Gradient Radar Backscatter	(X)			(X)	X		X		
Multi-Spectral Satellite Imagery	(X)	X	X		(X)		(X)		9.15
Total Bouguer Gravity Anomaly					(X)	X			9.16
Short-Wavelength Gravity Gradient					X		X		
Long-Wavelength Gravity Gradient					X			X	
Geological Map	(X)	(X)	X		(X)				9.17

distribution of volcanic rocks shown on the geological map is used to mask dark rocks on the satellite imagery to isolate dark non-volcanic basement and Paleozoic rocks.

- **Overlay of interpretation obtained from one data set onto another data set**—through this approach interpretation that is ambiguous on one data set can be supported by interpretation on another data set.

9.3.1 Interpreting Satellite Gravity Data for Basin Depth

At the start of the basin we infer the basin depth from satellite gravity measurements. The resulting free air gravity anomaly (Fig. 9.13a) gives some indications of the basin outline already. After elevation correction to the Bouguer gravity anomaly using the digital relief model and conversion to depth assuming a homogeneous density of the subsurface rocks, the relative basin depth can be estimated. The estimate for the absolute basin depth is calculated relative to the digital relief model. It is shown in Fig. 9.13b. The Ghadamis, Murzuq and Kufrah Basins stand out as basins clearly separated by basement highs of the Gargaf Arch, the ridge leading towards Tibesti and Gilf Kebir. The Sirt Basin in central Libya is well defined on the west and south but less certain on the eastern side, where we can assume its boundary to be described by the Ajdabiya Trough.

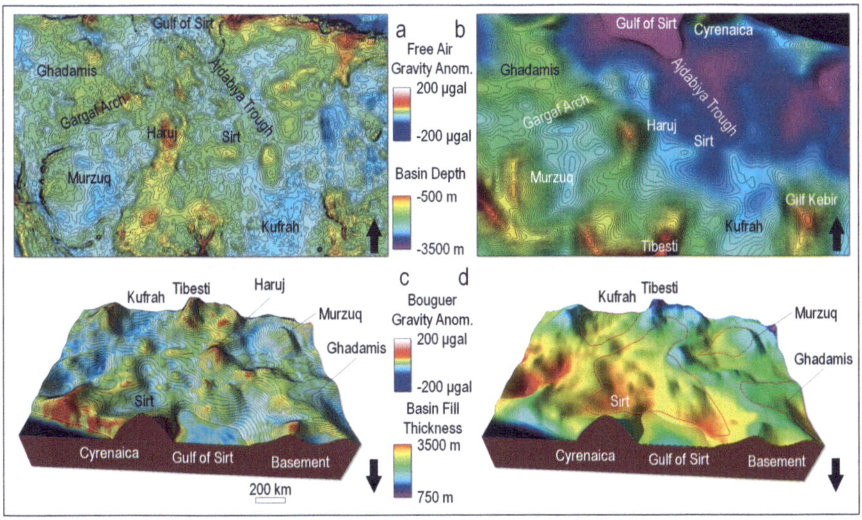

Fig. 9.13 Basin depth from gravity anomaly data derived from radar altimetry, (**a**) free air gravity anomaly. (Satellite gravity data courtesy Sandwell et al., University of California San Diego), (**b**) basin depth derived from gravity anomaly, (**c**) Bouguer gravity anomaly draped on top basement, (**d**) thickness of basin fill draped on top basement, basin outlines as re lineaments. (Satellite gravity data courtesy Sandwell et al., University of California San Diego)

The consistency of the basin depth estimate with the free air gravity input can be studied by draping the measurement onto the basement horizon and rendering it in 3D (c). For most basins the trends of free air gravity and estimated basin depth are aligned, only the Sirt Basin does not show a smooth basin floor, which is related to its intense internal structuration.

Finally, we correlate the thickness of the basin fill, i.e. the estimated thickness of the sediments filling the basins, with the topography of top of basement. When rendering these data in 3D (d), also the Sirt Basin stands out as a basin through its consistent thick sediment cover. Using the different contributions of gravity anomaly data we can now delineate the outline of the basins. They are shown as red lines in Fig. 9.13d.

9.3.2 *Interpreting the Digital Relief Model Data*

The digital relief model (DRM) characterizes the top layer of the basin model (Fig. 9.14). The input DRM (a) reveals two half-moon shaped elevated rims in the southwest. They form the southern boundaries of the plains of the Marzuq and Sirt Basins. To their east, the isolated highland of Gilf Kebir marks the eastern boundary of the plains of the Kufrah Basin. The north west is characterized by the mountainous terrain of the Gargaf Arch and the ridges bordering the Ghadamis Basin. In the center, the Haruj volcanic complex stands out as an isolated feature. Towards the

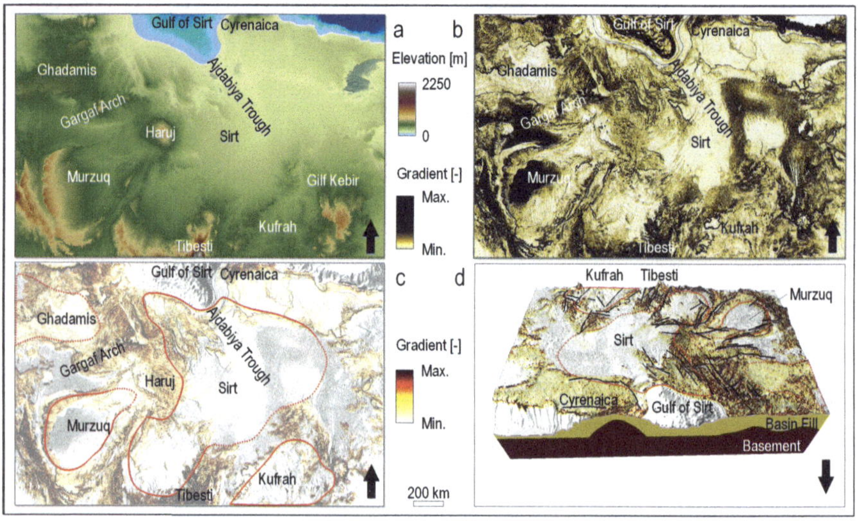

Fig. 9.14 Interpretation of the digital relief model (DRM) and its derivatives, (**a**) input DRM, (**b**) vertical gradient of DRM, (**c**) interpretation of the DRM texture for basin outline (red lines) and surface faults (black lines), (**d**) 3D view of the interpreted DRM. (SRTM digital elevation data courtesy of USGS)

east of the Haruj complex, the plains of the Sirt Basin stretch up to the Gulf of Sirt of the Mediterranean. Only the Cyrenaica platform at the Mediterranean coast exhibits mountainous character.

The topographic structure becomes evident in the vertical surface gradient or surface dip (b). The escarpments at the Murzuq and Sirt rims and the escarpment of Gilf Kebir stand out as clear lineaments. The structure of the mountainous area northeast of the Gargaf Arch reveals northwest-southeast trending structural horsts and grabens. The structural heterogeneity of the Haruj and Tibesti volcanic complexes is highlighted by multiple craters and lineaments associated with lava flows. There is, however, also noise that masks part of the Murzuq and Sirt Basins as well as the area west of the Gargaf Arch. These dark areas are the result of massive dune fields with dunes several 100 m high.

The surface expression of the basins follows the interpretation of the DRM and its gradient. When rendering the gradient as a surface attribute on the DRM, vertically exaggerating the DRM 100 times and applying slant illumination (c) the regional structure and the local texture become visible. For this display the noise related to the sand dunes has been removed using the mask derived from radar backscatter data, since radar backscatter in rough dunes areas is minimal.

The basin outline is masked by dune fields in the south and east of the Sirt Basin and not well expressed in the Ghadames Basin due to the rough terrain. The surface gradient guides most of the surface structural delineation, which is then supported with subtle elevation features better revealed on the DRM. In the Kufrah Basin the approach was swapped, starting the interpretation at the DRM and validating the lineaments on the surface gradient.

The 3D rendering (d) also employs 100 times vertical exaggeration and slant illumination to highlight the regional outlines of the basins and the local texture of horsts and grabens between Sirt and Ghadames Basins, the complex texture of the Haruj and Tibesti volcanoes and the sand-covered Sirt and Murzuq Basins. In the 3D rendering, the outline of the Kufrah Basin becomes more evident.

9.3.3 Interpreting Radar Backscatter Data

Radar backscatter data show a clear distinction between the sand-covered basins with minimal backscatter intensity and almost total absorption (Fig. 9.15a). The higher intensity of the surface backscatter from rocky areas clearly reveals the texture of the horst and graben terrain north of Haruj. It also shows structural details of the escarpment bordering the Murzuq Basin as well as the Haruj and Tibesti volcanoes. The Kufrah Basin is very well distinct from the Sirt Basin by its strong texture of lineaments, which are associated with shallow buried paleo-drainage. The Kufrah Delta stands out in the Sirt Basin just southeast of the southern end of the Ajdabiya Trough. Some noise from long ridges of high dunes on the southwest Sirt Basin distract from the two ridges of outcropping rock that connects the Haruj and Tibesti volcanoes.

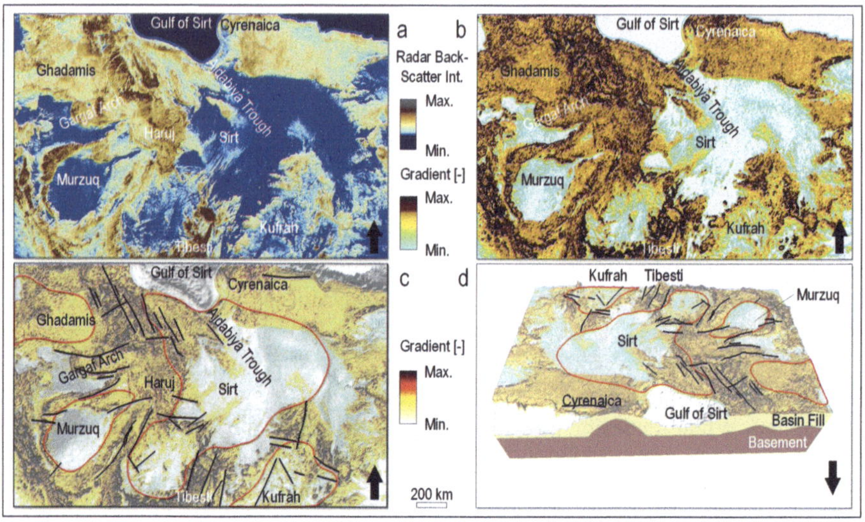

Fig. 9.15 Interpretation of the radar backscatter intensity and its derivatives, (**a**) input radar intensity, (**b**) gradient of radar intensity, (**c**) interpretation of the radar intensity gradient texture draped on the DRM for basin outline (red lines) and surface faults (black lines), (**d**) 3D view of the interpreted radar intensity data draped on the DRM. (DRM data courtesy of USGS, Radarsat-1 data courtesy of CSA)

The gradient of the radar backscatter intensity (b) highlights the lineaments already discovered on the backscatter data itself. The lineaments in the horst and graben terrain in the north, the eastern boundary of the Haruj volcano as well as strong north-south trending lineaments in the Tibesti volcanic system enhance the structural insight.

The rendering as a surface attribute on the DRM with slant illumination (c) emphasis the delineation of both the basis (red lines) and shallow structural lineaments (black lines).

The 3D rendering (d) shows more structural detail than the DRM in the northern horst and graben area between the Murzuq Basin and the Gulf of Sirt. The structural lineaments in the Tibesti and Kufrah areas are also enhanced.

9.3.4 Interpreting Multi-spectral Optical and Infrared Satellite Data

The satellite image of the study area shows vivid patterns of color corresponding to different surface rock lithology (Fig. 9.16a). The Murzuq Basin stands out as a sand-covered basin with a rim of dark rocks. The edges of the Kufrah Basin are sharply delineated as dark rock outcrops contrasting against the sand that is blown

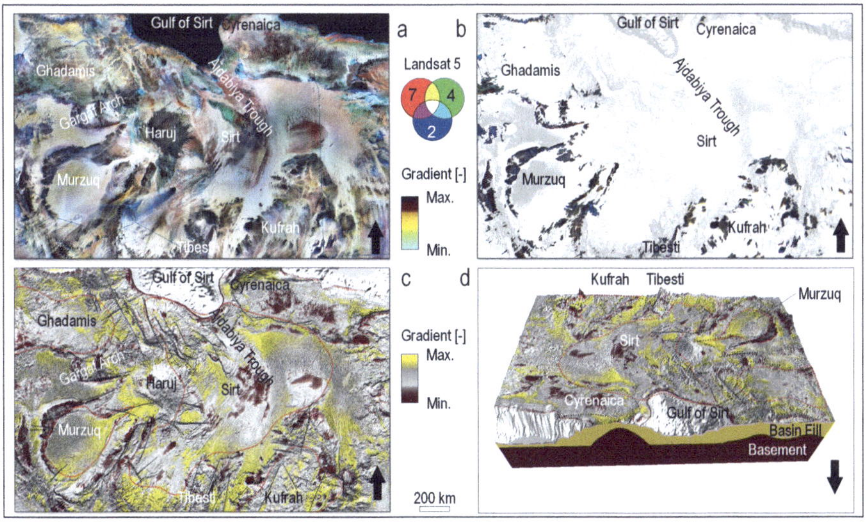

Fig. 9.16 Interpretation of multispectral satellite imagery and its derivatives, (**a**) input multispectral satellite image Landsat 5 742 RGB, (**b**) extract of non-volcanic dark rock outcrops, (**c**) interpretation of the Landsat image color intensity draped on the DRM for basin outline (red lines) and surface faults (black lines), (**d**) 3D view of the interpreted non-volcanic dark rock outcrops draped on the DRM. ((DRM and Landsat 5 mosaic data courtesy of USGS)

in from the Sirt Basin in the north. The Sirt Basin itself reveals a very heterogeneous surface lithological character. Sands (bright yellow) contrast with iron-rich deposits (reddish) and sabkha (cyan) as well as carbonates (greenish and brown). The basalts of the volcanoes of Haruj and Tibesti stand out as dark features. North of Haruj the horst and graben terrain is emphasized due to its contrasting lithology of sand cover in the grabens and dark rocks on the horsts.

For basin delineation the non-volcanic dark rock outcrops are extracted using a mask of Paleozoic and Mesozoic formations from the surface geological map whilst removing the dark volcanic rocks (b). The gradient of the DRM is plotted faintly in the background for orientation. Murzuq, Kufrah and Ghadamis basin outlines become evident, whereas the outline of the Sirt Basin can only be defined in the south.

The color intensity of the satellite image can be rendered as a surface attribute on the vertically exaggerated DRM with slant illumination (c) to highlight the basin outline. The horst and graben system north of Haruj is the only area where the multi-spectral satellite image adds to the surface structural mapping.

The 3D rendering on the vertically exaggerated DRM (d) confirms the observation that the main contribution of multispectral satellite imagery to basin model is to the basin outline.

9.3.5 Interpreting Satellite Gravity Data for Basin Structure

Bouguer gravity anomaly data are used as input for the interpretation of satellite gravity data (Fig. 9.17a). As described above, the Bouguer gravity anomaly data are computed from the free air gravity anomaly data that have been derived from satellite altimetry (Sandwell et al., 2001). The unfiltered Bouguer anomaly data show a combined density effect of elevation and shallow high-density rocks with the bulk gravity response of the basin fill. The unfiltered Bouguer anomaly is, therefore, a good indicator for tectonic structure between basement and surface.

The structural lineaments shown in (a) complement the structural lineaments derived from the interpretation of DRM and radar backscatter intensity, particularly the internal structuration of the Sirt Basin. The lateral gradient of the Bouguer gradient emphasizes structural lineaments. To obtain this, the Bouguer gravity anomaly has been smoothed at 5 and 10 km, which corresponds to spatial low-pass filtering at 5 and 10 km wavelength, respectively. The difference between the two filtered data sets corresponds to the texture of the Bouguer gravity anomaly in the range of 5–10 km wavelength, which gives an insight into the structural framework of the basin fill (b). The overlay of the lateral gradient on the unfiltered Bouguer gravity anomaly (c) enhances basin structural features, which have been delineated as black lineaments. The basin outlines, however, are still rather difficult to determine. This will be discussed later, when the map of the thickness of the basin fill is available.

The visualization of these data in 3D eases the interpretation for basin structure even more (d): unfiltered Bouguer gravity anomaly and lateral gradient are co-rendered on the top of basement surface that is estimated from satellite gravity data.

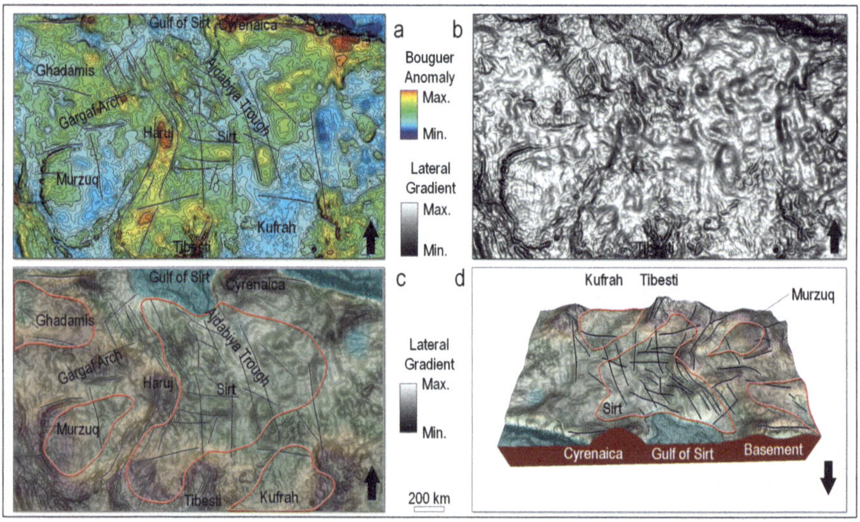

Fig. 9.17 Interpretation of satellite gravity data and its derivatives, (**a**) input Bouguer gravity anomaly, (**b**) Lateral gradient of Bouguer gravity anomaly, (**c**) structural interpretation of lateral gradient overlaid on the unfiltered Bouguer anomaly for basin outline (red lines) and surface faults (black lines), (**d**) 3D view data shown in (**c**). (Satellite gravity data courtesy Sandwell et al., University of California San Diego)

This approach also provides the possibility of validating the structural lineaments delineated in map view in the previous step.

9.3.6 Interpreting Surface Geologic Data

In the next step we amend the geologic age to the basin model. The surface geologic map (Fig. 9.18a) has been simplified to assist with outlining the basins.

Tertiary and Quaternary sediments cover most of the basin interior, whereas Paleozoic and volcanics outcrops determine the outline of the basins. Mesozoic formations assist in basin delineation of the Ghadames, Murzuq and Kufrah basins where outcrops exist (b). When overlaid on the digital relief model (c), the combination provides better clues for the basin outline than each data set alone. The basin outline from outcrop geology is shown as red lined. Rendering the data in 3D (d) supports this approach.

9.3.7 Integration of Basin Outline, Depth and Structure

The individual contributions of the individual remote sensing methods can now be integrated into on basin model (Fig. 9.19). Starting at the surface, we use the photogeologic map (a), i.e. the co-rendering of satellite imagery and surface geologic map, to identify basin outlines at outcrop. Volcanic rocks at the surface indicate that the significance of the gravity data needs to be studied carefully (see Sect. 9.6).

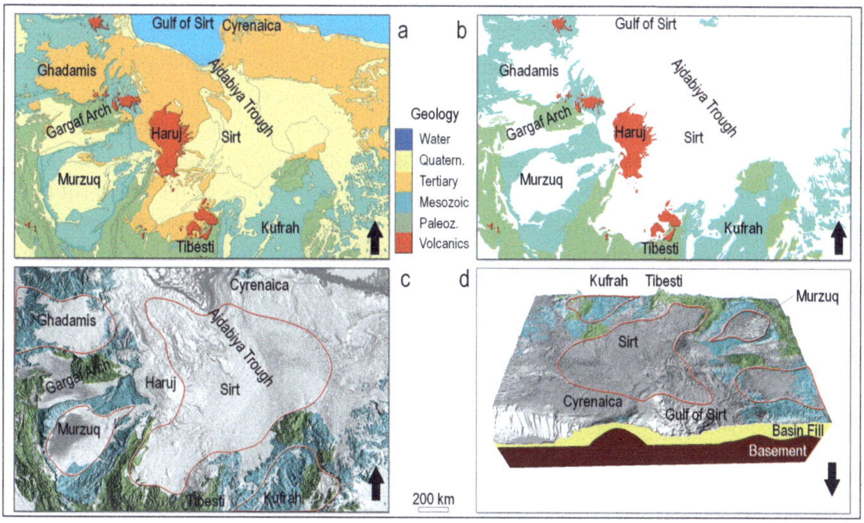

Fig. 9.18 Interpretation of surface geologic map in connection with the digital elevation mode, (**a**) simplified surface geologic map, (**b**) extract of Mesozoic and Paleozoic formation outcrops and surface volcanics, (**c**) idem on DRM, (**d**) idem in 3D view. Geologic map data and DRM based on USGS data

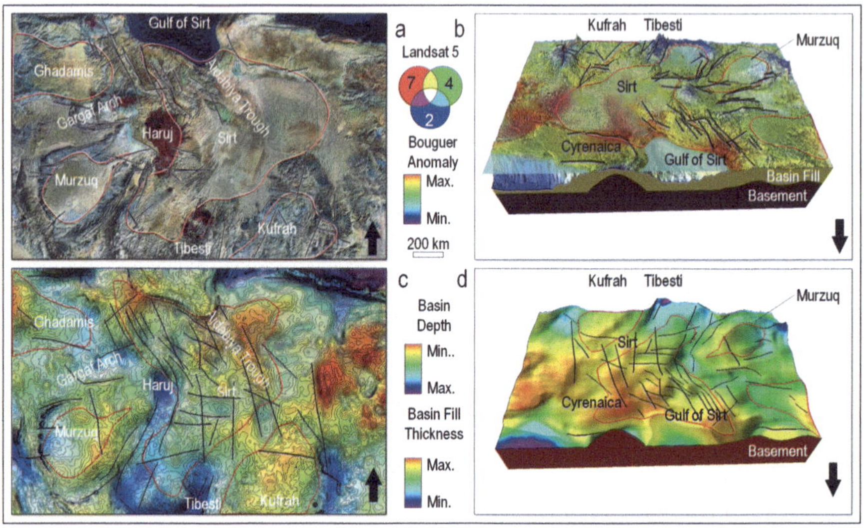

Fig. 9.19 Integration of interpretation from surface to basement, (**a**) photogeologic map, (**b**) Bouguer anomaly draped on DRM, (**c**) c-blending of basin depth and thickness of basin fill, (**d**) thickness of basin fill draped on top of basement horizon. Basin outlines are shown in red, black lineaments indicate inferred regional faults. (Satellite gravity data courtesy Sandwell et al., University of California San Diego, surface geologic map and DRM courtesy of USGS)

Rendering the Bouguer gravity anomaly on the DRM (b) gives a first indication of the location of the basins. The lateral gradient of DRM and radar backscatter provides the delineation of outcropping and shallow regional faults.

The basin structure is enhanced when co-rendering basin depth and thickness of sedimentary fill. For the thickness of the basin fill we use the standard color code from purple for minimum to red for maximum thickness. To achieve the enhancement, we invert the color scale for basin depth showing minimum depth as red and maximum depth as purple (c). The basin outline becomes easier to interpret. This rendering serves also for validation of the structural lineaments that had been obtained from the gradient of spatially filtered gravity data. To understand the basin structuration and compartmentalization we drape the thickness of the basin fill onto the top basement horizon and co-render the basin outline and the regional faults (d).

The basin model for the Libyan basins shows that Murzuq and Kufrah Basins are very well delineated with relatively simple internal structuration, whereas the Sirt Basin is extensively structured especially in the northwestern part north of Haruj and the Gargaf Arch.

9.4 Petroleum System Modeling Central Libya

Petroleum system modeling is a workflow that integrates in a basin the subsurface geological setting with depositional reconstruction and thermal models to estimate the maturation of kerogen to hydrocarbons and to predict their expulsion and migration. Hantschel and Kauerauf (2009), Magoon and Beaumont (1999), and Waples (1994) give an introduction into the methodology. The key ingredients required for petroleum systems modeling is the litho-structural basin model to which remote sensing can contribute through providing an estimate of the basin-wide surface to basement horizons as well as indications for regional structuration (Fig. 9.20). When integrating remote sensing data with stratigraphic charts, surface geologic maps and well logs from adjacent basins the structural basin model can be refined by basin strata. Heat flow and thermal models provide the information required to assess transformation from kerogen to hydrocarbons as well as expulsion and migration in the reservoirs.

We illustrate the workflow at the southwestern Sirt Basin in Libya (Laake et al., 2012). Rusk (2001), Hallett (2002), and Hassan and Kendall (2014) provide a comprehensive description of the petroleum geology of Libya. Gumati and Schamel (1988), Gumati et al. (1996), and Ahlbrandt (2001b) focus on the petroleum systems of the Sirt Basin. The simplified stratigraphic column for the Sirt Basin (Fig. 9.21) gives an overview of the subsurface lithology as well as the source and reservoir rocks after Rusk (2001) amended with the late Neogene and Quaternary strata. The Mesozoic petroleum system comprises multiple Upper Cretaceous source rocks and Cretaceous to Lower Paleocene reservoir rocks. Seals are composed of Upper Cretaceous evaporites, Paleocene shales and Eocene shales. We notice that the Paleozoic petroleum is absent in the Sirt Basin because of a major hiatus from Cambrian to Late Lower Cretaceous. In western and southern Libya, however, these formations are present and outcrop where they are accessible for remote sensing delineation of the basin outlines.

The remote sensing data contributing to the petroleum system modeling are summarized in Fig. 9.22. We overlay the surface geologic map and the satellite image to create the photo-geologic map and drape it over the vertically exaggerated digital relief model (a) to reveal the outcropping formation outlining the basins (red lines).

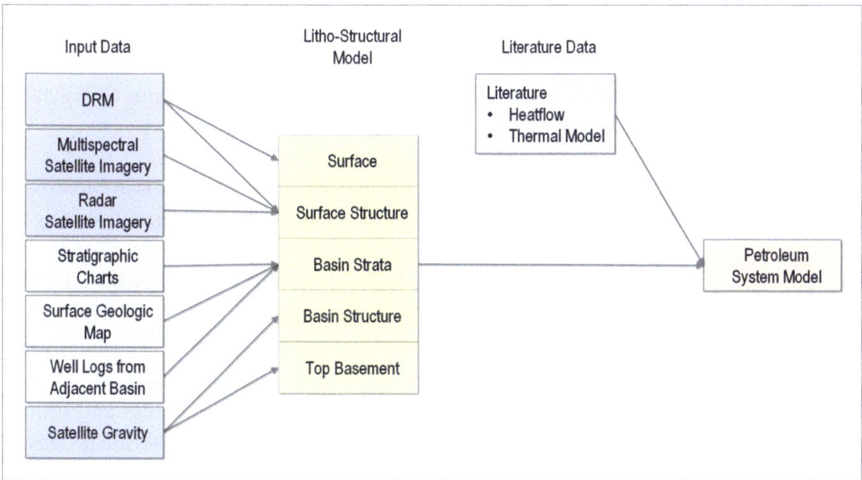

Fig. 9.20 Workflow of remote sensing-assisted petroleum system modeling

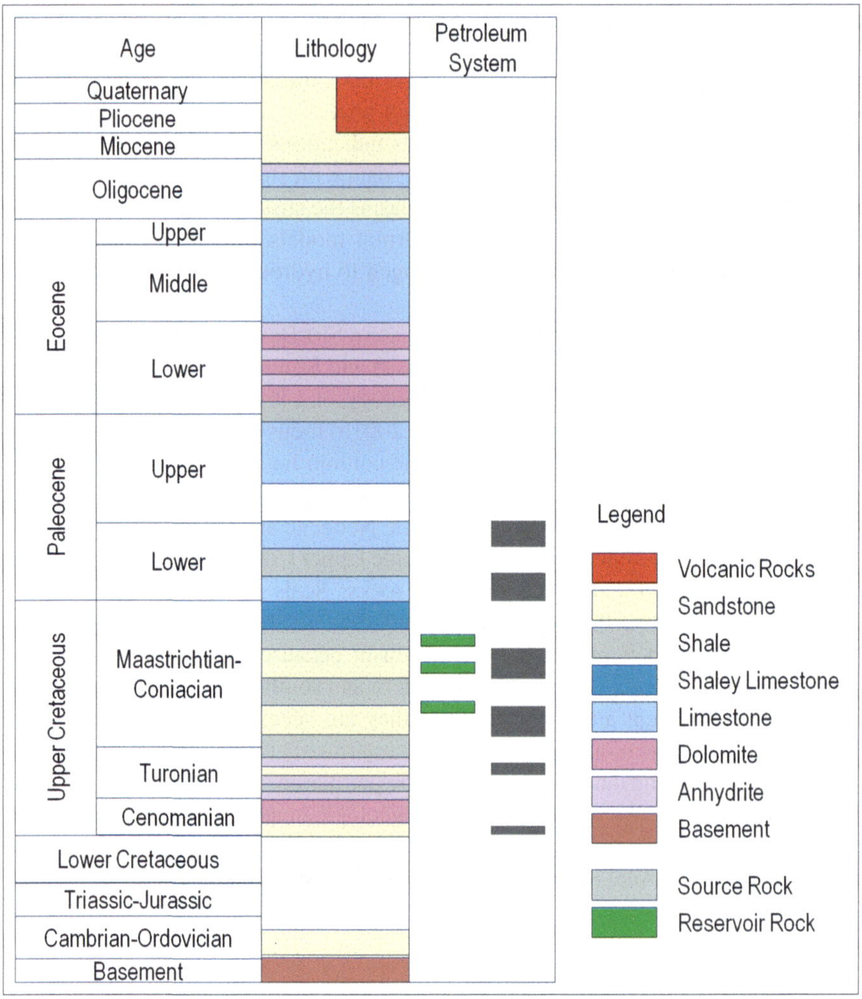

Fig. 9.21 Simplified stratigraphic column for the Sirt Basin, Libya. (Modified after Rusk (2001) with data from Elshaafi and Gudmundsson (2016))

Outcrops of regional faults are shown as black lineaments. We observe that Mesozoic and Paleozoic outcrops form the boundaries of the Ghadamis and Murzuq Basins as well as the Bu Tamaym Trough in the south western portion of the Sirt Basin whereas the western boundary of the Sirt Basin is formed by outcrops of Late Cretaceous and Paleogene strata as well as by volcanic rocks.

For the 2D petroleum system modeling we have chosen an arbitrary line (cyan) along the Bu Tamaym Trough around the Haruj Volcano into the western Sirt Basin, where a detailed petroleum system models exist. We take the thickness and lithology from wells in the Sirt Basin for the calibration of the southwestern Sirt Basin model that is derived from remote sensing data.

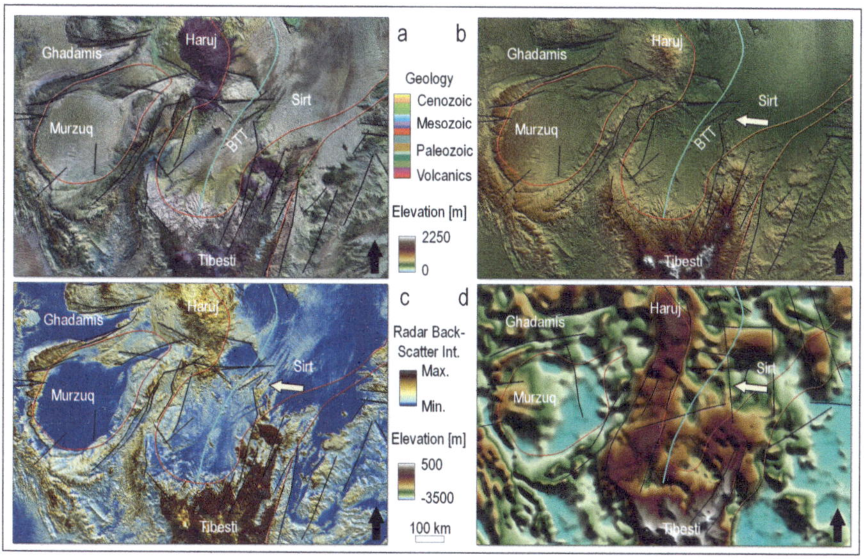

Fig. 9.22 Remote sensing data and interpretation used in remote sensing-assisted petroleum system modeling, (**a**) photogeologic map, (**b**) digital relief model, (**c**) radar back-scattering intensity, (**d**) basement depth derived from satellite gravity. *BTT* Bu Tamaym Trough. Landsat 5 TM satellite imagery. (SRTM DEM and surface geological map courtesy of USGS, Radarsat-1 data courtesy of CSA, satellite gravity data courtesy Sandwell et al., University of California San Diego)

The digital relief model (b) provides the surface topography delineating the basin outline, which is supported by radar back-scatter data (c). Regional fault outcrops are evident at the basin boundaries. In addition to that, the Bu Tamaym Trough appears segmented by faults into a northeastern and a southwestern part (see arrow).

The basin depth from satellite gravity (d) indicates that the Bu Tamaym Trough is less deep than the Murzuq Basin. The segmentation that had been observed at the surface appears to reach down to basement. We will, therefore, include it into the 2D petroleum system model.

From DRM and satellite gravity derived basement we can create the outline of the 3D basin model. Figure 9.23a shows the photogeologic map in 3D. The boundary of the Murzuq basin is clearly delineated by outcropping Paleozoic ridges. The Bu Tamaym Trough is located between the Haruj and the norther part of the Tibesti volcanic systems. Overlaying the free air gravity anomaly on the top basement surface (b) enhances the basement basin structures. The Bu Tamaym Trough (BTT) shows a clear segmentation into a northern portion near the Haruj Volcano and a southern portion. The location of the 2D section along which the petroleum system has been modeled is shown as a cyan lines.

For the creation of the 2D petroleum system model (Fig. 9.23b) we start in the north with existing wells in the western Sirt Basin to establish the lithostratigraphic sequence. The reference for the model is the top of the Lower Cretaceous Nubian Sandstone. In the south we use the formation outcrops the lithology. Satellite

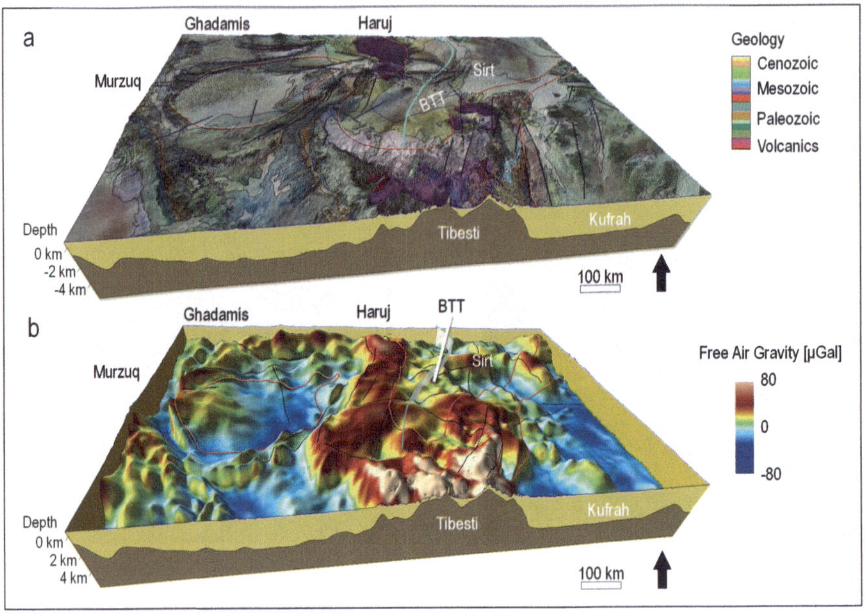

Fig. 9.23 3D basin model, (**a**) surface characterization from DRM overlaid with photogeologic map, (**b**) top basement horizon overlaid with free air gravity data, red lines—basin outlines, black lines—regional faults. (Landsat 5 TM satellite imagery, SRTM DEM and surface geological map courtesy of USGS, satellite gravity data courtesy Sandwell et al., University of California San Diego)

imagery provides clues for the outcrop lithology and the surface geologic map allows converting the outcrops to geologic time. The trend in between is inferred from satellite gravity. Satellite radar and gravity provide hints for the location of segmenting faults. The location of the 2D section as well as of the segmenting faults is shown in map view in (a) with an overlay of surface radar back-scattering intensity and basin depth from gravity (Fig. 9.24).

In the next step we introduce regional heat flow models to simulate the kerogen to hydrocarbon transformation ratio within the formations and apply them to the litho-stratigraphic section (Fig. 9.25c). We assume different heat flow curves for the Paleozoic time (b) and Paleogene time intervals (a). The thermal modeling focuses on the source in the Upper Cretaceous Sirt and Rachmat Formations (upper green formations in (c)). The resulting total hydrocarbon transformation ratio for these formations is shown in (d). For regional slowly varying heat flow, the transformation ration depends mainly on the burial depth. Therefore, lower parts of the interval of interest show higher transformation ratios than the shallower portions like the horst in the segmented area and the southern slope.

In the final step we estimate the chance of success for a working petroleum system generating hydrocarbons that could migrate into the subsurface structural trap (Fig. 9.26). For this computation we include the litho-structural horizon model, the structural segmentation and the transformation ratio. For the southern portion we

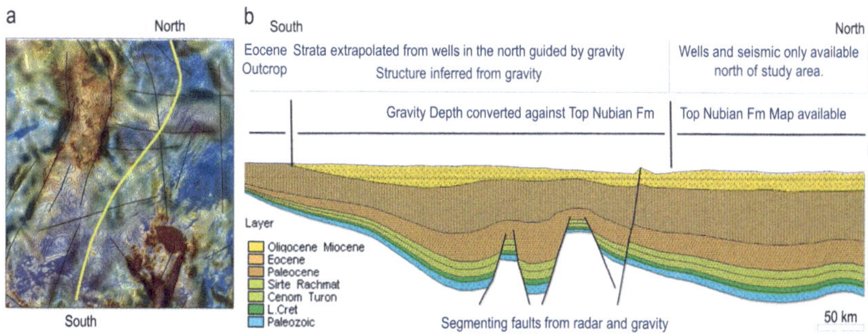

Fig. 9.24 Proposed South Sirt basin structure and geometry derived from remote sensing data and extrapolated from known basin model, (**a**) basin structural map, (**b**) structural section. (after Laake et al. 2012, Landsat 5TM satellite imagery, SRTM DEM and surface geological map courtesy of USGS, Radarsat-1 data courtesy of CSA, satellite gravity data courtesy Sandwell et al., University of California San Diego)

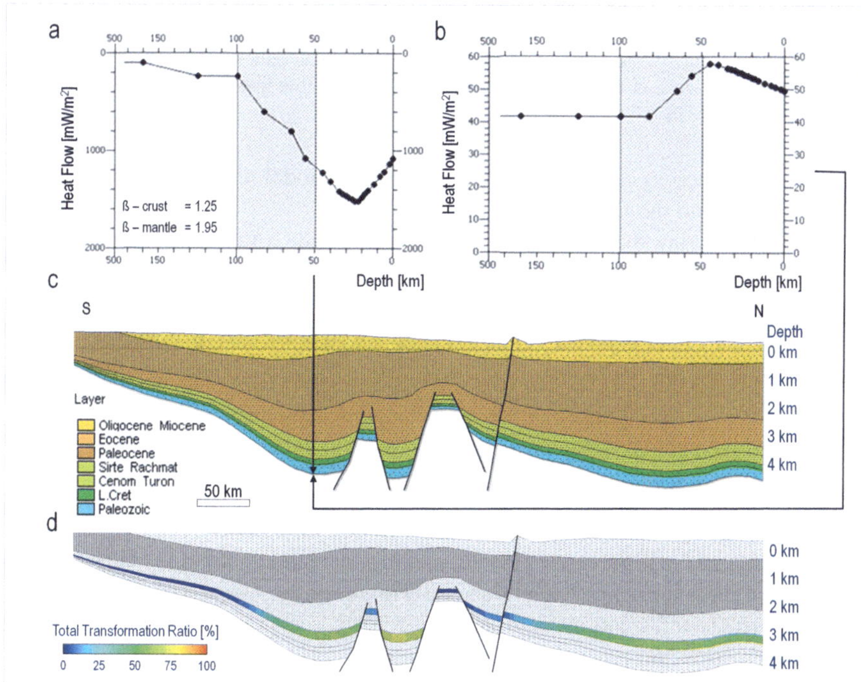

Fig. 9.25 Petroleum system model for South Sirt Basin, (**a**) regional heat flow curves for basin fill, (**b**) idem for basement, (**c**) proposed lithostratigraphic section, (**d**) transformation ration model for Upper Cretaceous Sirte and Rachmat formations (after Laake et al. 2012)

also include the risk of leakage due to potentially missing migration barriers. The chance of the success is then color coded from low (red) to high (green) along the geologic section. To complete the integration we co-render the geologic section and

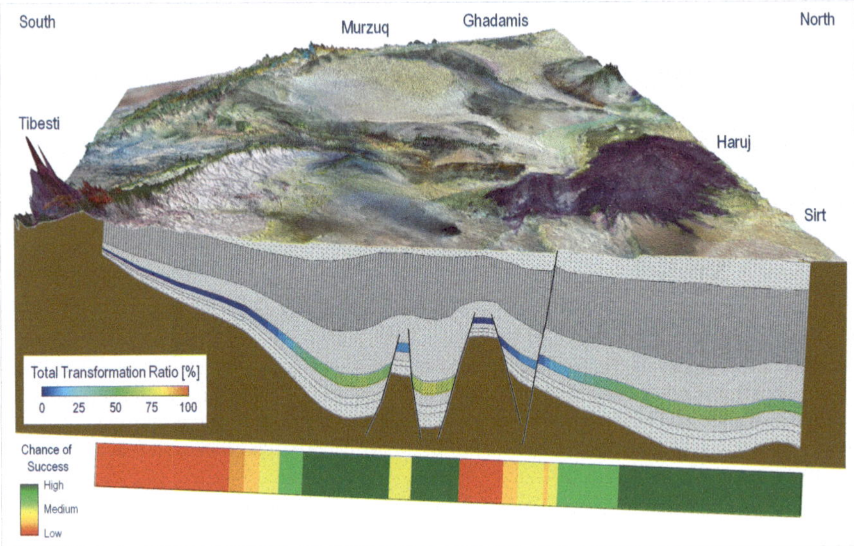

Fig. 9.26 Petroleum system model derived from remote sensing data in the context of the south Sirt Basin model. Laake et al. 2012. DRM, Landsat 7 TM satellite imagery and surface geologic map courtesy of USGS

the chance of success coding with the 3D geologic model with draping the photogeologic map onto the DRM surface.

It is worth noting that this methodology can provide a first approximation of a basin and petroleum system model. Unconformities, however, cannot be detected by the method and hence are not included in the model. The mapping of unconformities requires the availability of 3D geophysical data such as airborne full tensor gravity gradiometry (Veryaskin, 2018; Bell et al., 1997) or 3D surface seismic.

9.5 Surface-Subsurface Integration

In this chapter we will integrate remote sensing information about the surface with geophysical measurements that produce a 3D image of the subsurface. The first case study integrates information from satellite remote sensing with shallow seismic from shear waves and refraction and deep penetration reflection seismic. The second case study concerns the integration of regional 2D share wave seismic with optical and radar satellite imagery aiming at an estimation of the maximum penetration depth of satellite radar.

9.5.1 *Integration from Surface to Deep Formations Using Geologic Analogs*

The area for the first case study is located on the western coastal plains of the central Gulf of Suez in Egypt (Laake, 2011; Laake et al., 2011a, b, c). The study is set in the Tertiary rift environment of the Gulf of Suez/Red Sea rift (yellow box in Fig. 9.27). In the central part the asymmetric rift shows a broad coastal plain in the foreland of the western rift shoulder that is formed by the Red Sea Mountains (Alsharhan & Salah, 1995; Abd El-Naby et al., 2010; Moustafa & Khalil, 2017; Bosworth & McClay, 2001). The basement rocks of the Red Sea Mountains show southwest-northeast trending faults that pre-date the Gulf of Suez rift. These faults have created drainage passages from the Wadi Qena area to the west of the Red Sea Mountains allowing the transport of clastic sediments across the Red Sea Mountains into the Gulf of Suez Basin, where they form important clastic reservoirs in the subsurface.

The integration of surface information from remote sensing data is used to delineate the extent of the Gulf of Suez rift. Considering the geologic history of the fault-related sediment fairways across the Red Sea Mountains, present day channel and fan delta features are then used as geologic analogs for the interpretation of subsurface seismic data for fossil channels and fan deltas thus giving clues for fossil depositional environments and paleo-coastlines.

The workflow for the integration of interpreted satellite imagery and of shallow and deep seismic data is shown in Fig. 9.28. Geologic literature provides background information for the geological setting prior, during and post-rifting and, thereby, indicates how remote sensing of the present-day surface can assist in the subsurface interpretation.

Satellite remote sensing provides a digital relief model allowing to map the expression of pre-existing fault-related sediment fairways across the Red Sea

Fig. 9.27 3D rendering of the central Gulf of Suez and the study area (yellow box)

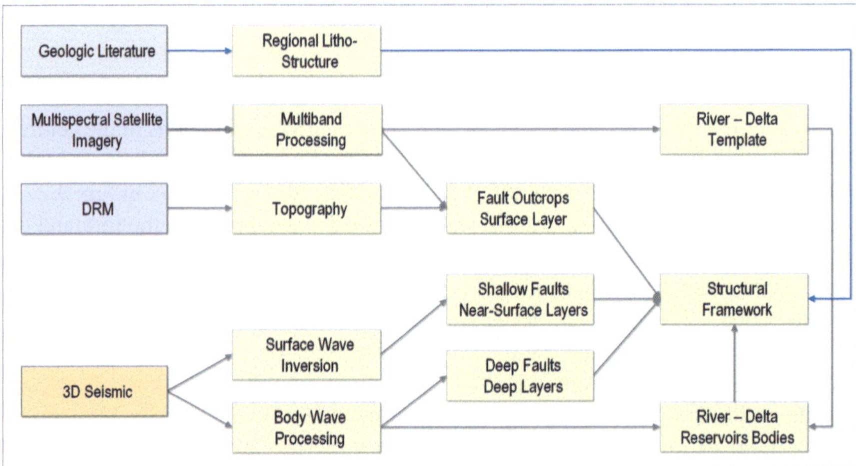

Fig. 9.28 Workflow for integration of satellite imagery, shallow and deep seismic imaging and the use of surface wadis as geologic for ancient channels and fan deltas

Mountains. The delineation of rift-parallel and transform faults in the coastal plains is obtained from multiband processing of multispectral satellite imagery because the faults impact the deposition of weathered debris from the Red Sea Mountains as well as the transport of clastic sediments across the Red Sea Mountains.

3D seismic data are processed separately for shallow horizons—from the surface to about 150 m below surface—and for deep horizons. For the shallow interval, the inversion of surface waves and the processing for shallow refracted waves provides a high-resolution insight into the elastic structure which is then correlated with the structural features observed in the processed satellite imagery. Horizons deeper than 150 m below surface are mapped from the processing of body waves.

Based on the tectonic and deposition history described in the literature, the river channel and fan delta template of the present-day Wadi Hawashia is used for detecting similar features in the subsurface when interpreting the seismic body wave data.

The litho-stratigraphic summary (Fig. 9.29) provides the stratigraphic context for the workflow. Crystalline and metamorphic basement rocks outcrop on either shoulder of the Gulf of Suez rift in the Red Sea Mountains and in the Sinai Peninsula. Cambrian to Mid Cretaceous formations are buried deep in the pre-rift deposits in the Gulf of Suez basin, but they outcrop in the Wadi Qena area where they represent a clastic source for Gulf of Suez clastic reservoirs. The Upper Cretaceous to Lower Miocene carbonate and clastic rocks formed reservoirs. We will study the clastic reservoirs in more detail when interpreting the deep seismic data. Mid to Upper Miocene salts that originate from the evaporation of the seas in the Eastern Mediterranean and the Red Sea during the Miocene salt crisis form the seals for the reservoir rocks. The ingression of the seawater through the opening of the Strait of Aden resulted in a major discontinuity after which clastic deposition mainly from the Sinai Peninsula and across the Red Sea Mountains filled Gulf of Suez basin almost completely.

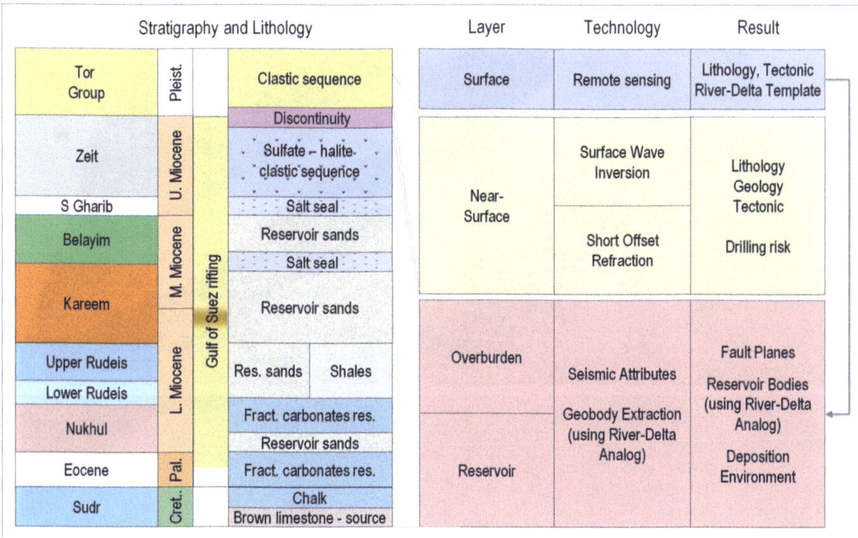

Fig. 9.29 Litho-stratigraphic summary of the case study area in the western Gulf of Suez coastal plain

When studying the elevations of the present-day setting of the fault-related valleys using the digital relief model, we notice that between the western entrance at the Wadi Qena area to the eastern mouth onto the Gulf of Suez coastal plains the elevation changes by a mere 20 m (Fig. 9.30a). This means that even today flash floods may transport sediments across the Red Sea Mountains into the Gulf of Suez Basin. The interpretation of the deep seismic data using the present-day Wadi Hawashia channel-fan template yields four fossil deltas from Mid Miocene Kareem and Upper Miocene Zeit to Pleistocene Wardan formations (b). For each stage the front of the fan delta gives clues for the location of the paleo-coastline (dashed lines in b), which divides the depositional environment into an onshore environment west of the coastline and an offshore environment east of it. The location of the deltas within the present-day Wadi Hawashia channel indicates that the transform fault that guides the course of the present-day wadi has been active and in the same place at least since the Lower Miocene, i.e. the last 20 Mio. years.

We will now look closer into the satellite remote sensing data. First, we build a surface geological model from the digital relief model and the multispectral satellite imagery. To assess the quality of the surface model we compare the digital surface geological model with an aerial photo from the same area (Fig. 9.31). The aerial photo (a) is affected by haze and relatively low contrast, which makes the interpretation for structural features and even the detection of the Wadi Hawashia delta difficult. The digital surface model (b) shows significantly more contrast, which can even be enhanced when introducing vertical exaggeration and slant illumination. The details of the wadi channels, the contours of its banks and the shape of the channel and the details are much clearer on the digital image than on the aerial photo.

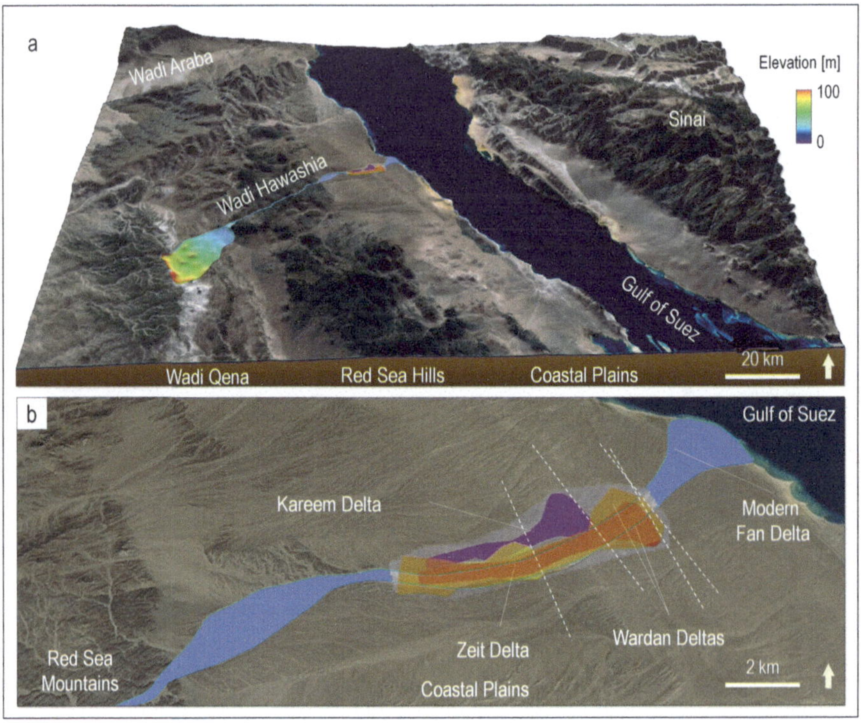

Fig. 9.30 Wadi alleys along transform fault crossing the Red Sea Mountains, (**a**) elevation from Wadi Hawashia catchment to Gulf of Suez, (**b**) fossil subsurface deltas from seismic interpretation projected to surface. (DRM and Landsat 5 satellite imagery courtesy of USGS)

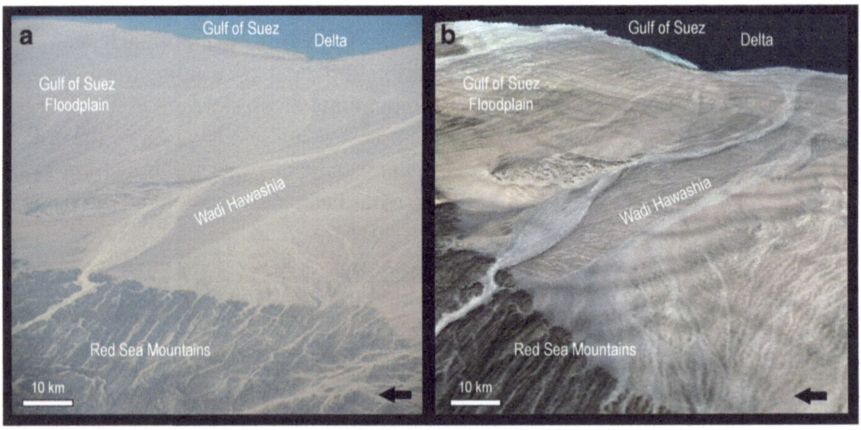

Fig. 9.31 Comparison of aerial photo (**a**) and digital surface geological model (**b**) for the study area. (Photo by author, DRM and Landsat 5 satellite imagery courtesy of USGS)

The digital surface geological model is, therefore, an appropriate representation of the natural environment.

The processing of the multispectral satellite imagery can highlight geologic features (Fig. 9.32). The three-band 321 RGB image of the optical bands (a) distinguishes between the dark basement rocks of the Red Sea Mountains and the light clastic sediments with wadis as well as the water of the Gulf of Suez.

The coastal plain appears largely featureless because the optical bands are not sensitive to lithologic details. Using six bands and forming a difference image of infrared and optical bands as 675-321 RGB enhances significantly the texture of the coastal plain (b). However, optical and infrared satellite imagery can only detect lithologic changes at the surface, i.e. changes in the mineral composition at the surface.

Let us study the multiband RGB a bit closer now (Fig. 9.33). One outstanding feature is the box structure around the course of Wadi Hawashia. The wadi bed shows purplish colors whereas most of the coastal plain is colored in red to orange tones. Also, the course of the wadi appears to bounce back from the straight boundary associated with the color change which suggests a structural boundary in the subsurface. We will use shallow seismic data to validate this hypothesis.

The Red Sea Mountain basement rocks appear now in two different colors. These are related to two different types of granite that are outcropping at the surface. The rather bluish signature of granite 1 outcrops upstream of the wadi. The drainage and the inferred faults guide the weathering debris from granite 1 outcrop into the wadi bed. The sand (light colors), however, originates from the western side of the Red Sea Mountains as no rock outcrops show this color. Granite 2 that outcrops to the north of granite 1 shows a distinct orange tone. The deposition of its weathering products across the coastal plain follows the channel network which in turn is guided by fault lineaments shown as dashed lines. Most of these fault lineaments are oriented WSW-ENE and thus perpendicular to the Gulf of Suez rift axis. We interpret these faults as transform faults which represent re-activated fault that pre-date

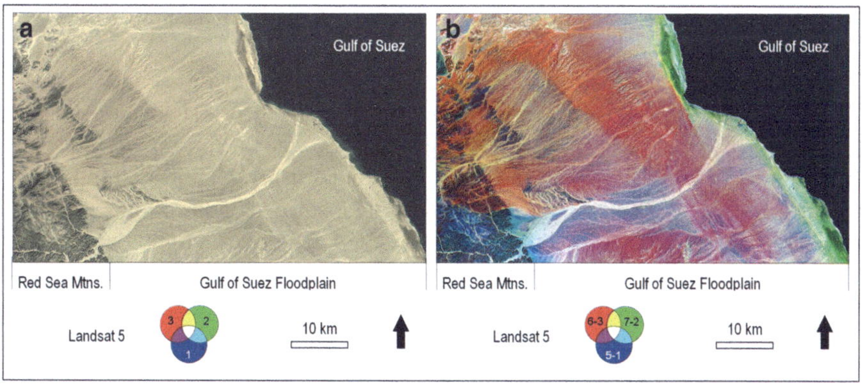

Fig. 9.32 Different methods of processing multispectral satellite imagery, (**a**) Landsat 5321 RGB, (**b**) Landsat 5675-321 RGB. (Landsat 5 satellite imagery courtesy of USGS)

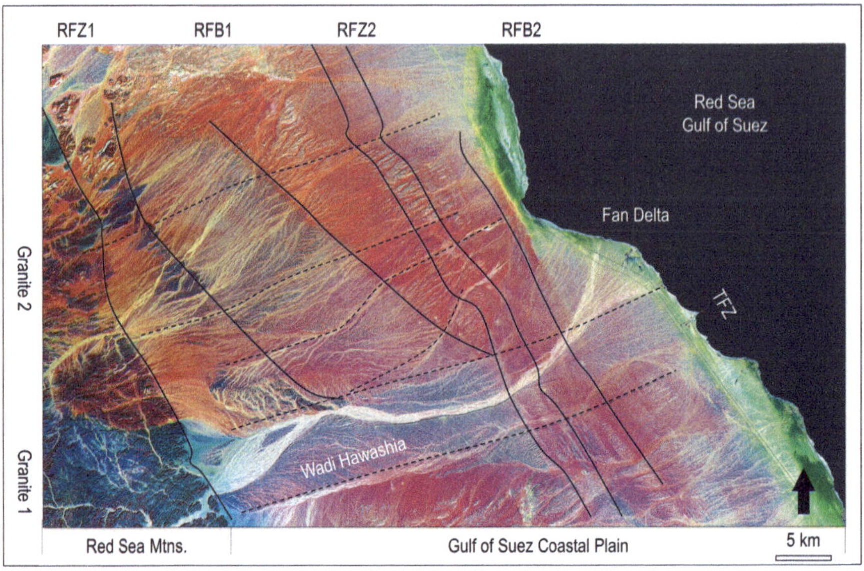

Fig. 9.33 Multiband difference image Landsat 5675-321 RGB with structural interpretation. (Landsat 5 satellite imagery courtesy of USGS)

the rifting. The course of Wadi Hawashia follows the most prominent of these transform fault zones (TFZ).

Rift parallel faults (continuous lines) are marked by patches of dark to bluish color, which mark outcrops of granitic gravels that had been deposited some time ago. The zones with rift faults (RFZ1 and RFZ2) delineate the zones with tilted fault blocks (RFB1 and RFB2).

Rifting breaks up the subsurface of the coastal plain in fault block that are tilted to the west due to the extensional rift forces. The edges of these fault blocks create very subtle edges in the terrain which have an impact on the drainage and hence on the distribution of the weathered debris. Therefore, we infer most changes in color are related to fault outcrops.

When analyzing the panchromatic band of Landsat 5 imagery (Fig. 9.34a) we can color code the reflection intensity such that it is directly correlated with geological features observed in the field. In brownish tones we see the lineaments related to the outcrop of rift parallel faults (white arrows) and transform faults delimiting the Wadi Hawashia block (blue arrows). On the ground, however, none of these features are visible (b). The coastal plain appears as a largely featureless slightly undulating plain covered with sand and gravel of very similar color. This observation confirms the featureless satellite image we saw in Fig. 9.32a.

Interestingly, the only tree on the entire coastal plain in the study area (c) is located right at the intersection of a rift parallel fault (blue arrow) with the northern transform fault of the Wadi Hawashia block (dashed line). Groundwater appears to be diverted by the faults and directed towards fault intersections where

Fig. 9.34 Satellite image processed for lithology and field photos. (**a**) Landsat 5 band 8 intensity image color coded for lithology, (**b**) location of the only tree on the coastal plain at the intersection of a rift parallel fault with the Wadi Hawashia block, (**c**) view of the featureless coastal plain north of Wadi Hawashia. (after Laake 2011, Photos by author, Landsat 5 satellite imagery courtesy of USGS)

a tree may find sufficient moisture to survive. The occurrence of this tree reminds of the Tree of Life in Bahrain that has survived in similarly hostile for almost 400 years. That tree is located at the intersection of anticlinal faults with a major rock fall.

We change our focus now to Wadi Hawashia itself. Its course and the fan delta are directly correlated with transform faults. In Fig. 9.35 we show two options to enhance drainage in optical satellite images. The band ratio of the infrared band 6 and the panchromatic band 8 (a) highlights the dendritic character of the drainage channels on the coastal plain. It also reveals the outcrops of rift parallel faults near the coast. The inverted multiband 321 RGB image clearly enhances the wadi courses without highlighting the structural blocks. For the extraction of the channel and fan delta geologic analog we chose the band ratio 6/8 since it revealed not only the wadi outline but also its internal channel pattern.

Wadi Hawashia consists of two distinct channel patterns: a braided stream pattern in the western and central portion of the coastal plain and a fan delta in the

Fig. 9.35 Processing of satellite imagery for drainage. (**a**) Band ratio Landsat 5 bands 6/8, (**b**) inverse RGB image of Landsat 321 RGB. (Landsat 5 satellite imagery courtesy of USGS)

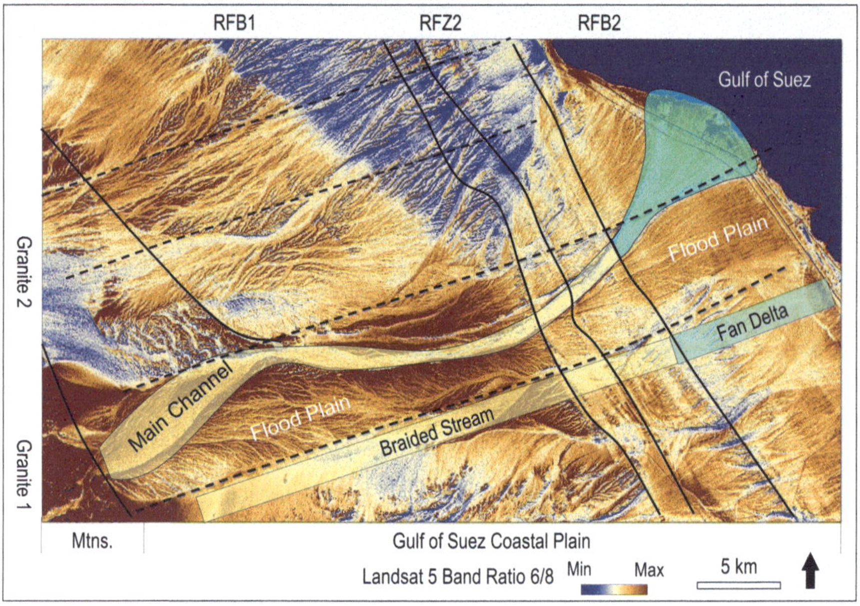

Fig. 9.36 Satellite imagery processed for channels. Landsat 5 band ratio 6/8 with structural and drainage interpretation. (Landsat 5 satellite imagery courtesy of USGS)

eastern portion near and at the coast (Fig. 9.36). The braided stream portion is confined in the north and south by transform faults, whereas the fan delta in the shallowest part of the coastal plain slope crosses the northern transform fault. The remainder of the rectangular area between the two transform faults is covered by the flood plain from older channels. We merge the braided stream and the fan delta patterns into one coastal channel-fan template. Later, when interpreting the seismic data, we will first search for channel-fan patterns in the shallow subsurface and then expand the search downwards. When scanning the present-day coastal plain we note that Wadi Hawashia is not an isolated phenomenon. All major fault-related valleys crossing the Red Sea Mountains show the same channel-fan delta pattern.

After extracting structural and depositional features from satellite imagery we study now the shallow seismic data. Diving surface waves provide insights into the shear properties of the near surface, which is more sensitive to subtle elastic property changes than compressional body waves (Strobbia et al., 2009a, b, 2010, 2011). The inversion of seismic surface waves provides a shear wave velocity cube in the present case for the first 150 m below the surface. This velocity cube is then processed using color processing of seismic data (Laake, 2015) to obtain a color image of shear wave impedance contrast that resembles a false color RGB satellite image. Shallow seismic refractions travel mainly along defined subsurface horizons. Their propagation velocity is an elastic characteristic of the rocks in which they travel. Figure 9.37 shows satellite imagery, color processed shear wave velocity and seismic refraction velocity side by side to show the correlation between features observed at the surface and in the subsurface. The main course of Wadi Hawashia that is clearly visible on the satellite imagery is also visible on shear waves and refracted waves since the material of the wadi channel has different mechanic properties than the material in the flood plain.

The shear wave image (b) highlights depositional features such as channels of paleo-wadis in the subsurface due to their different lithology to which shear waves are much more sensitive than compressional waves at about 5 m lateral and vertical resolution. It also reveals fault zones due to their lower shear wave velocity as a result of mylonitization, i.e. grinding of the rock matrix within fault zones.

The refracted wave image (c) generally shows the same picture as the shear waves except that the vertical resolution is significantly poorer—we estimate a resolution of about 50 m which does not warrant color processing of the refraction data. Consequently, the courses of the paleo-wadis are less well defined. The rift parallel

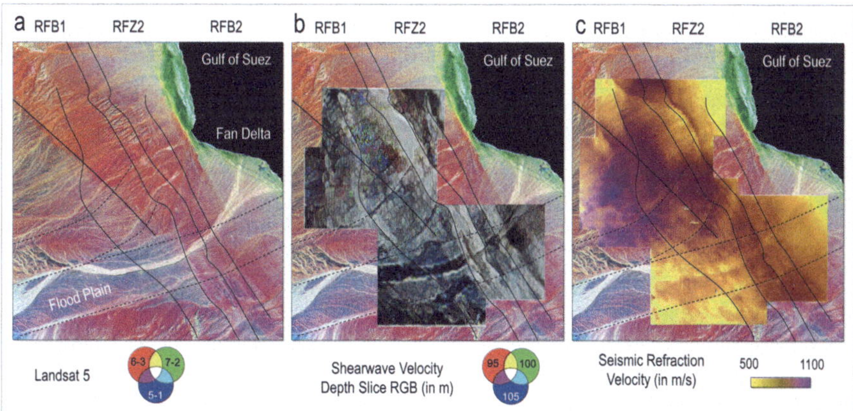

Fig. 9.37 Correlation of surface satellite imagery with near surface seismic data. (**a**) Landsat 5 multiband difference image 675-321 RGB, (**b**) color processed depth slice of the shear wave velocity cube at 100 m depth, (**c**) depth slice of seismic refraction velocity at 100 m depth. (after Laake 2011, Landsat 5 satellite imagery courtesy USGS, seismic data courtesy of TransGlobe Energy, Dara Petroleum Co. and WesternGeco)

faults are poorly imaged whereas the transform faults cannot be imaged at all because of the lack of vertical throw associated with these shear faults.

For the detailed study of the near surface geology we select a shallow horizon at the top of the first consolidated sandstone and texture it with color processed shear wave velocities (Fig. 9.38). We interpret the box structure of the Wadi Hawashia flood plain is actually a structural horst—whereas the surface shows the inversion of this setting. Towards the coast, rift parallel faults orthogonally intersect the structure and separate it from a tilted fault block, the structure of which could also be confirmed in deeper body wave data. The rift parallel fault zones are very obvious across the entire survey area. Under the coastal floodplain we discover an area of anomalous high velocity which we interpret as an isolated reef that developed in very shallow water. Today, clastic deposition by the wadis bringing sediments from across the Red Sea Mountains buried this structure completely leaving no detectable trace at the surface. Therefore, remote sensing cannot detect this feature.

We can now integrate the surface information from satellite remote sensing and shallow subsurface information from seismic data (Fig. 9.39). The surface satellite image processed to highlight lithology is Landsat 7 742 RGB draped onto the digital relief model (a), Landsat 7 band 8 sharpened to delineate wadis and draped also on the DRM is shown in (b) and (c) renders the color processed shear wave velocity data at the shallow top sandstone horizon. The box-shaped Wadi Hawashia horst block is evident in (b) and (c) delineated by transform faults on either side. The rift parallel faults near the coast are also evident in (b) and (c) and have only a faint expression on (a). The tilted horst block in the subsurface (c) appears to leave a subtle surface expression in the shape of a slightly high elevation, which deviates the Wadi Hawashia's fan delta towards north out of the horst block (dashed line in a). The rift parallel faults are only detectable in (b) and (c).

Fig. 9.38 3D view of a shallow horizon of color processed shear wave velocity cube with structural and lithologic interpretation. (after Laake 2011, Seismic data courtesy of TransGlobe Energy, Dara Petroleum Co. and WesternGeco)

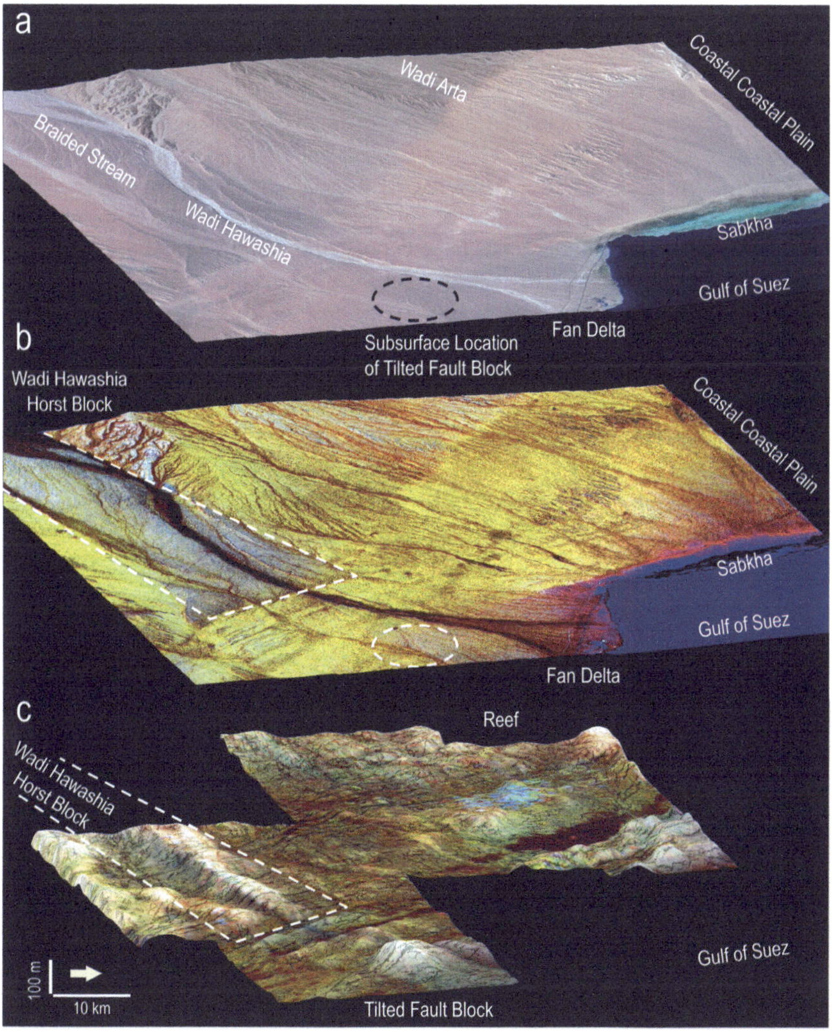

Fig. 9.39 3D view satellite data at the surface and shear wave velocity data at a shallow horizon. (**a**) Landsat 5742 RGB image, (**b**) Landsat 5 band 8 intensity image, (**c**) shallow horizon textured with color processed shear velocity data. (after Laake et al. 2011, Seismic data DRM and Landsat 5 satellite imagery courtesy USGS, seismic data courtesy of TransGlobe Energy, Dara Petroleum Co. and WesternGeco)

One striking difference between the shallow subsurface image obtained from seismic and the geologic expression at the surface is the structure of the shallow sandstone horizon. The elevation changes in this horizon as obtained from inverted surface waves reaches up to 150 m.

From these observations we conclude that remote sensing data need to be processed with a geologic objective in mind to complement existing data. For structural mapping we can either process the satellite such that they reveal directly subtle lithologic boundaries. This can be achieved by multiband differences for example using infrared and visible bands in a combination like 675-321 RGB. Alternatively, drainage patterns can be highlighted since drainage is often related to outcropping faults. Anomalous straight channel patterns are often an indicator for fault-guided drainage.

Another form of integration is the use of present-day depositional features as geologic analogs for fossil objects in the subsurface. For example, it is important to understand whether a sand body has been deposited onshore or offshore, and in the latter case at what water depth. The depositional environment plays a significant role for reservoir quality and seal quality. In the case of the Gulf of Suez study the litho-stratigraphic column (Fig. 9.29) indicates that sand bodies originating from deltaic and shallow water deposition are favorable exploration targets.

Therefore, it is important to understand the location of the paleo-coastline at the target formation. The paleo-coastline can directly be obtained from the front of the fan delta. We use the Landsat 5 band ratio 6/8 to create the channel-delta template (Fig. 9.40a). Then we take this template and use for identification of subsurface geobodies of similar shape in different attributes of the seismic data. We were successful to identify four distinct channel patterns starting in the mid Miocene

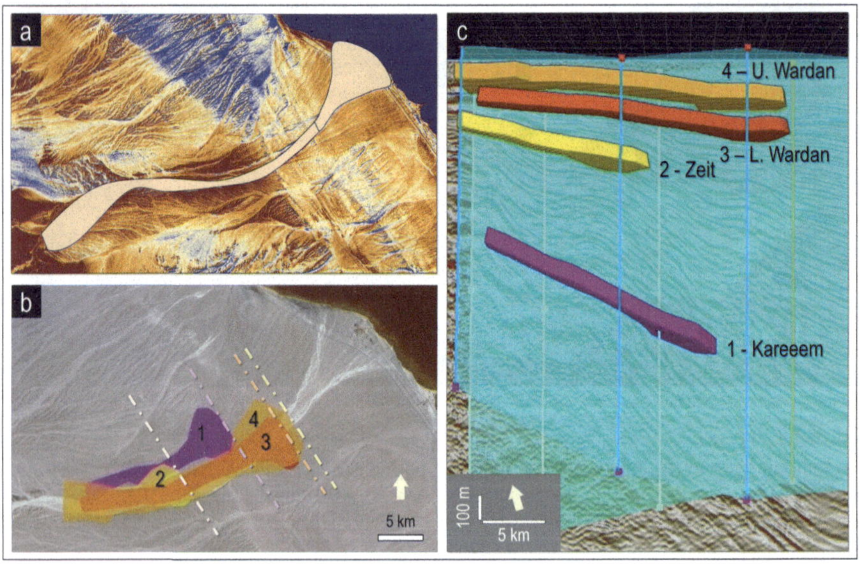

Fig. 9.40 Application of surface channel-delta template as a geologic analog for subsurface interpretation. (**a**) Surface template created from satellite imagery, (**b**) projection of subsurface channel-delta geobodies using the surface template in subsurface interpretation, (**c**) approximate location of channel-delta geobodies in the subsurface from interpreted seismic data. (after Laake & Strobbia 2012, Landsat 5 satellite imagery courtesy of USGS, seismic data courtesy of TransGlobe Energy, Dara Petroleum Co. and WesternGeco)

Kareem Fm at the time of the main Gulf of Suez rifting (b). The coastline at that time was approximately 20 km further inland compared to today. During the Zeit Fm, the coastline had receded another 10 km to return to about 15 km inland compared to today's coastline during the Pleistocene Wardan Fm. This implies that reservoir sand bodies can be associated with their depositional environment depending on their age. For example, most Kareem reservoirs are located either within the 20 km zone from the coast or offshore. These would have been deposited most likely in an offshore environment. The shelf slope can be obtained from the seismic data to further specify at what depth on the shelf the reservoir sands had been deposited. The relative location of the channel-delta geobodies in space is shown in (c).

Following the observation that the braided channel appears to be constrained by the northern and southern transform fault, the channel-delta geobodies may also be used to delineate the transform faults since these are difficult to map on seismic directly due to the small vertical fault throw associated with the transform faults.

In summary, the integration of remote sensing and subsurface seismic data provides comprehensive clues for the interpretation for structural, stratigraphic and depositional features in the subsurface (Fig. 9.41). Remote sensing data provide lineaments from outcropping faults and depositional templates for channel-delta analogs. The faults are propagated first into the near surface through correlation with shallow seismic surface wave and refraction data and then propagated down deep using seismic body waves. Channel-delta templates are employed when interpreting with the help of seismic attributes to search for subsurface depositional features matching the characteristics of the surface geologic analogs. Using this approach paleo-coastlines can be interpreted in subsurface formations, which in turn allows deeper insight into the depositional environment at the time of deposition.

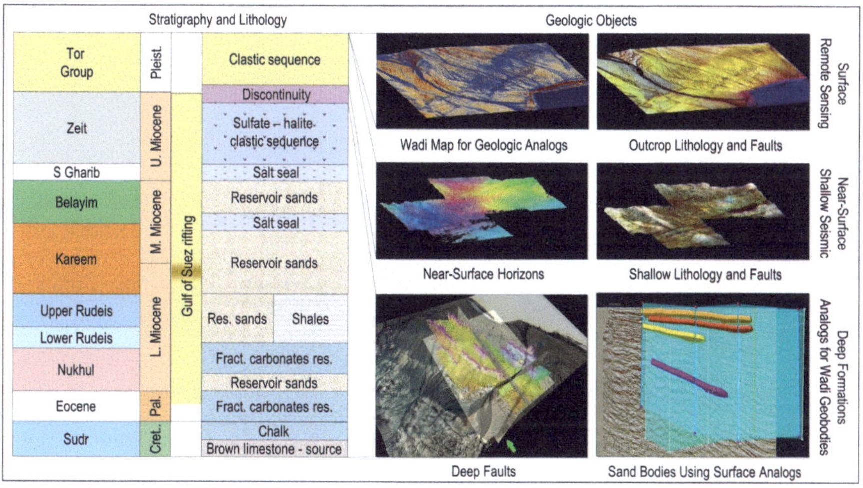

Fig. 9.41 Integration of surface remote sensing, near surface shallow seismic and deep seismic using channel-delta analogs from surface templates. (after Laake 2011, DRM and satellite imagery courtesy of USGS)

9.5.2 *Estimation of the Penetration of Satellite Radar from Integration with Seismic Data*

The second case study demonstrating the integration of remote sensing data and subsurface seismic data focuses on the estimation of the penetration depth of satellite radar data (Laake & Strobbia, 2013a, b). We characterize the surface for topography and hardness using the combination of digital relief model and multi-spectral satellite imagery to discriminate the satellite radar back scatter intensity for surface back scatter from rocky surfaces and subsurface back scatter from soft sand and clay surfaces (Fig. 9.42). The concept to estimate the penetration depth of satellite radar data into sand uses a flat surface with sand dunes of variable height rising above the flat surface. From hydrodynamic considerations we consider a floodplain with braided channels to be the closest to a natural flat plain in a desert environment. The surface outcrop of such a floodplain can be detected from multispectral and radar satellite data. We then follow the signature of braided channels from the radar back scatter response along the floodplain under the sand dunes until we loose the channel signature on the radar data. This point is equivalent to the maximum penetration depth of the radar data into the sand.

For an independent validation of the remote sensing approach, we use seismic data from which surface waves have been extracted, that in turn have been inverted for surface wave velocity. Since sand dunes and channel fill have a distinguishably different surface wave velocity, we can locate the depth of the channels under the dunes on the surface wave section. The vertical accuracy of this method is approximately 5 m.

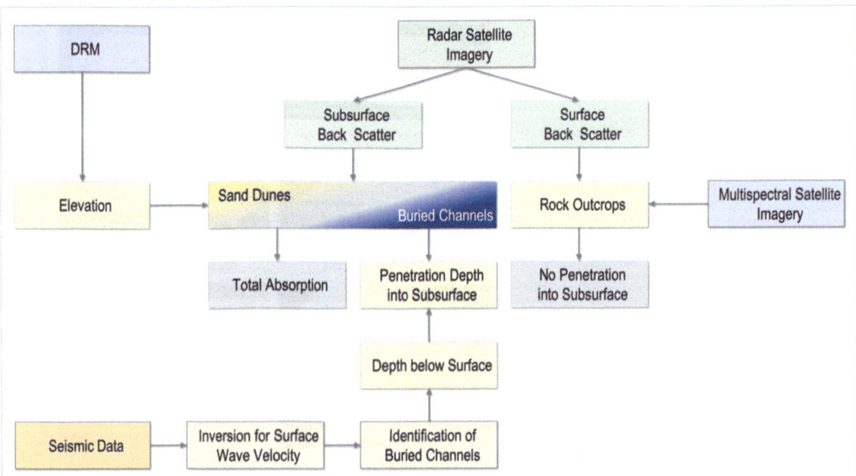

Fig. 9.42 Workflow to estimate the depth of penetration into the subsurface for satellite radar data using the integration of remote sensing and seismic

In the final step we determine the level of back scatter intensity at which the delineation of buried channels vanishes and associate the depth of the channels below the thickness of sand overburden the penetration depth for satellite radar data. We compare our findings for very dry soil in our test area with the modeling results by Redadaa et al. (2007).

Input data are digital elevation model, multispectral satellite imagery and radar imagery (Fig. 9.43). The overlay of the Landsat 5 multispectral composite 742 RGB image on the digital elevation model (a) provides an insight into the geomorphology by highlighting areas where hard rocks outcrop along an escarpment (red arrows) at the edge of the Gilf Kebir plateau and at the highlands bordering the El Mesaha plain to the north. The eastern half of the image is dominated by long stretches of sand dunes (yellow arrows), whereas in the western part outside the Gilf Kebir plateau channels of paleo-rivers are visible (light blue arrows). Towards the east, these channels are buried under sand dunes and become invisible for optical surface satellite imagery.

The radar imagery (b) allows tracing the channels further under the sand to establish a more continuous mapping of the braided channel system covering the wide floodplain. However, where the sand cover is too thick, the radar beams are absorbed before they reach the floodplain and, therefore, no back scatter can be detected (dark blue areas). Towards the west where rocky outcrops are intersected by sand-filled wadis, the radar imagery enhances the expression of these valleys through the strong contrast between strong surface back scatter intensity on the rocky outcrops and very weak back scatter in the sand and clay filled valleys. We use this strong contrast to classify the radar back scatter intensity into five classes: the strongest back scatter intensity comes from hard rock outcrops (brown), slightly lower intensity is associated with softer rocks and weathered rock debris (light brown). Combined surface and subsurface back scatter occurs in areas with thin sand cover (green), whereas subsurface back scatter correlates with sand-covered areas where the sand is thin enough to allow radar penetration and back scatter from clays in buried channels (light blue). Total absorption, i.e. neither surface nor subsurface back scatter of detectable intensity is associated with deep dry sand in high sand dunes (yellow). Figure 9.44 shows surface features illustrating the manifestation of the radar back scatter classes in the field. Surface back scatter from hard rock occurs at granite (a) and ferruginous sandstone (b). At escarpments where the sandstone is weathered to boulders, eolian sand is blown in (c). Soft sandstone is weathered to soft sand partially covering the sandstone outcrop (d). Large areas are covered with weathered boulders and pebbles of the ferruginated sandstone (e), that may also be covered by thin eolian sand (f). Valleys cutting into the sandstone plateau often show dry channels in their thalweg (g). These channels contain clay and are the features we use to determine the penetration of radar into the ground, for example when they are covered by medium to thick sand (h, i).

We will now determine the depth of radar penetration into dry sand using the concept of sand dunes rolling onto to a channelized floodplain (Fig. 9.45). Our attention is on an area south of the sandstone escarpments (golden brown features in a). We zoom into the area on the Landsat 5 multispectral composite 742 RGB image

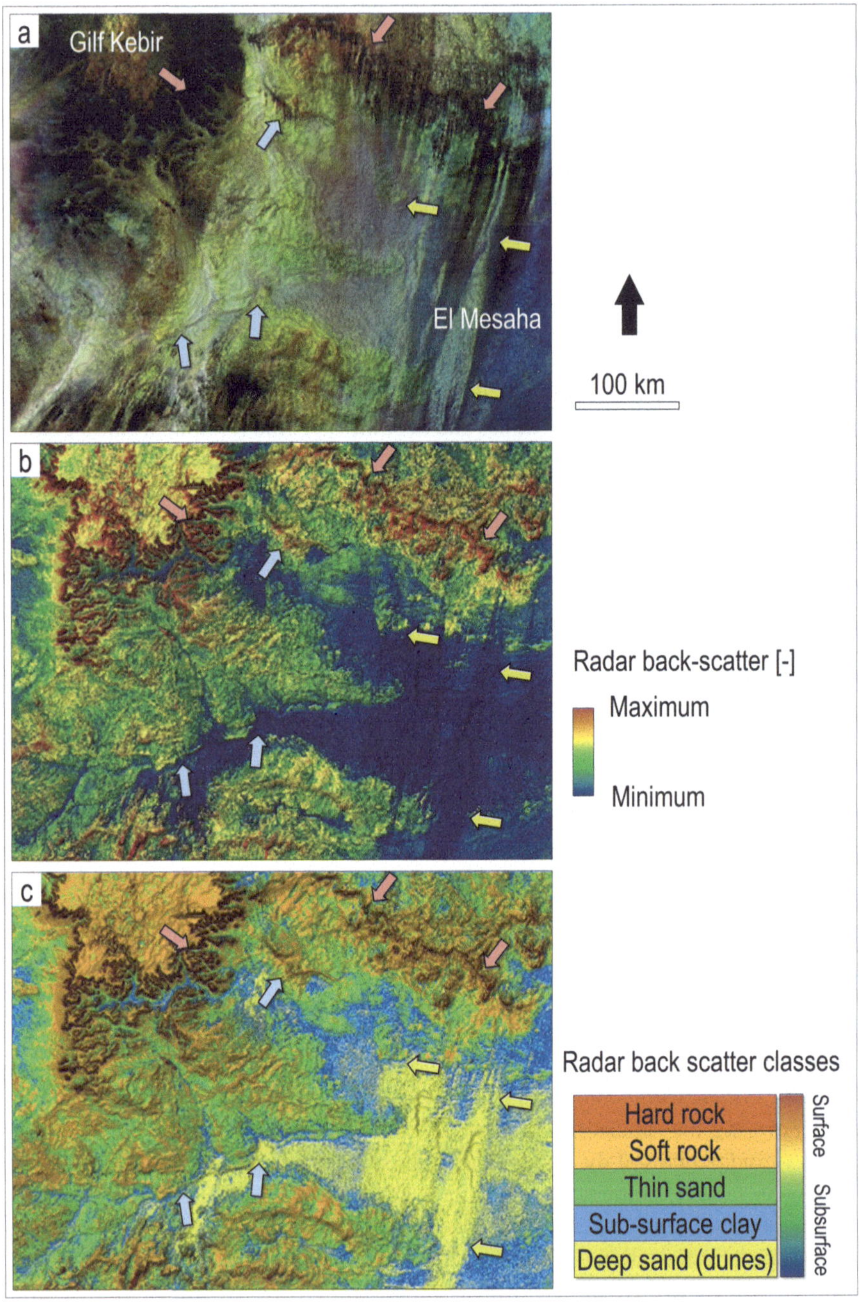

Fig. 9.43 Satellite imagery characterizing the surface and near surface in Gilf Kebir, southwest Egypt. (**a**) Composite of Landsat 5 742 RGB and DRM, (**b**) radar backscatter intensity data, (**c**) classification of radar backscatter intensity. (Satellite imagery courtesy of USGS, Radarsat-1 imagery courtesy of CSA)

Fig. 9.44 Surface photos illustrating the radar back scatter classes related to surface cover and rock type. Explanations in the text

which highlights lithological changes (b). The dark green to purple colors indicate ferruginated sandstone, yellow-brown areas with a ripple texture show sand dunes and light lineaments indicate the location of the channelized floodplain.

We observe that the channels are most distinct near the sandstone outcrops where the sand cover is either absent or intermittent so that the averaging over the sample size of 30 m ×30 m still allows the detection of a channel. The radar image (c) allows tracing the braided channel system much further down even below the dune cover. The image is slightly more noisy than the multispectral image because rocky debris at the surface and the texture of the dune surface causes local variations in the radar back scatter energy and in the distribution of the intensity between surface and subsurface back scatter.

So far, we have only been able to obtain a qualitative measure for the penetration of radar energy into the subsurface. We can use the shallow propagation and

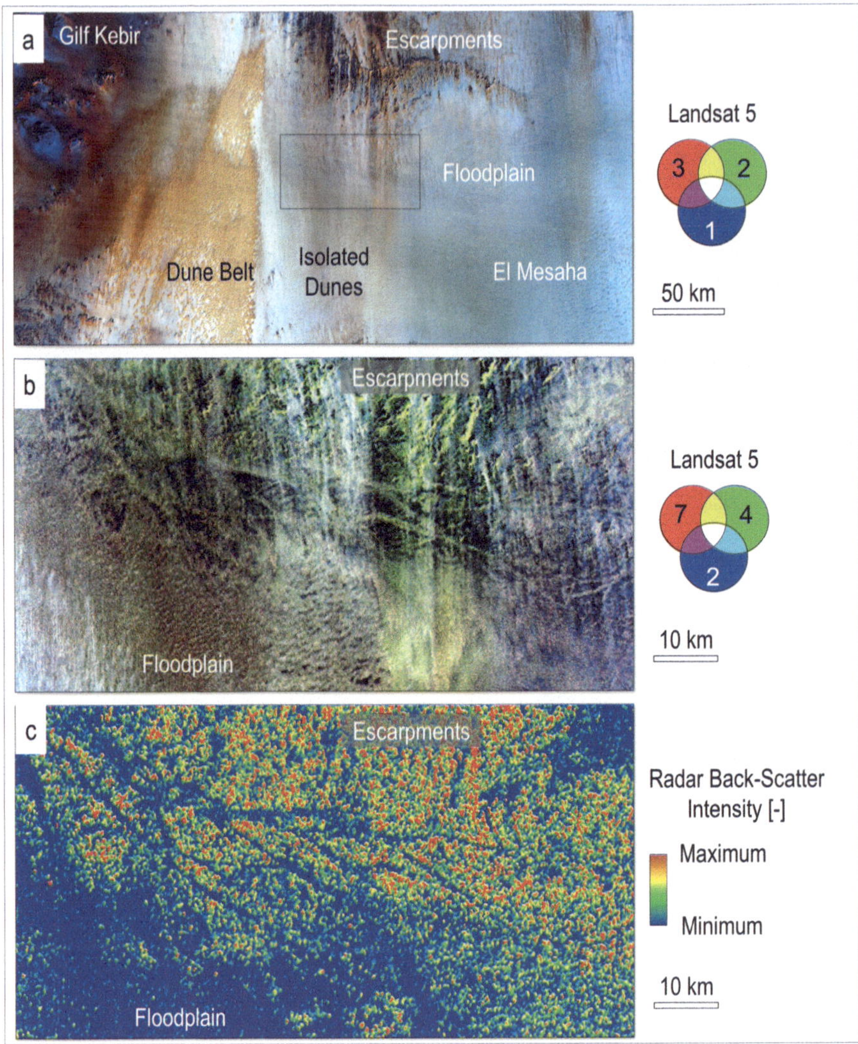

Fig. 9.45 Representation of channelized floodplain on satellite imagery. (**a**) Optical satellite image Landsat 5 321 RGB, (**b**) zoom into floodplain, multi-spectral composite Landsat 5 742 RGB, (**c**) same zoom, radar back scatter intensity

absorption characteristics of seismic surface waves to quantify this observation (Fig. 9.46). Across the El Mesaha Basin (Fig. 9.43a) we have access to a regional seismic section that runs north-south from the sandstone escarpments across the basin into the southern hilly terrain that delimits the basin in the south. For the channelized floodplain we have studied in the previous paragraph, we can create a schematic shallow geologic model comprising the sand-filled part of the basin with Pleistocene channels and the partial cover by sand dunes (a). Below, we display the measured radar back scatter intensity classes (b) that we showed in Fig. 9.43c. And at the bottom (c) we plot a velocity section from seismic surface waves computed from surface wave inversion (Laake & Strobbia, 2013a, b). We color code the velocity ranges for sand (white to gray), clay (dark green), ferruginated sandstone (brown) and hard rocks representing metamorphic and crystalline rocks in various stage of weathering. The velocity section has a vertical resolution of better than 5 m.

At the channel outcrop the surface wave section shows a thin clay layer (green) that widens as we go to the left in Fig. 9.46c. The surface wave section shows some undulation which is most likely correlated with further channels, whereas the radar intensity class just indicates sand, meaning that the sand is too thick to reveal any buried channels in the back scatter intensity. The zone where the radar back scatter intensity allows the delineation of channels (center of b) correlates with an area in the surface wave section (c) where the weathered ferruginated sandstone (yellow) is at about 20 m depth.

We repeat this approach across the entire length of the north-south seismic section. In Fig. 9.47 we show the integration of the radar back scatter image draped on the digital relief model (top) and the seismic surface wave section (bottom). The shallow surface wave velocity profile is available for a total of 400 km length and runs from the sandstone escarpments north of Wadi Maftuh across the wadi floodplain through a dune field into a ridge of ferruginated sandstone and further into a wide floodplain that is complete covered by a dune field before intersecting the weathered basement granite outcrops at its southern end.

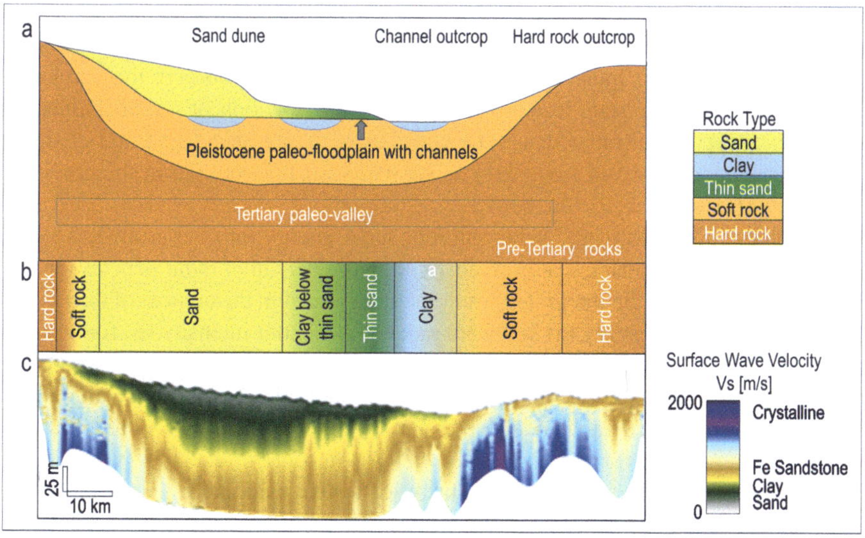

Fig. 9.46 Context of shallow geologic setting, radar backscatter intensity and seismic surface wave velocity. (**a**) Schematic of shallow geology, (**b**) dominant near surface lithology, (**c**) seismic surface wave section. (after Laake & Strobbia 2013, Seismic data courtesy of Melrose Resources Egypt and WesternGeco)

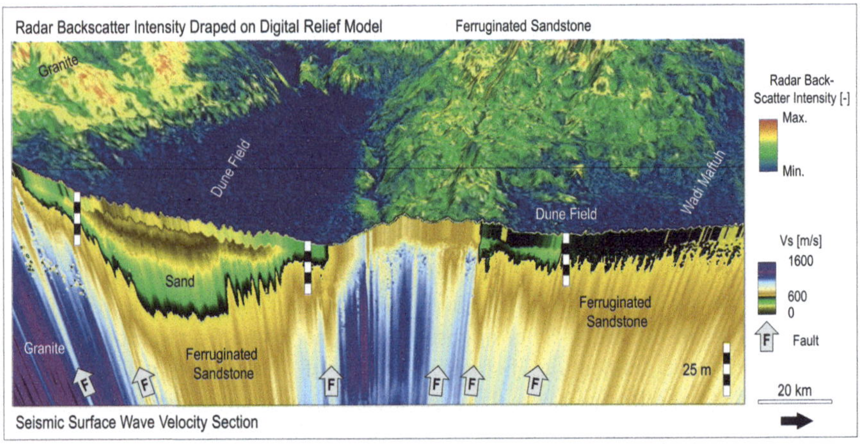

Fig. 9.47 Surface-subsurface correlation of radar backscatter intensity and seismic surface wave shear-wave velocity. (after Laake & Strobbia 2013, Radarsat-1 satellite data courtesy of CSA, DRM courtesy of USGS, seismic data courtesy of Melrose Resources Egypt and WesternGeco)

Along the seismic section, we have placed measuring bars with 5 m vertical intervals to indicate the depth of buried channel clays on the seismic section for correlation with the radar back scatter intensity map. We confirm that the mapping of braided channels is possible down to approximately 20 m depth below which radar back scatter intensity fails to map buried channels.

On the seismic section we also observe some near vertical low velocity zones (yellow-brown) reaching into the hard basement rocks (blue to purple). These zones indicate shallow faults most of which appear to outcrop. We observe that the location of these faults appears to be correlated with the location of regional channel trends as observed on radar data. The correlation of regional faults and channel courses has been observed frequently as we had seen for example in the previous case study (Laake et al., 2011a, b, c).

With the calibration of the penetration of radar energy into the ground, we can now convert the radar back scatter intensity map into a map of radar penetration. In Fig. 9.48 we compile the integrated result (c) and show it in the context of the inputs from multispectral imagery (a) and from radar back scatter intensity (b) for an area of about 100,000 sq km. The radar penetration is absent to minimal where hard rock outcrops in the highland of Gilf Kebir, in the granite outcrops in the south and in areas with ferruginated sandstone at the surface. Intermediate back scatter values occur on the perimeter of the sand dune fields with buried channels (wadis in a) being mapped under the sand. Minimal back scatter intensity indicates total absorption of the radar energy in the dry sand without any mapping capability.

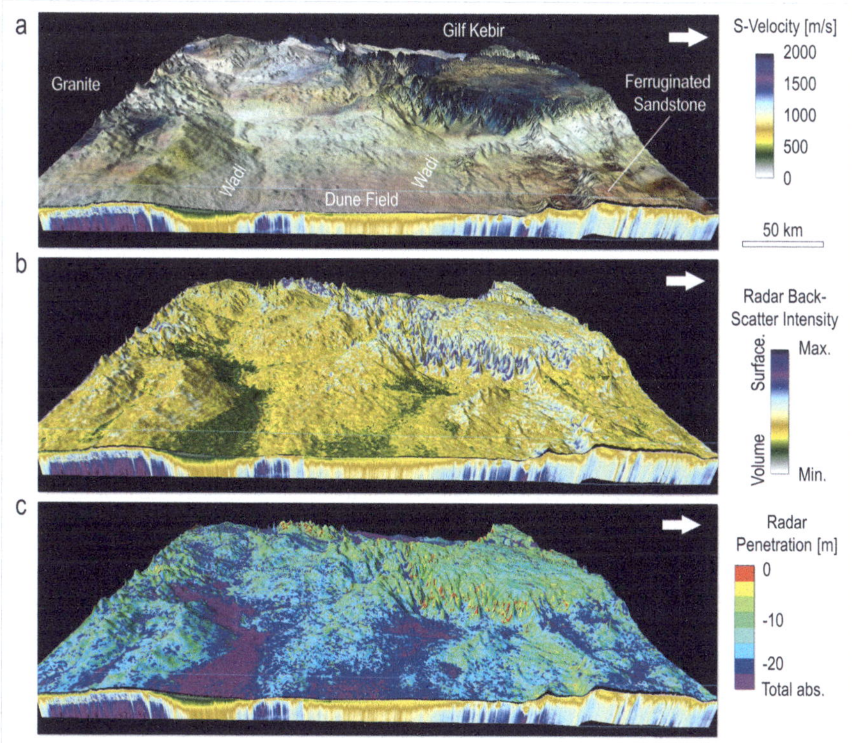

Fig. 9.48 Surface-subsurface integration of radar satellite imagery and seismic surface waves to determine the penetration of satellite radar into the surface. (**a**) Landsat 5 742 RGB satellite imagery draped over DRM, (**b**) radar backscatter intensity draped over DRM, (**c**) computed radar penetration into the ground draped over DRM. Vertical section in all images shows the shear-wave velocity of seismic surface waves. (after Laake & Strobbia 2013, DRM and Landsat 5 imagery courtesy of USGS, Radarsat-1 imagery courtesy of CSA, seismic data courtesy of Melrose Resources Egypt and WesternGeco)

9.6 Estimating Impact of Volcanic Rocks

The impact of igneous intrusions on petroleum systems have been described in general by Schofield et al. (2017) and Senger et al. (2017). Iyer et al. (2017), Luo et al. (2017), Nuzzo et al. (2012), and Peace et al. (2017) have studied the impact on maturation whereas Rateau et al. (2013) focused on the impact on hydrocarbon migration. The detection, delineation and characterization of volcanic rocks in petroleum basins is therefore an important challenge.

Volcanic rocks often show dark colour which makes them detectable from optical satellite imagery. Their deposition occurs as volcanoes and surface lava flows like flood basalts the shape of which can be detected on topography data. And, finally, volcanic rocks such as basalt have a significantly higher density than sedimentary rocks, which we can detect on satellite gravity data. For additional support to outline areas that are covered by volcanic rocks, geologic maps provide good source for calibration.

The case study chosen for the characterization of volcanic rocks is the extensive Haruj volcanic system in central Libya. It provides a variety of different geologic settings to explain how to detect and compensate for the effect of volcanic systems. The Haruj volcanic system is an intra-continental plate shield volcano which created huge flood basalt fields and lava flows (Elshaafi & Gudmundsson, 2016; Brugge, n.d.) resulting in volcanic rock deposits of widely varying thickness.

Figure 9.49 shows a sketch of the geological setting found at and near the Haruj volcanic system. A particularity of the Haruj volcano is that is set next to a graben, the surface expression of which is covered by the flood basalt. We expect this graben to leave a signature on the Bouguer gravity anomaly data.

The workflow for the remote characterization of the Haruj volcanic system and the estimation of the its impact on the petroleum system of the western Sirt Basin (Fig. 9.50) uses the topography from the digital relief model and the surface geologic map and multispectral satellite imagery for the delineation of the extent of the volcanic rocks at the surface. For the mapping of the approximate location of the main vent, satellite gravity data are used.

The photogeologic map is created from the overlay of surface geologic map and satellite image on the DRM, which yields the extent of the surface volcanic rocks. Where these consist of thin layers of flood basalt or lava flows, the Bouguer gravity anomaly is only affected insignificantly and still provides reliable information about the basin depth. In this case, the impact on the maturity of the hydrocarbons in the subsurface is rather small (Senger et al., 2017). The main vent and the magma chamber are major heat sources and are expected to have a significant impact on the maturation of hydrocarbons in the Sirt Basin.

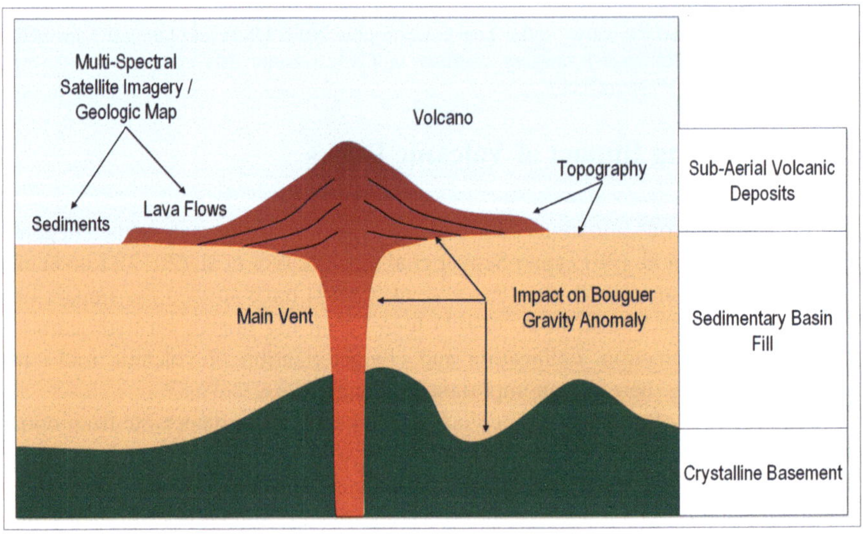

Fig. 9.49 Schematic of the geological setting of the Haruj volcanic system and indications for its impact on remote sensing data

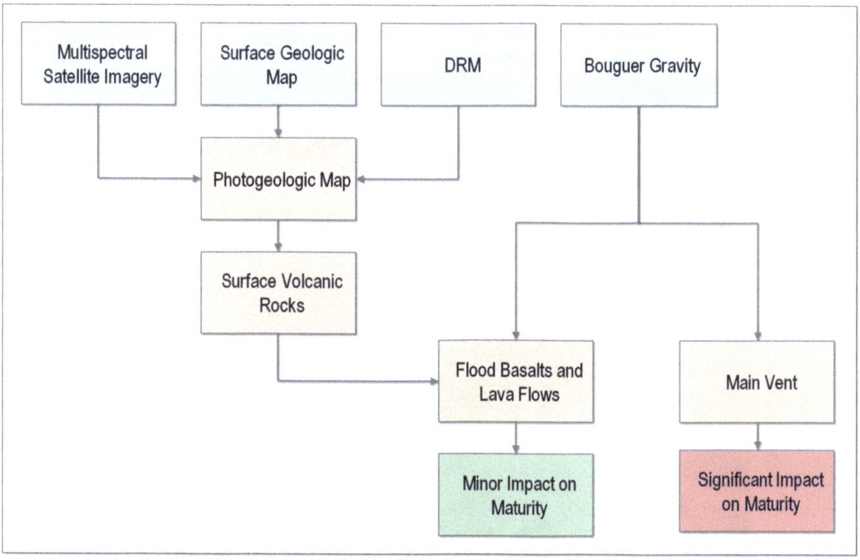

Fig. 9.50 Workflow for the characterization of volcanic systems from remote sensing data

The geologic literature indicates that the Haruj volcanic system is located next to the intersection of Hun Graben in the north, Dor El Abid Trough to the east and Bu Tamaym Trough to the southeast and south (Abadi et al., 2008; Elshaafi & Gudmundsson, 2016; Ahlbrandt, 2001a). Figure 9.51 shows the satellite image from Landsat 5 742 RGB with the outline of these grabens and horsts. The vertical exaggeration and slant illumination form the northeast shows the texture of the Haruj volcanic system. Very dark colors indicate basaltic lavas and flood basalt fields of the Haruj and Sawda volcanoes, light colors are related to sediments of different composition. In the east towards the center of the Sirt Basin two platforms separate the Dor El Abid Trough from the interior of the Sirt Basin. In our analysis we will use remote sensing data to assess the thickness of the volcanic rocks covering the trough. We will also attempt to map the main vent of the Haruj volcano.

We start the assessment by draping the input data onto the digital relief model (Fig. 9.52). The digital relief model (a) indicates the peak of Haruj is about 800 m higher than the surrounding terrain. The relatively smooth elevation contours confirm prior information (Brugge, n.d.) that Haruj is a shield volcano. More than 400 vents primarily on the northern, eastern and southern flanks dot the relatively smooth surface of the volcano. The surface geologic map (b) shows an elongated outcrop of volcanic rocks consisting of a large part in the north, Al Haruj al Aswad (AHAS), and a smaller part in the south, Al Haruj al Abyied (AHBS). The geologic age of the surface rocks varies from Paleozoic in the southwest and Mesozoic in the west to Tertiary and Quaternary elsewhere indicates the location of the Haruj volcano at a boundary between the Mesozoic basins of western and southern Libya and the Tertiary and Quaternary cover of the Sirt Basin in central Libya. The Landsat 5 mosaic (c) shows multiple flood basalt and lava flows mostly in a dominant

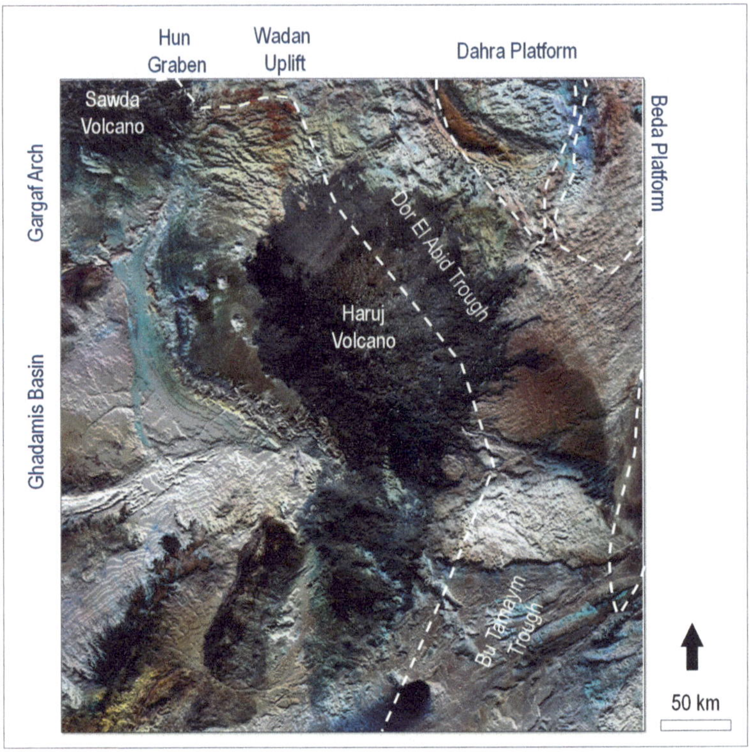

Fig. 9.51 Overview map of Haruj volcano. Structural units after Abadi et al. (2008) and Elshaafi and Gudmundsson (2016). (Landsat 5 TM 742 RGB mosaic, satellite imagery courtesy USGS)

northeast-southwest orientation. The overlay of the Bouguer gravity anomaly on the digital satellite image and the DRM (d) shows a high positive gravity anomaly on the central and southern part of the AHAS sub-province. At the location of the Dor El Abid Trough the Bouguer gravity anomaly shows a clear negative feature (arrow) indicating the presence of a basin below the basalt. This feature is only detected by the gravity anomaly since the gravity signal is determined by the rock density across the entire rock column between surface and basement.

In the next step we integrate the findings from the individual data (Fig. 9.53). The Landsat 5 satellite image draped over the digital relief model (a) serves as a reference. The total reflected energy in the visible and near infrared bands (b) not only reveals the presence of volcanic rocks (shades of purple), it also allows distinguishing individual lava flows with color shades ranging from cyan for older deposits to purple for the youngest flows. Since the satellite imagery only maps the surface occurrence of volcanic rocks, we employ the satellite measurement of the gravity anomaly to estimate the thickness of the high-density volcanic rocks (c). We observe larger gravity values in the center and on the southern flank of the volcanic system indicating thicker layers of high-density volcanic rocks.

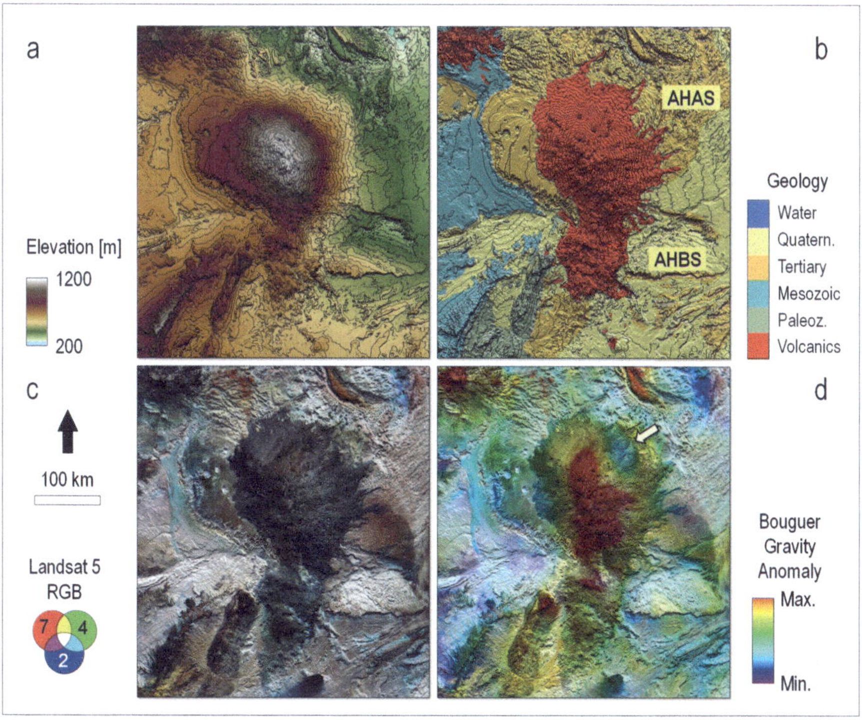

Fig. 9.52 Input from remote sensing data, (**a**) DRM, (**b**) surface geologic map draped on DRM, (**c**) multispectral satellite imagery, Landsat 5 742 RGB mosaic, (**d**) overlay of Bouguer gravity anomaly. AHAS Al Haruj al Aswad, AHBS Al Haruj al Abyied. (DRM, geologic map and satellite imagery courtesy USGS, Satellite gravity data courtesy Sandwell et al., University of California San Diego)

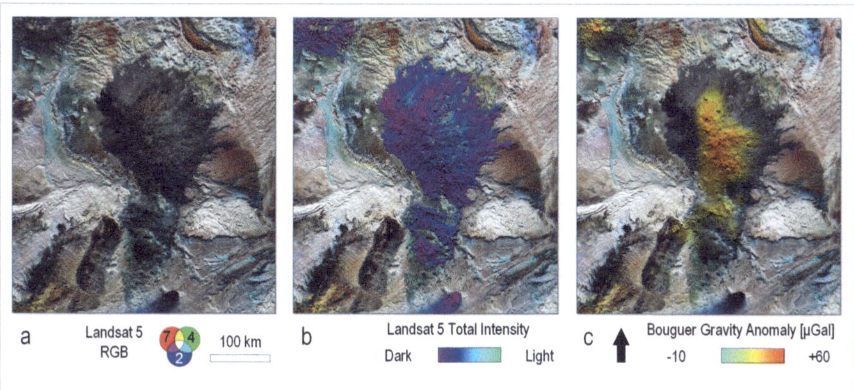

Fig. 9.53 Identification of lava from optical satellite imagery and location of main vent from Bouguer gravity anomaly, (**a**) multispectral satellite imagery Landsat 5 742 RGB mosaic, (**b**) total reflection intensity of satellite image from a overlaid on Landsat image, (**c**) Bouguer gravity anomaly for the interval of −10 to +60 μgal overlaid on Landsat image. (Satellite imagery courtesy USGS, Satellite gravity data courtesy Sandwell et al., University of California San Diego)

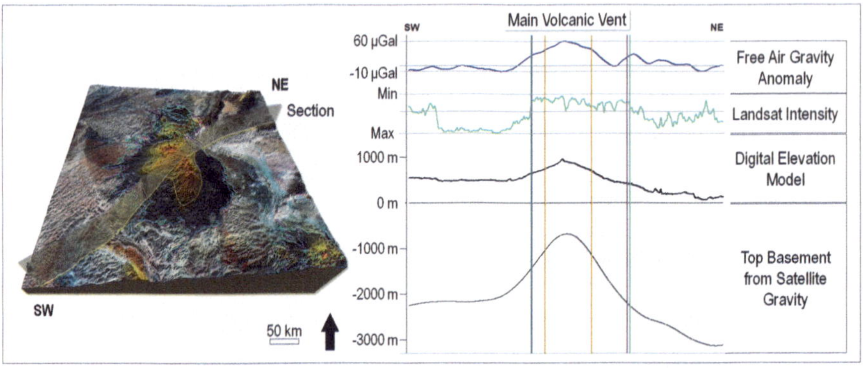

Fig. 9.54 Correlation of remote sensing data along a section across the Haruj volcano system. (DRM, geologic map and satellite imagery courtesy USGS, Satellite gravity data courtesy Sandwell et al., University of California San Diego)

Using the basin modelling technique described in Sect 9.3 we can build a geologic model for the area of the Haruj volcanic system. When we intersect the model from northeast to southwest across the highest point of Haruj, we can co-render the different measurements (Fig. 9.54). The boundaries of outline of dark rocks from the Landsat intensity is shown as blue bars whereas the outline of high values of the gravity anomaly are shown in light brown.

The digital elevation model shows the typical contour of a shield volcano. The deposition of volcanic rocks as represented by very low Landsat intensity values is well correlated with the contour of the volcano. Please observe that the intensity scale is inverted to better show the correlation with the other measurements. The free air gravity from satellite altimetry mainly follows the contour of the Haruj volcano except for area in the northeast, a negative anomaly indicates the presence of a basin below the volcano.

The depth of top basement does not provide this level of granularity because of the 40 km low-pass filter that is required to obtain the depth without inversion. The data for top basement are, however, still indicative of the strong positive gravity anomaly that results from the high-density column of the main volcanic vent. From the co-rendering of all remote sensing measurements we conclude that the is located below the area of highest value of the gravity anomaly.

We can now merge all the observations to present the remote characterization of the Haruj volcanic system (Fig. 9.55). As the basis for the rendering we use the vertically exaggerated digital elevation model with the Landsat 5 satellite image draped over the surface. The expression of the surface texture is emphasized through slant illumination from the southeast and northeast.

Low values of the total Landsat 5 intensity mark the extent of the surface volcanic rocks (shades of purple) which is overlaid by Bouguer gravity anomaly data above 30 µGal (yellow to reddish tones) indicating the large thickness of volcanic rocks at the main volcanic vent. The areas showing lower gravity anomaly values indicate that the volcanic rocks are less thick which is indicative of thinner layers of

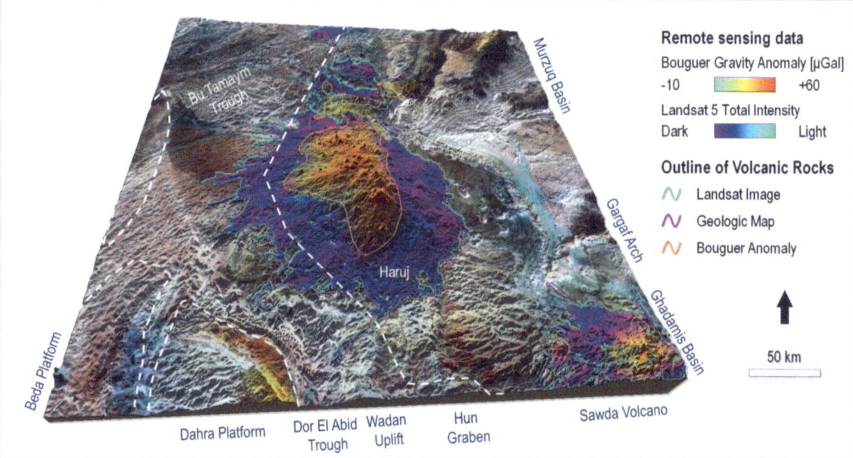

Fig. 9.55 Geologic characterization of the Haruj volcano system from remote sensing data

flood basalt and lava flows. For orientation, the outlines of the main basins and uplifts are indicated by dashed white lines.

The characterization of the Haruj volcanic systems implies that only the area of the main vent is likely to have a major impact on the maturation of the hydrocarbons in the petroleum systems of the Sirt Basin. The thin flood basalt and lava flows appear constrained to the surface overlaying unaffected sedimentary rocks. This observation is important since the activity of the Haruj volcano started in the Pliocene (Elshaafi & Gudmundsson, 2016) whereas the main hydrocarbon reservoirs have been filled latest during Paleogene times (Ahlbrandt, 2001a). Therefore, the Haruj volcanism itself did not contribute to the maturation of the hydrocarbons. It could, however, introduce over-maturation in the basin of the Dor El Abid Trough if the volcanic rocks had been more extensive than can be estimated from this remote sensing study.

References

Interaction with Earth Surface

Braitenberg, C., Wienecke, S. & Wang, Y. (2006). Basement structures from satellite-derived gravity field: South China Sea ridge. *Journal of Geophysical Research 111*, B05407, https://doi.org/10.1029/2001GC000252.

Bird, P. (2003). An updated digital model of plate boundaries. *Geochemistry, Geophysics and Geosystems 4(3)*, https://doi.org/10.1029/2001GC000252.

Cande, S.C., Stock, J.M., Müller, R.D., & Ishihara, T. (2000). Cenozoic motion between East and West Antarctica, *Nature 404,* 145–150.

Consortium for Ocean Leadership Inc. (2007). Ocean drilling program, Final Technical Report 1983–2007, National Science Foundation contracts ODP83-17349 and OCE93-08410, Washington DC, 2007.

Crosby, A.G., McKenzie, D. & Sclater, J.G. (2006). The relationship between depth, age and gravity in the oceans, *Geophysics Journal International 166*, 553–573. https://doi.org/10.1111/j.1365-246X.2006.03015.x.

Gibbs, P.B., Brush, E.R. & Fiduk, J.C. (2003). The evolution of the syn rift and transition phases of the central/southern Brazilian and W. African conjugate margins: the implications for source rock distribution in time and space, and their recognition on seismic data, Proceedings 8th International Congress of the Brazilian Geophysical Society.

Fairhead, J.D., Green, C.M. & Odegard, M.E. (2001). Satellite-derived gravity having an impact on marine exploration, The LEadging Edge, 873-876.

Heine, C., Zoethout, J. & Müller, R.D. (2013). Kinematics of the South Atlantic rift. *Solid Earth 4*, 215–253, https://doi.org/10.5194/se-4-215-2013.

Laake, A. (2010). Integration of surface and subsurface data – From satellites to reservoir, EAGE/SEG Summer Research Workshop - Towards a Full Integration from Geosciences to Reservoir Simulation, paper 22461, https://doi.org/10.3997/2214-4609.201402470.

Laake, A., & Strobbia, C. (2013a). Mapping paleorivers from satellite radar and surface-wave velocities. In *Proceedings 75th EAGE Conference & Exhibition Incorporating SPE EUROPEC 2013, London, UK.*

Lentini, M.R., Fraser, S.I., Sumner, H.S. & Davies, R.J. (2010). Geodynamics of the central South Atlantic conjugate margins: implications for hydrocarbon potential. *Petroleum Geoscience 16,* 217–229. https://doi.org/10.1144/1354-079309-909

MacLeod, S.J., Williams, K.J., Müller, R.D. & MacLeod, S.J., Williams, S.E., Matthews, K.J., Müller, R.D., and Qin, X.D. (2017). A global review and digital database of large-scale extinct spreading centers: Geosphere. 13(3), 911–949, https://doi.org/10.1130/GES01379.1.

Mohole: http://nasonline.org/about-nas/history/archives/milestones-in-NAS-history/project-mohole.htm. [accessed 16 June 2020]

Mohriak, W.U., Danforth, A., Post, P.J., Brown, D.E., Tari, G.C., Nemčok, M. & Sinha, S.T. (2013). Conjugate divergent margins: an introduction, in: Mohriak, W. U., Danforth, A., Post, P. J., Brown, D. E., Tari, G. C., Nemčok, M. & Sinha, S. T. (eds), Conjugate Divergent Margins. *Geological Society, London, Special Publications, 369*, 1–10, https://doi.org/10.1144/SP369.26.

Müller, R.D., Roest, W.R., Royer, J.-Y., Gahagan, L.M. & Sclater, J.G. (1997). Digital isochrons of the world's ocean floor. *Journal of Geophysical Researc 102(B2)*, 3211–3214.

Smith, W.H.F. & Sandwell, D.T. (1994). Bathymetry prediction from dense satellite altimetry and sparse shipboard bathymetry. *Journal of Geophysical Research 99(B11)*, 21803–21824.

Smith, W. H. F. & Sandwell, D. T.(1997). Global seafloor topography from satellite altimetry and ship depth soundings. *Science 277*, 1957-1962.

Zanella, E., Doran, H. & Collard, J. (2018). Is a conjugate an analog ?, GeoExPro Sept 2018, 14–17.

Multi-Band Composite Images

Laake, A., Sheneshen, M. S., Strobbia, C., Velasco, L., & Cutts, A. (2011a). Integration of surface/subsurface techniques reveals faults in Gulf of Suez oilfields. *Petroleum Geoscience, 17*, 165–179. https://doi.org/10.1144/1354-079310-014

Multiband Differences and Ratios

Laake, A., Sheneshen, M. S., Strobbia, C., Velasco, L., & Cutts, A. (2011b). Integration of surface/subsurface techniques reveals faults in Gulf of Suez oilfields. *Petroleum Geoscience, 17*, 165–179. https://doi.org/10.1144/1354-079310-014

ETOPO1

Amante, C., & Eakins, B. W. (2009). *ETOPO1 1 arc-minute global relief model: Procedures, data sources and analysis, NOAA Technical Memorandum NESDIS NGDC-24*. National Geophysical Data Center, Marine Geology and Geophysics Division.

Free Air Gravity Anomaly from Satellite Altimetry

Sandwell, D. T., Müller, R. D., Smith, W. H. F., Garcia, E., & Francis, R. (2014). New global marine gravity model from CryoSat-2 and Jason-1 reveals buried tectonic structure. *Science, 346*(6205), 65–67. https://doi.org/10.1126/science.1258213

Magnetic Anomaly

Dyment, J., Lesur, V., Hamoudi, M., Choi, Y., Thebault, E., Catalan, M., the WDMAM Task Force, the WDMAM Evaluators, the WDMAM Data Providers, & World Digital Magnetic Anomaly Map version 2.0. (2015). Map available at http://www.wdmam.org. The WDMAM Task Force: J. Dyment (chair), M. Catalan (co-chair), A. de Santis, M. Hamoudi, T. Ishihara, J. Korhonen, V. Lesur, T. Litvinova, J. Luis, B. Meyer, P. Milligan, M. Nakanishi, S. Okuma, M. Pilkington, M. Purucker, D. Ravat, E. Thébault. (alphabetical order), accepted at IUGG General Assembly of Prag in June 2015 as WDMAM version 2.0.

Crustal Age NOAA

Muller, R. D., Sdrolias, M., Gaina, C., & Roest, W. R. (2008). Age, spreading rates, and spreading asymmetry of the world's ocean crust. *Geochemistry, Geophysics, Geosystems, 9*, Q04006. https://doi.org/10.1029/2007GC001743

Sediment Thickness NOAA

Straume, E. O., Gaina, C., Medvedev, S., Hochmuth, K., Gohl, K., Whittaker, J. M., et al. (2019). GlobSed: Updated total sediment thickness in the world's oceans. *Geochemistry, Geophysics, Geosystems, 20*, 1756–1772. https://doi.org/10.1029/2018GC008115

Deep Sea Drilling Programs

DSDP. (n.d.). http://deepseadrilling.org/
IODP. (n.d.). http://iodp.org/

Global Modeling Software Platform

GPlates. (n.d.). https://www.gplates.org/

Müller, R. D., Qin, X., Sandwell, D. T., Dutkiewicz, A., Williams, S. E., Flament, N., Maus, S., & Seton, M. (2016). The GPlates Portal: Cloud-based interactive 3D visualization of global geophysical and geological data in a web browser. *PLoS One, 11*(3), e0150883. https://doi.org/10.1371/journal.pone.0150883

Müller, R. D., Cannon, J., Qin, X., Watson, R. J., Gurnis, M., Williams, S., Pfaffelmoser, T., Seton, M., Russell, S. H. J., & Zahirovic, S. (2018). GPlates: Building a virtual earth through deep time. *Geochemistry, Geophysics, Geosystems, 19*(7), 2243–2261. https://doi.org/10.1029/2018GC007584

Exploration Concepts and Techniques

Laake, A., Kornpihl, K., & Bishop, A. (2012). *Integrated Workflow for Exploration in Frontier Areas, SEG Las Vegas 2012 Annual Meeting, 1-5*. https://doi.org/10.1190/segam2012-1037.1

Gravity and Magnetic Inversion

Camacho, A. G., Montesinos, F. G., & da Silva Vieira, R. (2002). A 3-D gravity inversion tool based on exploration of model possibilities. *Computers & Geosciences, 28*(2), 191–204. https://doi.org/10.1016/S0098-3004(01)00039-5

Li, Y., & Oldenburg, D. W. (1998). 3-D inversion of gravity data. *Geophysics, 63*(1), 109–119.

Sjöberg, L. E., & Bagherbandi, M. (2017). Gravity inversion. In L. E. Sjöberg & M. Bagherbandi (Eds.), *Gravity inversion and integration* (pp. 247–373). Springer. https://doi.org/10.1007/978-3-319-50298-4_8

Wu, G. J., Liu, H., Zou, Z. B., Yang, G. L., & Shen, C. Y. (2014). 3-Dimensional inversion for gravity anomaly calculation in complex geologic region. *Advanced Materials Research, 962–965*, 238–241. https://doi.org/10.4028/www.scientific.net/amr.962-965.238

Petroleum System Modeling

Ahlbrandt, T. S. (2001a). *The Sirte Basin Province of Libya—Sirte-Zelten total petroleum system*, USGS Bulletin 2202-F.

Bell, R. E., Anderson, R., & Pratson, L. (1997). Gravity gradiometry resurfaces. *The Leading Edge, 16*(1), 55–58.

Gumati, Y. D., & Schamel, S. (1988). Thermal maturation history of the Sirt basin. *Libya: Journal of Petroleum Geology, 11*, 205–218.

Gumati, Y. D., Kanes, W. H., & Schamel, S. (1996). An evaluation of the hydrocarbon potential of the sedimentary basins of Libya. *Journal of Petroleum Geology, 19*(1), 95–112.

Hallett, D. (2002). *Petroleum geology of Libya*. Elsevier.

Hantschel, T., & Kauerauf, A. I. (2009). *Fundamentals of basin and petroleum systems modeling*. Springer.

Hassan, H. S., & Kendall, C. C. G. (2014). Hydrocarbon provinces of Libya—A petroleum system study, chap. 4. In L. Marlow & L. Yose (Eds.), *Petroleum systems of the Tethyan region: American Association of Petroleum Geologists Memoir No. 106* (pp. 101–141).

Magoon, L. B., & Beaumont, E. A. (1999). Petroleum systems. In E. A. Beaumont & N. H. Foster (Eds.), *Treatise of petroleum geology/handbook of petroleum geology: Exploring for oil and gas traps, Chapter 3, AAPG Special Volumes, 3-1-3-34.*

Rusk, D. C. (2001). Libya: Petroleum potential of the underexplored basin centers—A twenty-first century challenge. In M. W. Downey, J. C. Threet, & W. A. Morgan (Eds.), *Petroleum provinces of the twenty-first century: AAPG Memoir 74* (pp. 429–452).

Sandwell, D. T., & Smith, W. H. F. (2009). Global marine gravity from retracked Geosat and ERS-1 altimetry: Ridge sedimentation versus spreading rate. *Journal of Geophysical Research, 114*, B01411.

Veryaskin, A. V. (2018). *Gravity, magnetic and electromagnetic gradiometry: Strategic technologies in the 21st century (IoP Concise Physics)*. Morgan & Claypool. https://doi.org/10.1088/978-1-6817-4700-2. ISBN 978-1-6817-4701-9.

Waples, D. W. (1994). Modeling of sedimentary basins and petroleum systems. In L. B. Magoon & W. G. Dow (Eds.), *The petroleum system—From source to trap: AAPC Memoir 60 chapter 18* (pp. 307–322).

Gulf of Suez Integration

Abd El-Naby, A. I. M., Ghanem, H., Boukhary, M., Abd El-Aal, M., Luning, S., & Kuss, J. (2010). Sequence-stratigraphic interpretation of structurally controlled deposition: Middle Miocene Kareem Formation southwestern Gulf of Suez, Egypt. *GeoArabia, 15*(3), 129–150.

Alsharhan, A. S., & Salah, M. G. (1995). Geology and hydrocarbon habitat in rift setting: Northern and central Gulf of Suez, Egypt. *Bulletin of Canadian Petroleum Geology, 43*, 156–176.

Bosworth, W., & McClay, K. (2001). Structural and stratigraphic evolution of the Gulf of Suez Rift, Egypt: A synthesis. In P. A. Ziegler, W. Cavazza, A. H. F. Robertson, & S. Crasquin-Soleau (Eds.), *Peri-Tethys Memoir 6: Peri-Tethyan Rift/Wrench Basins and Passive Margins* (Mem. Natn. Hist. nat. 186, pp. 567–606), Paris. ISBN: 2-85653-528-3.

Laake, A. (2011). Integration of surface and subsurface data—From satellites to reservoir. In *EAGE/SEG Summer Research Workshop—Towards a Full Integration from Geosciences to Reservoir Simulation, Trieste, 2011*. https://doi.org/10.3997/2214-4609.201402470

Laake, A. (2015). Structural interpretation in color—A new RGB processing application for seismic data. *Interpretation, 3*(1), SC1–SC8.

Laake, A., Sheneshen, M. S., Strobbia, C., Velasco, L., & Cutts, A. (2011c). Integration of surface/subsurface techniques reveals faults in Gulf of Suez oilfields. *Petroleum Geoscience, 17*, 165–179. https://doi.org/10.1144/1354-079310-014

Moustafa, A. R., & Khalil, S. M. (2017). Control of extensional transfer zones on syntectonic and posttectonic sedimentation: Implications for hydrocarbon exploration. *Journal of the Geological Society, 174*, 318–335. https://doi.org/10.1144/jgs2015-138

Strobbia, C., Laake, A., Vermeer, P. L., & Glushchenko, A. (2009a) Surface waves—Use them then lose them. In *EAGE 71st Conference and Exhibition, 8–11 June 2009, Amsterdam, paper T 015.*

Strobbia, C., Vermeer, P., Glushchenko, A., & Laake, A. (2009b). Advances in surface-wave processing for nearsurface characterisation in land seismic. In *71st EAGE Conference & Exhibition 2009.*

Strobbia, C., Vermeer, P., Laake, A., Glushchenko, A., & Re, S. (2010). Surface waves: Processing, inversion and removal. *First Break, 28*, 85–91.

Strobbia, C., Laake, A., Vermeer, P., & Glushchenko, A. (2011). Surface waves: Use them then lose them. Surface-wave analysis, inversion and attenuation in land reflection seismic surveying. *Near Surface Geophysics, 9*, 503–513. https://doi.org/10.3997/1873-0604.2011022

Radarsat Integration

Laake, A., & Strobbia, C. (2013b). Mapping Paleorivers from satellite radar and surface-wave velocities. In *75th EAGE Conference and Exhibition 2013, Workshop Paper G06*.

Redadaa, S., Boualleg, A., Merabtine, N., & Bensama, M. (2007). A study of P-band penetration capabilities: Application in sub-surface remote sensing. *Romanian Journal of Physics, 52*(5–7), 575–591.

Estimating Impact of Volcanic Rocks

Abadi, A. M., van Wees, J.-D., van Dijk, P. M., & Cloetingh, S. A. P. L. (2008). Tectonics and subsidence evolution of the Sirt Basin, Libya. *AAPG Bulletin, 92*(8), 993–1027.

Ahlbrandt, T. S. (2001b). *The Sirte Basin Province of Libya—Sirte-Zelten total petroleum system*, USGS Bulletin 2202-F.

Brugge, N. (n.d.). *Crater-like structures in context with the large flood basalt field of Al-Haruj al-Aswad (Libya)*.

Elshaafi, A., & Gudmundsson, A. (2016). Volcano-tectonics of the Al Haruj Volcanic Province, Central Libya. *Journal of Volcanology and Geothermal Research, 325*, 189–202. https://doi.org/10.1016/j.jvolgeores.2016.06.025

Iyer, K., Schmid, D. W., Planke, S., & Millett, J. (2017). Modelling hydrothermal venting in volcanic sedimentary basins: Impact on hydrocarbon maturation and paleoclimate. *Earth and Planetary Science Letters, 467*(1), 30–42. https://doi.org/10.1016/j.epsl.2017.03.023

Luo, X., Gong, S., Jun, F. J., Wang, Z. H., & Qi, J. S. (2017). Effect of volcanic activity on hydrocarbon generation: Examples in Songliao, Qinshui, and Bohai Bay Basins in China. *Journal of Natural Gas Science and Engineering, 38*, 218–234. https://doi.org/10.1016/j.jngse.2016.12.022

Nuzzo, M., Elvert, M., Schimdt, M., Scholz, F., Reitz, A., Hinrichs, K.-U., & Hensen, C. (2012). Impact of hot fluid advection on hydrocarbon gas production and seepage in mud volcano sediments of thick Cenozoic deltas. *Earth and Planetary Science Letters, 341–344*, 139–157. https://doi.org/10.1016/j.epsl.2012.05.009

Peace, A., McCaffrey, K., Imber, J., Hobbs, R., van Hunen, J., & Gerdes, K. (2017). Quantifying the influence of sill intrusion on the thermal evolution of organic-rich sedimentary rocks in nonvolcanic passive margins: An example from ODP 210-1276, offshore Newfoundland. *Canada, Basin Research, 29*, 249–265. https://doi.org/10.1111/bre.12131

Rateau, R., Schofield, N., & Smith, M. (2013). The potential role of igneous intrusions on hydrocarbon migration, West of Shetland. *Petroleum Geoscience, 19*, 259–272. https://doi.org/10.1144/petgeo2012-035

Schofield, N., Holford, S., Millett, J., Brown, D., Jolley, D., Passey, S. R., Muirhead, D., Grove, C., Magee, C., Murray, J., Hole, M., Jackson, C. A.-L., & Stevenson, C. (2017). Regional magma plumbing and emplacement mechanisms of the Faroe-Shetland Sill Complex: Implications for magma transport and petroleum systems within sedimentary basins. *Basin Research, 29*, 41–63. https://doi.org/10.1111/bre.12164

Senger, K., Millett, J., Planke, S., Ogata, K., Eide, C. H., Festoy, M., Galland, O., & Jerram, D. A. (2017). Effects of igneous intrusions on the petroleum system: A review. *First Break, 35*, 47–56.

Chapter 10
Seismic Logistics and Planning

Abstract The success of hydrocarbon exploration depends on the quality of the three-dimensional information of the subsurface. Seismic data play an important role in generating high quality subsurface models. The planning of onshore seismic surveys aims at acquiring the best possible data quality with the least effort. In flat open terrain without obstacles the planning of seismic surveys is a straight-forward task and the seismic acquisition is mostly optimized to fast acquisition. In complex terrain, however, significant attention is paid to mitigating the risks related to people, equipment and data quality in addition to fast acquisition with the goal to optimize all parameters for the most efficient operation. This part of the planning is where remote sensing data prove to be most useful. Typically the following remote sensing data and derivatives are used for the task:

- **Digital relief model and surface gradient**: these data provide the basis for the topography and terrain dip which have a major impact on logistics and data quality.
- **Multispectral satellite imagery**: the optical bands which provide the best spatial resolution are used for terrain reconnaissance to enhance the terrain information obtained from the DRM. The shortwave infrared bands enable the interpretation for lithology to discriminate different rock types and vegetation.
- **Radar back scatter imagery**: is sensitive to terrain roughness at a resolution well below the wavelength of the microwaves. The back scatter intensity allows the discrimination of soft sandy surfaces from clay or gypsum flat and hard rock blocks to the texture of slopes and escarpments.

The three aspects covered in this section, i.e. logistics, data quality and near surface modeling, are closely linked. Therefore in each chapter, we will have a major focus on one aspects whilst mentioning aspects of the other two aspects as well.

Keywords Logistic risks: steep slopes, roll-over, sliding, break through surface, rocky blocks, wet sabkha, manoeuvreability · Source data quality risks: point loading, baseplate sliding, baseplate break through surface, ground support, compliant surface · Receiver data quality risks: hard surface point coupling, sensor

A. Laake, *Remote Sensing for Hydrocarbon Exploration*, Springer Remote Sensing/Photogrammetry, https://doi.org/10.1007/978-3-030-73319-3_10

burying, soft surface resonance, (wind exposure) · Near surface velocity modeling: flat layers, surface karsting, escarpments, sand dunes, scattering, surface multiples, near surface interbed multiples, subsurface scattering · Geological aspects: surface lithology, evaporite, gypsum pan, basalt, basalt blocks, lava ridge, escarpment, vegetation, grassland, rock-to-vegetation ratio, sand dunes, star dune, barchan dune · Remote sensing aspects: digital elevation model, surface gradient, surface dip, terrain classification, lithology interpretation, vegetation index, basalt index

10.1 Introduction

The process of planning and conducting seismic acquisition operations and processing and interpreting the resulting data is an integrated process that ensures the optimal result, i.e. the best possible subsurface geologic image which is suitable for the exploration for hydrocarbons. Seismic acquisition comprises the following components:

- **Seismic source**: hydraulic vibrators, shot-hole drilling rig, dynamite source
- **Seismic receivers**: geophones as single sensors or in arrays, line equipment connecting the geophones (unless the operation is cable-less, Crice et al., 2015)
- **Recording truck**: nerve center of the seismic operation which commands the operation of the seismic source, the operation of the receiver spread and the collection of the data
- **Seismic crew**: workforce that plants and recovers the geophones
- **Support equipment**: vehicles that transport the geophones and line equipment to move the receiver spread in the field, in case of heliportable operation this task is performed by helicopters

Not considered in this list is the infrastructure that is required to operate the seismic crew, such as camp and transportation of supplies.

Under ideal conditions, seismic operation would be carried in flat and open terrain without infrastructure or vegetation. The terrain provides a compliant surface that supports the weight of the vehicles involved in the operation and supports the planting of the receivers with perfect coupling. However, seismic operation is not carried out under such ideal conditions. Seismic acquisition operation is, therefore, subject to risk related to operational logistics and data quality (Laake, 2011, Laake & Insley, 2004, Laake & Cutts, 2007). We will study the following aspects:

- **Logistic risks**: vehicle roll-over, sliding, breaking through soft surface, obstruction by hard rocky blocks at the surface
- **Source data quality risks**: hard surface point loading, baseplate sliding on steep slopes, baseplate brake through surface
- **Receiver data quality risk**: hard surface point coupling, soft surface resonance

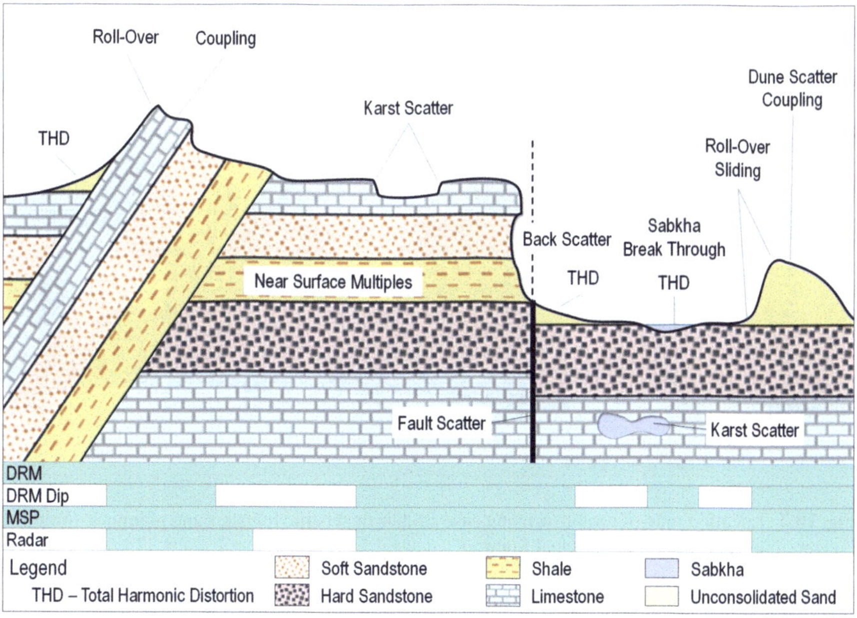

Fig. 10.1 Overview some seismic operations risks related to the surface and near surface geology

Table 10.1 Overview of the remote sensing examples for seismic logistics planning

Section	Geologic setting	Area of interest	Remote sensing input for
11.2	Volcanic plateau	Payun, Argentina	Seismic logistic planning
11.2	Coastal sabkha	Abu Dhabi, UAE	Operation in dry and wet sabkha
11.2	Sand dunes	Berkine, Algeria	Operations in sand dunes
11.3	Mesas and gravel plains	Tademait Plateau, Algeria	Seismic source and receiver coupling
11.3	Surface limestone karst	Ghazalat Plateau, Egypt	Mapping of surface karst scatterers
11.4	Mesa erosion	Ghazalat Plateau, Egypt	Near-surface velocity modeling

Figure 10.1 shows a schematic overview of the above-mentioned seismic operations risks that are related to the geology at the surface and in the near surface. The colored bars below the geologic section indicate for which features what remote sensing data are helpful. These examples are discussed in the following paragraphs.

The remote sensing examples discussed in this section are listed in Table 10.1 and their locations shown in the map in Fig. 10.2.

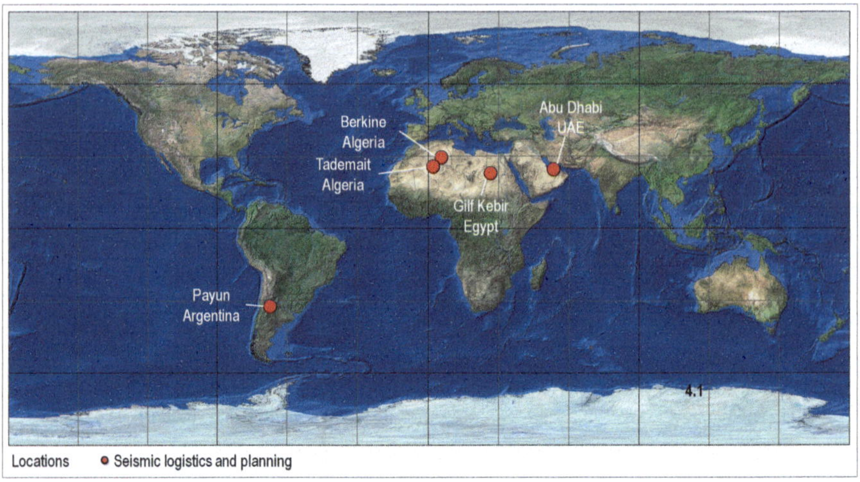

Fig. 10.2 Location map of remote sensing case studies fir seismic logistics and planning. (Satellite imagery courtesy of U.S. Geological Survey)

10.2 Seismic Logistics Planning

The success of hydrocarbon exploration depends on the quality of the three-dimensional information of the subsurface. Seismic data play an important role in generating high quality subsurface models. Therefore, the planning of seismic surveys aims at acquiring the best possible data quality with the least effort. In flat open terrain without obstacles the planning of seismic surveys is a straight-forward task and the seismic acquisition is mostly optimized to fast acquisition.

In this chapter we will cover complex settings that challenge seismic acquisition from a logistics point of view. We will see that often logistics challenges are coupled with challenges for the data quality. Hence, we will try to optimize both aspects if possible. The challenges we cover in the chapter comprise:

1. **Rough volcanic terrain** with steep slopes with the risk for vehicle roll-over as well as lava flows and large blocks that impact the movement of vehicles. This case study is located on the Payún Plateau in the Province Mendoza, Argentina.
2. **Wet and dry sabkha terrain** with areas of very low ground support threatening vehicles to break through the surface crust and sink into liquified soil. We have chosen Sabkha Matti in western Abu Dhabi, U.A.E. for this study.
3. **Giant sand dunes** with steep slopes, unconsolidated sand and gypsum pans in the inter-dune space. The logistics threats in this terrain comprise roll-over and sliding of vehicles as well as poor ground support and high signal distortion in the gypsum pans. We have chosen the large dune field of the Berkine Basin in eastern Algeria for this case study.

For each study, we give a short introduction into the geological setting and define the challenges posed by the terrain. Then we asses which remote sensing data, processing and interpretation provides solutions. We conclude with a set of risk maps that can used when planning seismic surveys in challenging terrain.

10.2.1 Seismic Logistics Planning in Rough Volcanic Terrain

We start the study of how remote sensing imagery and their interpretation can assist seismic logistics planning in the rough volcanic terrain of the Andean foothills. The Payún Plateau is located in the foothills of the central Andes in Argentina. The subduction of the Pacific Plate under south America resulted in complex compressional tectonics which created the north-south trending Malargüe fold and thrust belt (Hernando et al., 2012, Fig. 10.3). The melting of the subducted crust initiated widespread volcanism. In the study area the shield volcano of Payún Matrú deposited the lavas of the Payún Plateau which are complemented with volcanic ashes from the stratovolcano Payún Liso. The erosion of the soft strata below the lavas lead to the collapse of the edges of the lava plateau leaving steep slopes of very rough terrain which poses threats particularly to vehicular operation. Whilst vibrators can manage steep terrain, the vehicles carrying the line equipment often cannot leaving costly heliportable operation as the only solution to move equipment efficiently. In our study area heliportable logistics has been used mostly in the areas of Cañado Amarillo in the south and Barda in the east.

Landsat 7 multispectral imagery and the SRTM DEM provided the basis for the surface model shown in Figs. 10.3 and 10.4a.

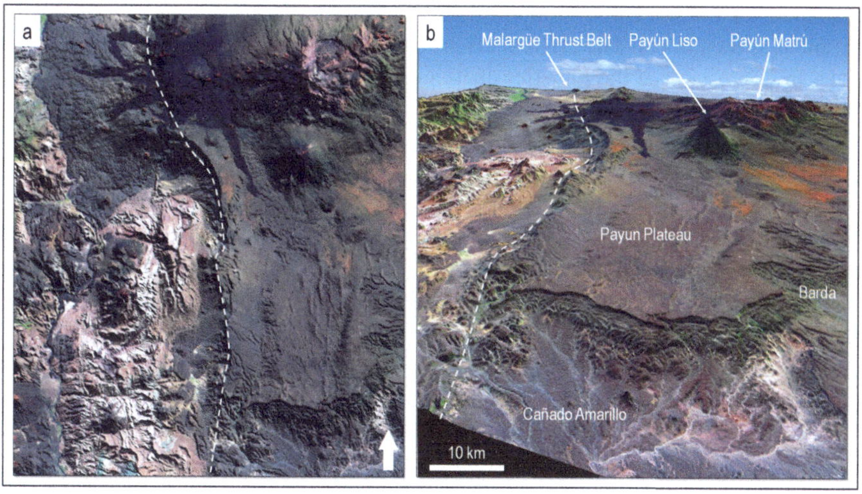

Fig. 10.3 Geologic setting of seismic operations planning in rough volcanic terrain in the foothills of the Argentinian Andes represented in multispectral satellite imagery. (**a**) Map view, (**b**) virtual 3D view with annotations of the main geographic features. (DRM and Landsat 7742 RGB image courtesy of U.S. Geological Survey)

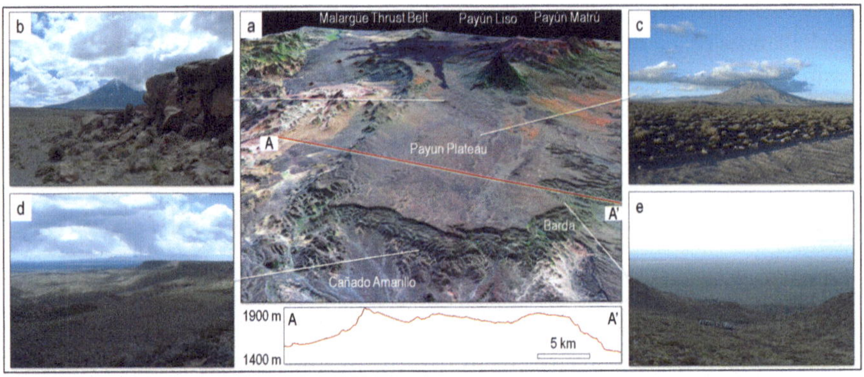

Fig. 10.4 Challenges for seismic logistics planning for operation on and near the Payún Plateau. (**a**) 3D view of the target area with a topographic section, (**b**) edge of lava flow, (**c**) grassland of Payún Plateau with Payún Lisso volcano in the background, (**d**) steep escarpments towards Cañado Amarillo, (**e**) steep ravines in the Barda area with vibrators in the foreground. (DRM and Landsat 7 imagery courtesy of U.S. Geological Survey)

The band combination of short wave infrared (band 7), near infrared (band 4) and visible red (band 2) allowed generating a composite RGB image that discriminated basalt (purple tones), tuff and volcano-clastics (reddish to white tones) and evaporites (orange) from vegetation (green). Closer inspection at the surface revealed that large areas of the plateau are covered in grass and basalt gravel which poses no threat to vehicular movement (Fig. 10.4c). However, in the north lava flows of several meters thickness create barriers that cannot be crossed by any vehicle (b). Detours for vibrators and line equipment carriers are unavoidable. The escarpments of Cañado Amarillo (d) and the steep ravines in the Barda area (e) pose significant challenges to the movement of line vehicles, whereas vibrators master this difficult terrain as shown in (e). The cross-section AA′ shown below the virtual 3D view (a) gives an impression of the logistics challenges at the edges of the plateau.

The operational challenges faced in rough volcanic terrain are shown in Fig. 10.5 for three different terrain types. Column 1 represents rather flat terrain (a1) covered with a mix of sand and pebbles. The vibrator baseplate can establish full surface contact with the ground (a2) which is represented in the padmark imprint (a3). The terrain shown in column 2 is still rather flat (b1) but covered with larger boulders which the baseplate cannot push into the ground during the sweep (b2). The consequence is point loading on individual boulders leaving no baseplate padmark (b3) and resulting in both severe signal distortion and damage to the baseplate. On steep slopes (c1) boulders are present but mostly spread sparsely with soft sediments in between. The baseplate has good contact to the ground but can slide of up to 50 cm during the sweep (c2) leaving a characteristic padmark pattern after the sweep (c3). The sliding causes the vibrator control electronics to compensate for the measured loss in ground force in a non-linear way which in turn leads to high distortion. Cases b and c are, therefore, the target for remote sensing support with logistics and data quality planning in steep and rocky terrain.

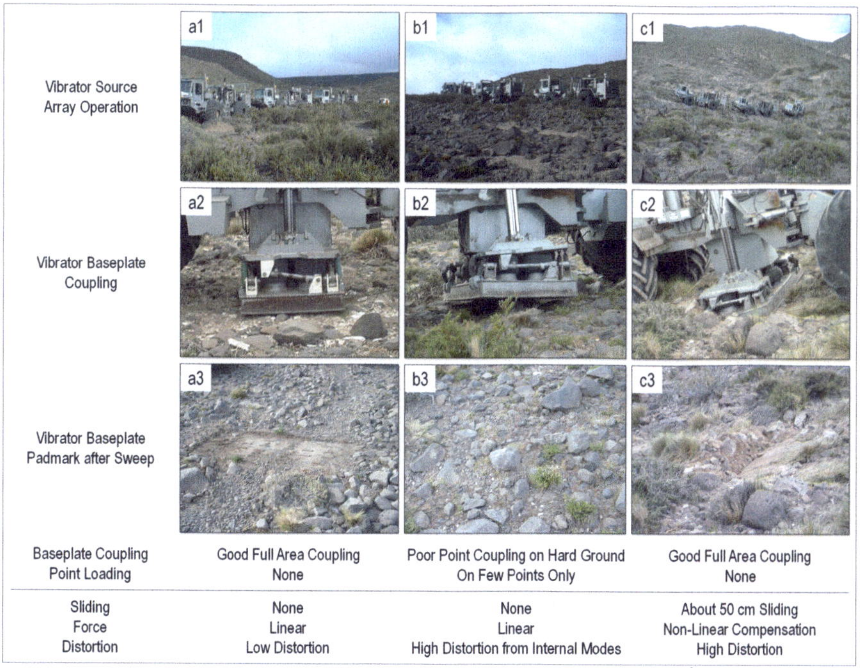

Fig. 10.5 Operation of vibroseis operation in rough volcanic terrain

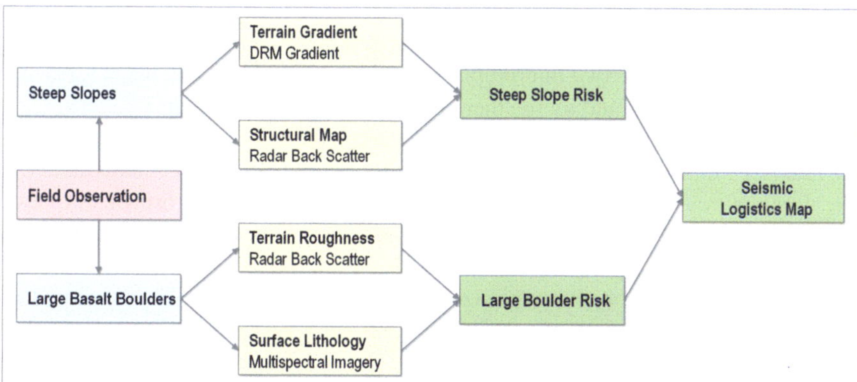

Fig. 10.6 Workflow for remote sensing-based logistics planning in steep and rocky terrain

With the definition of the challenges in place, we can now build a workflow to develop solutions based on remote sensing data (Fig. 10.6). The two key challenges are steep slopes posing the threat of vehicle roll-over and large basalt blocks which make any vehicle movement impossible. The first challenge can be addressed by the

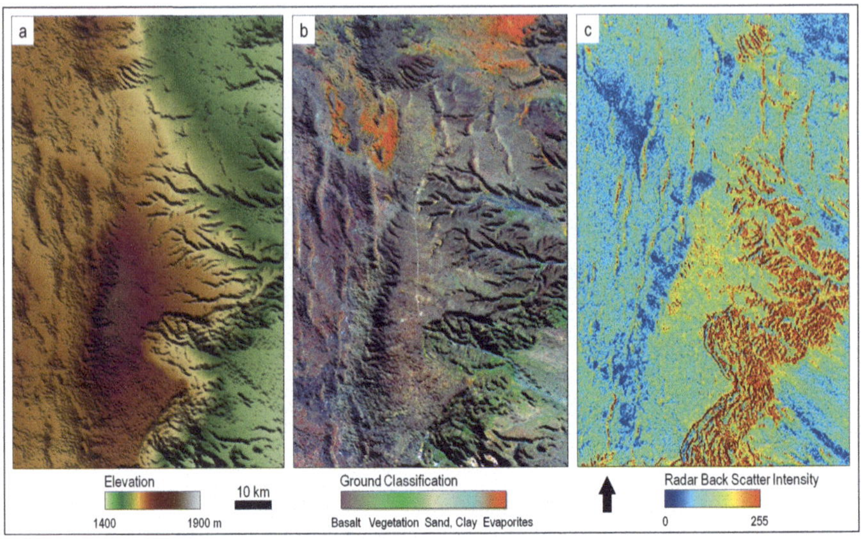

Fig. 10.7 Overview of remote sensing input data. (**a**) Digital elevation model, (**b**) multispectral satellite image composite Landsat 7 742 RGB, (**c**) radar back scatter intensity image. (DRM and Landsat satellite imagery courtesy of U.S. Geological Survey, Radarsat-1 imagery courtesy of CSA)

terrain gradient from the DRM. High amplitudes in radar back scatter data give clues where steep slopes are covered by hard rock. Additionally, radar data also provide a structural map showing features like lava flows. For the detection of large blocks, the surface roughness from radar back scatter highlights rough areas and multispectral satellite imagery interpreted for lithology complements information about basalt. From these two interpretation routes we can obtain risk maps for steep slopes and large blocks and merge them into a seismic logistics map.

The remote sensing input data (Fig. 10.7) comprise the digital elevation model (a), Landsat 7 multispectral 742 RGB composites (b) and radar back scatter data (c). The digital elevation model shows an undulating plateau surface (brownish tones) with some lava flows and lava ridges. Multispectral imagery shows as the dominant color on the plateau a purple tone that had been interpreted as basalt, however it does not give information about the size of individual rock bodies. This is provided by the radar back scatter intensity that highlights steep rocky slopes in the east and lava ridges (elongated red to yellow features).

For more detailed mapping of the vegetation, we use the band combination of very near infrared, shortwave infrared and green bands (Fig. 10.8a). Barren and rocky ground appears as dark red to purple and white tones whereas grassland shows dark to light green colors. Bush are represented by greyish-green colors. The green and purple areas match with the field photo showing grass and gravel cover (d). The operation of wheeled vehicles in this terrain is easy. Logistics can be planned for efficiency.

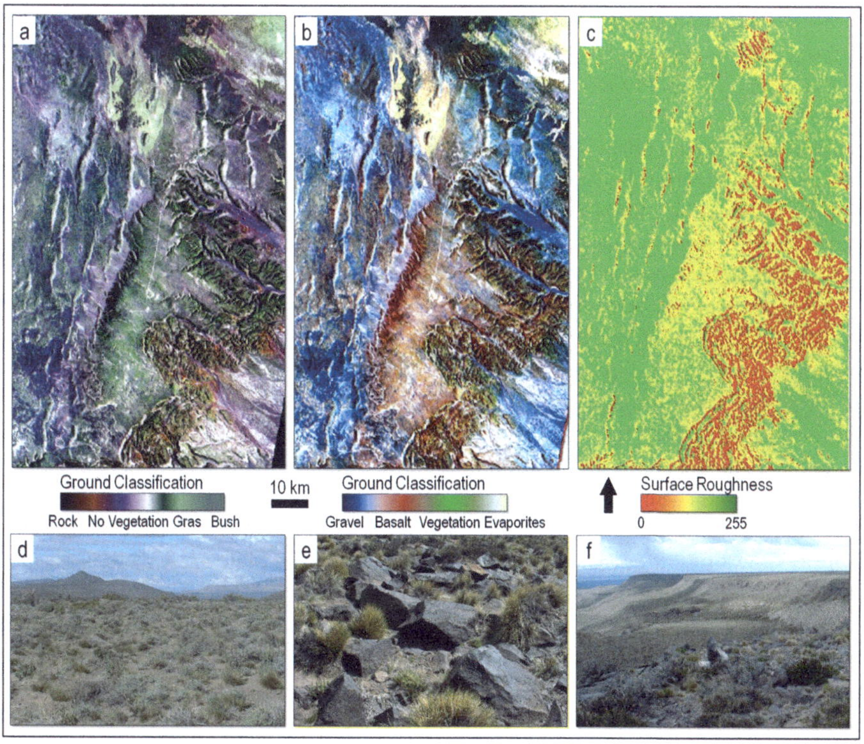

Fig. 10.8 Processing of remote sensing data for delineation of logistic risks. (**a**) Vegetation index from Landsat 7 453 RGB composite, (**b**) basalt index from Landsat 7 675 RGB composite, surface roughness from RADARSAT-1 back scatter intensity, (**d**) grassland on Payún plateau, (**e**) basalt blocks on ridges and slopes, (**f**) steep slopes. (DRM and Landsat 7 imagery courtesy of U.S. Geological Survey, Radarsat_1 imagery courtesy of CSA)

The lithological discrimination becomes available when using the infrared bands in a composite RGB image of thermal infrared (band 6) and shortwave infrared (bands 7 and 5). Figure 10.8b reveals sharply defined areas of gravel (blue tones), basalt blocks (red) vegetation (green) and evaporites like gypsum (c). The distinction between basalt blocks and basalt gravel becomes possible because lavas and large basalt blocks as shown in (e) differ in heat capacity.

The resulting heat radiation is then detected primarily by the thermal infrared band 6 (b). Terrain colored in reddish tones in (b) poses, therefore, considerable risk for the operation of wheeled vehicles. Multispectral imagery alone cannot distinguish terrain with predominantly large basalt blocks and scattered bush from terrain covered in basalt gravel with grass when the relative percentage of basalt and vegetation is similar. The reason is that multispectral satellite imagery generates images with a pixel size of of 100–600 m^2 and hence does not resolve individual large basalt blocks.

This is where radar back scatter imagery becomes important. The spatial resolution of radar imagery is similar to optical satellite imagery, but the averaged back scatter intensity, i.e. the terrain roughness, is directly related to the average size of scattering objects at the surface (c). This means that basalt gravel and grass show significantly lower values of surface back scatter (green in (c)) than large blocks with bush interspersed between the blocks (yellow in (c)). Escarpments with steeply dipping rocky slopes have a characteristic length of typically 10s of meters (f) and are, therefore, directly mapped by radar imagery. They are shown in (c) as red areas and indicate a high risk of roll-over for wheeled vehicles.

Table 10.2 summarizes the capabilities of mapping different terrain types from radar back scatter intensity imagery and direct imaging of surface objects from radar and multispectral imagery.

From the integration of radar image interpretation for surface roughness and multispectral imagery for surface lithology and vegetation we can generate maps showing surface cover classification and surface roughness (Fig. 10.9a, b). These maps are translated into a risk map for heavy vehicle roll-over (c) and into wheeled vehicle manoeuvre risk related to large basalt blocks. The surface cover classification (a) shows basalt cover (purple tones) on most of the plateau as well as on the block basalt ridge near the escarpment. The escarpment itself shows significant vegetation cover (green). In the north flats are covered by sand, clay and evaporites like gypsum (bright areas). The surface roughness map (b) identifies most of the plateau as smooth terrain that is only interrupted by local lava flows (orange-red). The block basalt ridge shows intermediate roughness corresponding to the coarse texture induced by the basal blocks. The escarpments are characterized by large-scale rough terrain (red).

The roll-over risk is linked mainly to high levels of terrain dip (red). Wheeled vehicles need approach these areas carefully and only from certain angle relative to the terrain dip to avoid roll-over. During field operation most of these areas have actually been served by heliportable operation to avoid the long detours for wheeled vehicles imposed to avoid roll-over risk. The vibrator source lines have been overlaid on this map to demonstrate the impact the terrain has on the source points during acquisition. Rough surface areas (yellow) are impassable for most wheeled and

Table 10.2 Mapping and imaging characteristics of optical and radar remote sensing imagery, *RBS* radar back scatter intensity

Terrain	Sand, clay pan	Grassland, gravel plain	Basalt block field	Lava ridge, forest	Escarpments
Lithology	Sand, clay	Basalt	Basalt	Basalt lava	Hard rock, hard soil
Vegetation	Absent	Grass	Bush	Trees	Absent
Typical length	0.01–1 mm	1–20 cm	0.5–5 m	5–50 m	50–500 m
RBS mapping	0–5%	10–30%	40–60%	70–100%	70–100%
Radar imaging [8–100 m Res.]	No	No	No	(Yes)	Yes
Optical imaging [10–90 m Res.]	No	No	No	(Yes)	Yes

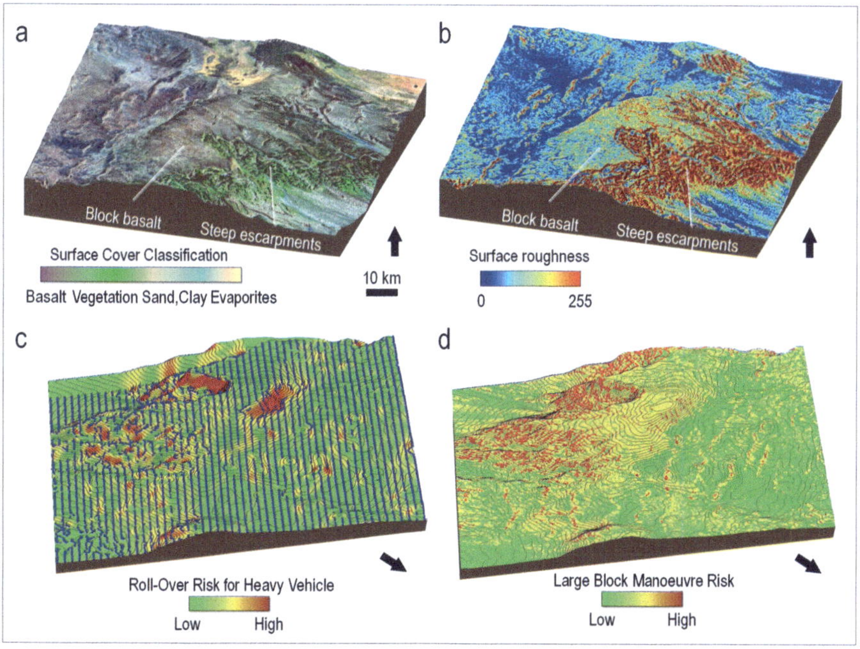

Fig. 10.9 Interpretation of multispectral and radar satellite imagery for logistics planning. (**a**) Surface cover classification from multispectral Landsat 7 742 RGB composite, (**b**) surface roughness interpretation of radar back scatter intensity, (**c**) map for heavy machinery roll-over risk, (**d**) map of manoeuvre risk related to large basalt blocks. (DRM and Landsat 7 imagery courtesy of U.S. Geological Survey, Radarsat-1 imagery courtesy of CSA)

have, therefore, been served by heliportable operation. The logistics map assisted in planning the heliportable operation efficiently to ensure seamless handover between wheeled vehicle and heliportable operation.

10.2.2 Operations in Wet and Dry Sabkha

Seismic operation often encounters salt flats called sabkha which pose specific challenges to operations. Kendall (1978) described the genesis of sabkha and evaporite flats in general, Goodall et al. (2006), Evans et al. (2002), and Sanford and Wood (2001) focused on the coastal sabkha in our case history. In the context of coastal sabkha, the Arabic expression describes a low lying terrain near mean sea level in hot and arid climate. Evaporation and complex processes of upward leakage of salty brines, downward recharge of water as well as lateral inflows and outflows are constantly changing the composition of the sabkha. Typical lithologies found in sabkha comprise clays, sand, halite, anhydrite and gypsum. The varying contents of water and brine changes the ratio between fluid-soaked formation, i.e. wet sabkha, and dried out crusts, i.e. dry sabkha. Dry sabkha cannot only convert to wet sabkha due a change in water and brine supply. Low frequency vibration for example from vehicles driving slowly across sabkha terrain can change duricrusts to wet sabkha when these overlay fluid-soaked formations.

This transformation can go as far as liquefaction as we will see below. The geologic setting of Sabkha Matti in the western UAE (Fig. 10.10) is defined by an ancient river delta that had been active during the Pleistocene. The elevation map (a) shows the location of Sabkha Matti between the limestone plateau associated with the Qatar Arch in the west and the sand dunes of the Rub Al Khali desert in the east and the south. The low lying sabkha terrain reaches up to 50 km inland. The terrain gradient reveals the sabkha terrain as a flat and smooth terrain which can be clearly distinguished from the rough topography of the Qatar Arch and the dune lineaments of Rub Al Khali (b). Multispectral imagery (c) allows further distinction between coastal sabkha (cyan) and inland sabkha (reddish brown) indicating slight lithologic differences between these two zones.

The 3D rendering (Fig. 10.11) of the elevation model (a) and the multispectral satellite image draped over the elevation model (b) shows the correlation between the occurrence of sabkha and the terrain elevation. Both images show a series of anomalous islands that are related to salt plugs, that are not related to the sabkha.

The workflow for mapping wet and dry sabkha for logistics planning (Fig. 10.12) considers digital elevation data for mapping the terrain elevation and multispectral satellite imagery for delineating the occurrence of sabkha minerals. The sabkha model is generated from the integration of the structural map from the terrain gradient, flat areas from the topography and evaporite lithologies such as halite, gypsum,

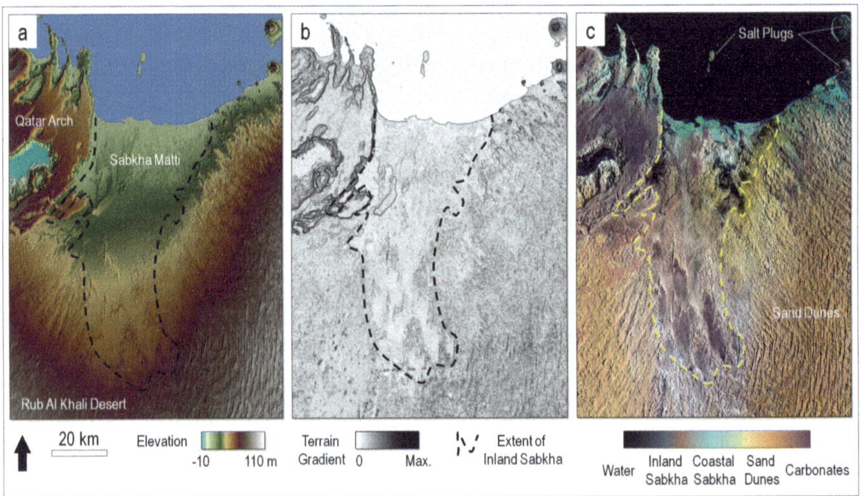

Fig. 10.10 Geologic setting of Sabkha Matti, UAE. (**a**) Digital elevation model, (**b**) Terrain gradient, (**c**) Landsat 7 742 RGB composite, extent of active sabkha indicated by dashed line. (DRM and Landsat 7 satellite imagery courtesy of U.S. Geological Survey)

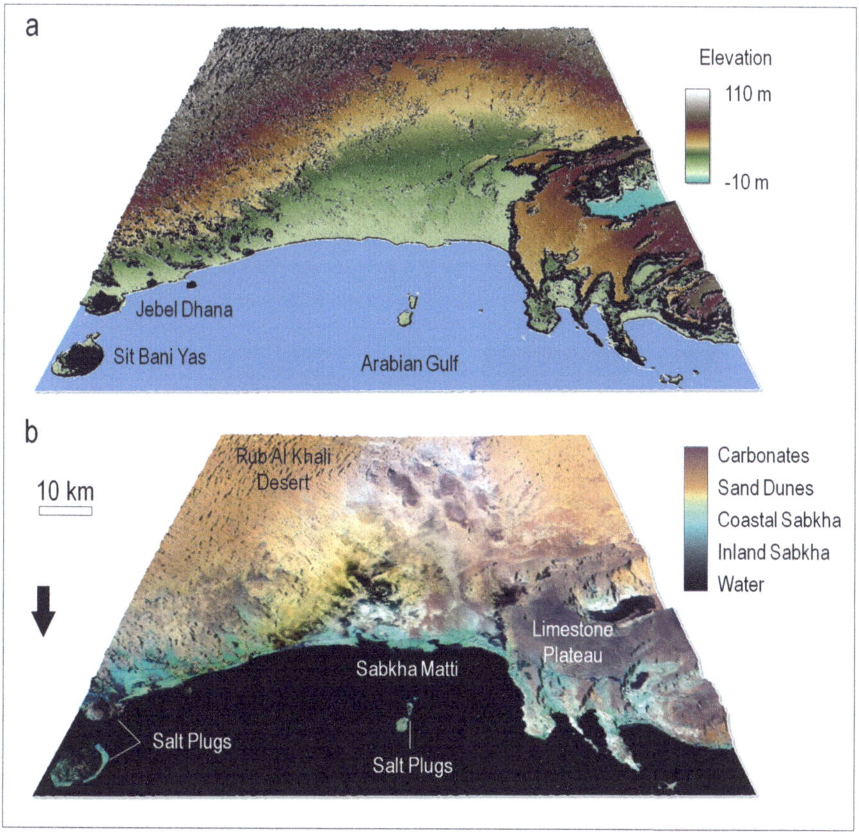

Fig. 10.11 Interpretation of remote sensing data for sabkha characterization. (**a**) DRM overlaid with surface gradient, (**b**) Landsat 7 742 RGB composite draped over DRM with interpretation of key geologic features. (DRM and Landsat 7 imagery courtesy of U.S. Geological Survey)

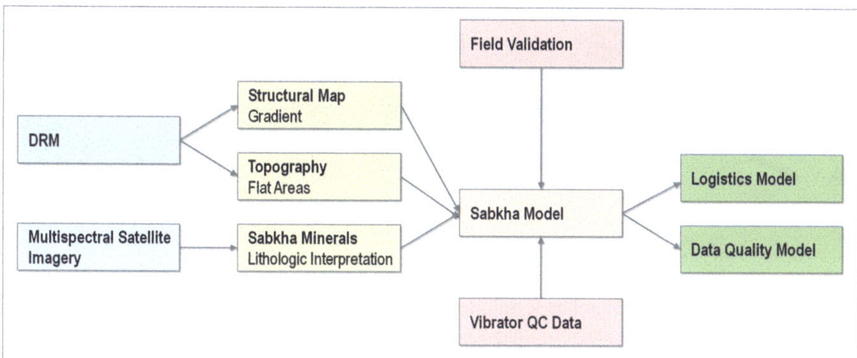

Fig. 10.12 Workflow for generating logistics and data quality prediction models from remote sensing data

anhydrite as well as clay and sand from multispectral imagery. Field photos and vibrator QC data from the area provide the validation and allow the generation of a logistics and a data quality model.

Field photos (Fig. 10.13) give an insight into the surface features as well as the challenges imposed on seismic operations in sabkha terrain. We observed that most areas with logistics problems also showed data quality problems due to degraded source coupling of the vibrator baseplate to the ground. We, therefore, include source-related data quality into our study.

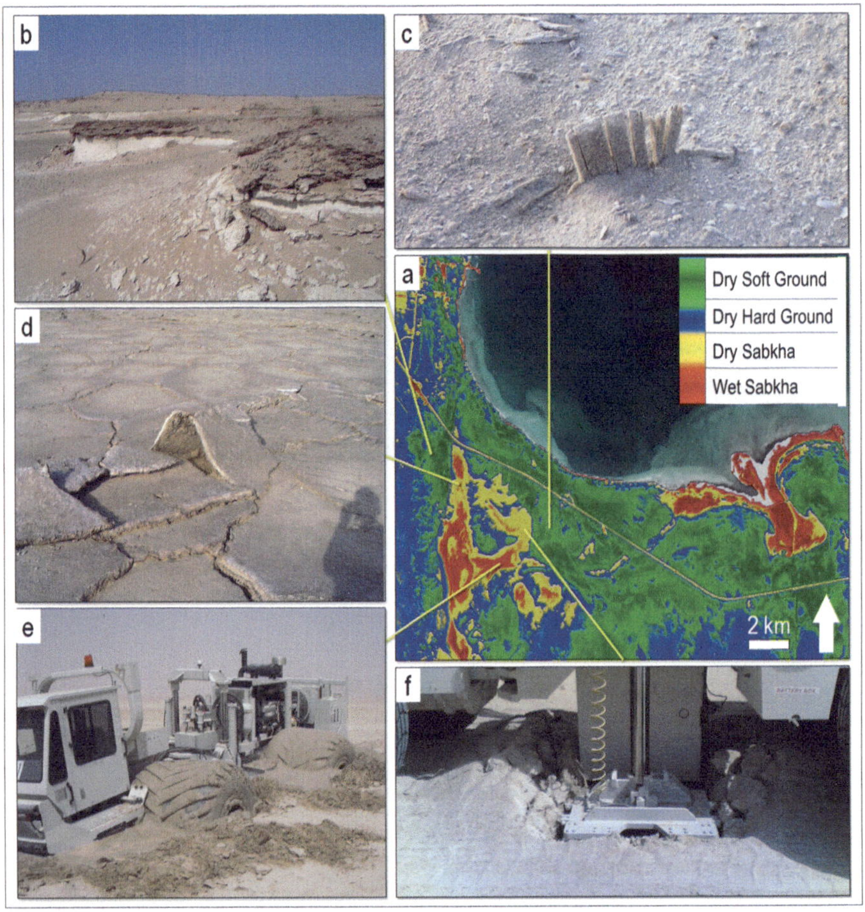

Fig. 10.13 Classification of sabkha terrain for logistics planning. (**a**) Terrain classification map obtained from multispectral satellite imagery. Field photos of logistic challenges: (**b**) limestone escarpments, (**c**) gypsum crystals indicative of shallow wet sabkha, (**d**) salt crusts, (**e**) vibrator breaking through crust of wet sabkha, (**f**) vibrator baseplate breaking into dry sabkha. (Landsat 7 imagery courtesy of U.S. Geological Survey, terrain classification map after Cutts & Laake (2008))

Supervised classification of multispectral images interpreted for lithology provides a map of ground properties. We discriminate dry sabkha represented mainly by salt and gypsum crusts (d) from wet sabkha which is characterized by smooth gypsum and clay minerals. Interestingly, weathered gypsum crystals that protrude vertically from the soft surface like teeth appear to be indicative of shallow wet sabkha (c). For non-sabkha areas we distinguish hard ground, i.e. the limestone plateau (brown formations at the top in (b)) from soft sandy ground (beige to gray formations in (b)).

The challenges for vibroseis operation in sabkha terrain are quite impressive. Low frequency vibrations from driving on wet sabkha leads to liquefaction (e). The ground can now longer support the heavy weight of the vibrator, which breaks through the thin dry crust at the surface and sinks into the liquid mix of brine and mud until it floats. Obviously, recovering the vibrator as quickly as possible from this situation is the first priority because the brine is very corrosive causing severe damage to the vibrator's mechanical and electrical components.

In dry sabkha, the baseplate may break the duricrust and sink in (f), which results in sever distortion of the signal emitted from the vibrator baseplate. Corrosion damage to the baseplate may occur as well.

From the observation we have made so far, it appears desirable to manage and mitigate the logistics and data quality risks of seismic operation in sabkha already in the planning stage. Remote sensing data provide topography and lithology information which in turn allows the interpretation for operations risk (Fig. 10.14a).

The operational risk is best assessed in 3D, for example through draping the operations risk map on the digital elevation model. We observe two distinct areas of wet sabkha, which poses the biggest operational risk. One area is located on the coast in what appears to be a channel of the ancient delta. The other area is located inland on the western channel of the ancient delta. As indicated earlier, areas of elevated wet sabkha-related risk are also correlated with high distortion of the vibrator signal. The map of the total harmonic distortion (THD) of the vibrator signal (b) shows this correlation. We do observe also that the THD is elevated on the slope of the Miocene limestone plateau due to problematic ground coupling of the baseplate in the rough terrain (red zone west on the slope west of Sabkha Matti).

We mentioned in the introduction to this chapter that the distribution of wet and dry sabkha depends on the migration of fluids in the subsurface. To assess the temporal change of the sabkha distribution, we compare two multispectral images (Fig. 10.15). We use the Landsat 8 image from November 2019 as a reference. The band 421 RGB composite comprises the green, blue and coastal blue bands, which allows mapping both, the onshore and offshore (a). To provide the reference for the sabkha distribution we overlay the operations risk map on the 432 RGB composite (b). Then we build the difference between the Landsat 8 image from 2019 and the Landsat 5 image from 2002, which had been used to create the operations risk map. We notice that the wet sabkha shows changes that are indicated by the yellow arrows. This implies that assessment of the sabkha risk should be done close to the planning of the seismic survey using the most recent multispectral satellite image.

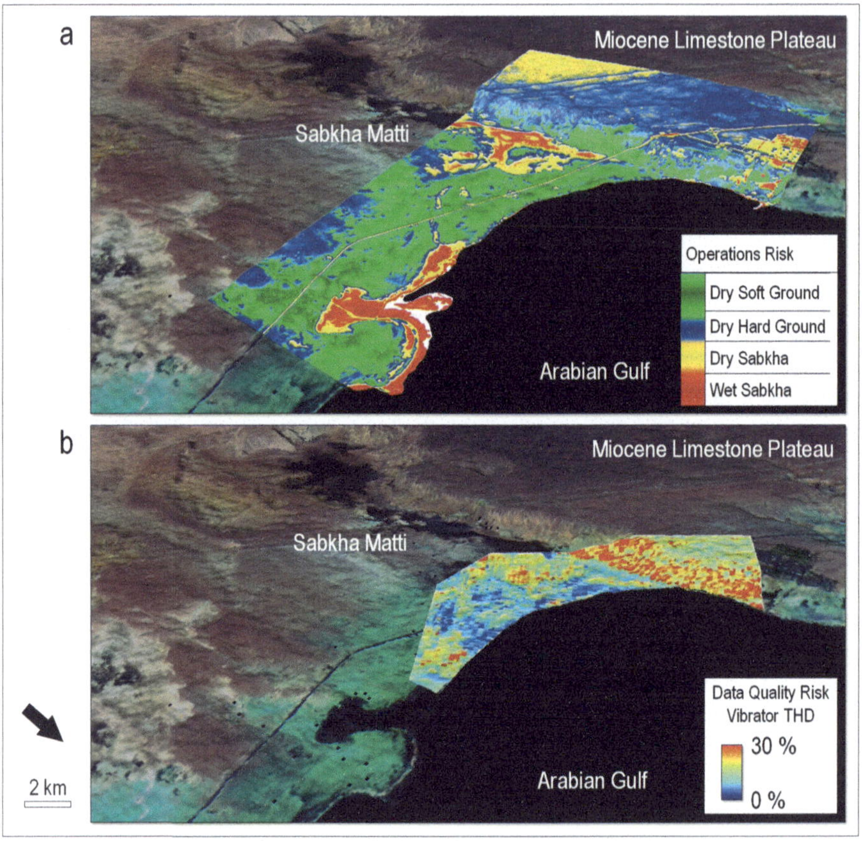

Fig. 10.14 Risk maps for seismic operation in sabkha terrain. (**a**) Operations risk related to wet and dry sabkha, (**b**) data quality risk related to harmonic distortion of the emitted vibrator signal. (DRM and Landsat 7 imagery courtesy of U.S. Geological Survey, operations risk map and vibrator THD map after Cutts & Laake (2008))

We also observe artifacts from construction. These are indicated by red arrows. For example, the major highway UAE11 linking Saudi Arabia with Abu Dhabi has been expanded significantly. The most obvious difference is visible at the intersection roughly in the middle of Sabkha Matti. It gives access to the coastal settlement, which are indicated by numerous red arrows. Further construction affected coastal protections near Jebel Dhanna. A sea dike has been built which shows as a strong red lineament. The wet sabkha west of it has been cut off its connection to the open sea by a dam (red arrow), which resulted in changing distribution of the sabkha on the landward side. Finally, the settlement Sila has expanded considerably which shows as yellow and red patches.

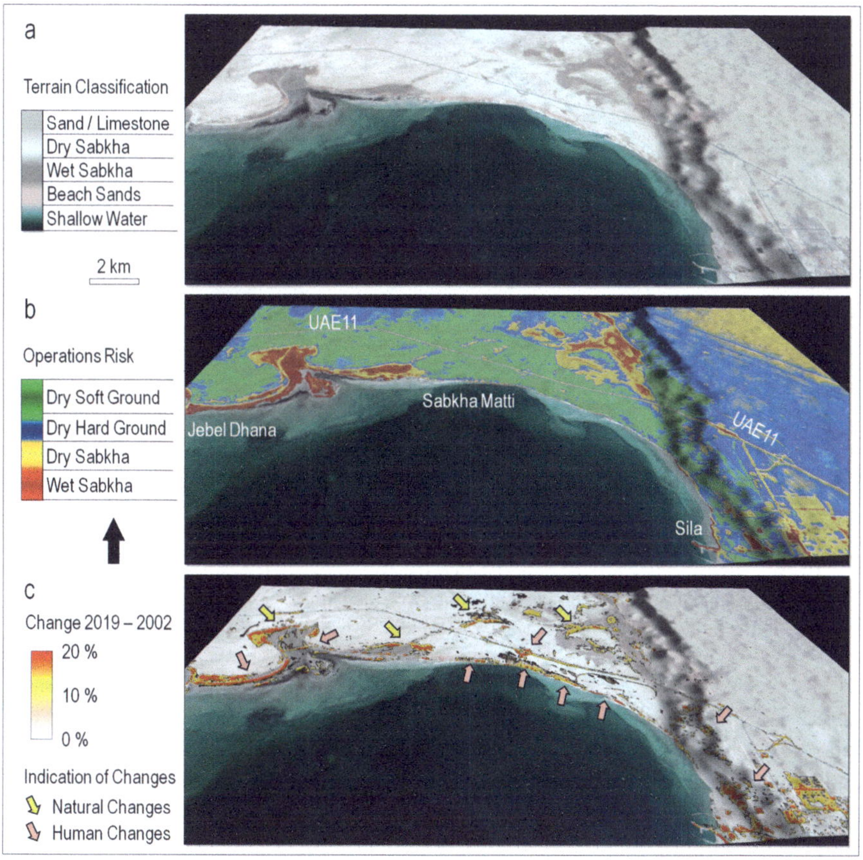

Fig. 10.15 Temporal change of sabkha character and distribution from multi-temporal Landsat composites. (**a**) Landsat 8 421 RGB image from November 2019 with terrain classification, (**b**) idem overlaid with operations risk obtained from Landsat 5 image from July 2002, (**c**) idem overlaid with change of sabkha character and distribution from 2002 to 2019. (DRM and Landsat imagery courtesy of U.S. Geological Survey, operations risk map after Cutts & Laake (2008))

10.2.3 Operations in Sand Dune Terrain

The last example in this section concerns sand dunes. Details on sand dune classification and genesis can be found in Tsoar (2008). Sand dunes expose seismic operations to a wide range of challenges from roll-over risk on steep slopes to sliding on unstable sand and finally to gypsum flat that frequently cover the inter-dune space. As with the previous examples, areas with high logistics risk mostly represent areas exposing poor data quality (Insley & Laake, 2004; Laake & Insley, 2004a, b). To cover a range of dune challenges as wide as possible we chose the study area in the giant sand dune field of the Berkine Basin in eastern Algeria (Fig. 10.16). Sand dunes of an impressive height of up to 300 m cover the area in a dense pattern (c). The digital relief model (a) indicates that these dunes are arranged in organized

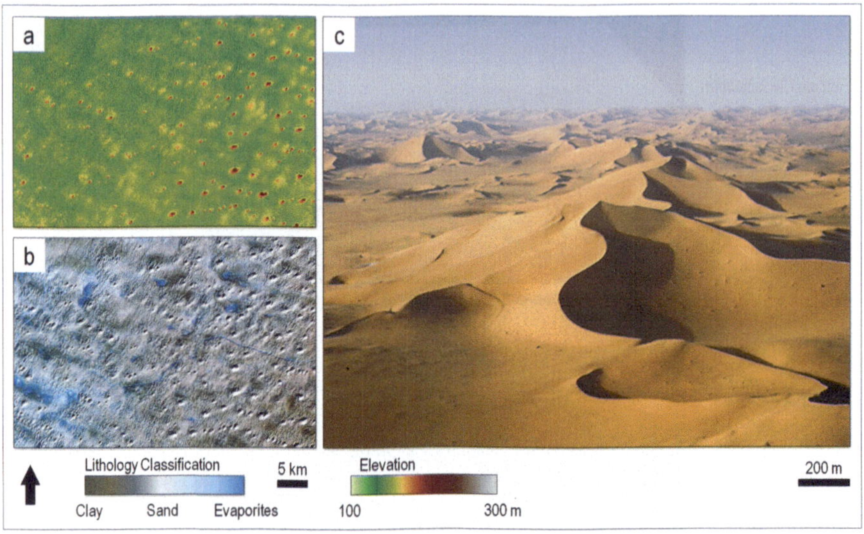

Fig. 10.16 Overview of sand dunes in the Berkine Basin, East Algeria. (**a**) DRM, (**b**) Landsat 7 742 RGB composite image showing lithology, (**c**) mega-dunes composed of star dunes and barchan dunes after Insley and Laake (2004). (DRM and Landsat imagery courtesy of U.S. Geological Survey)

patterns that are guided by the dominant wind direction. The distribution of the dunes also determines the distribution of surface lithologies (b). The dunes in the study area are composed mainly of quartz sand (light colors). The interdune space shows either the underlying clays (brownish colors) or in depressions accumulations of evaporites like gypsum (blue tones).

The challenges for seismic operation in sand dune terrain are illustrated in Fig. 10.17. Gypsum pans (a) may pose expose heavy vibrators to the risk of insufficient ground support. Typically, gypsum pans do not support the force excited by the vibrator baseplate, which results in the baseplate sinking into the gypsum during the sweep, which in turn results in significantly increased harmonic distortion of the vibrator source signal. Steep sand dunes (b) yield the risk of vehicle roll-over.

The best mitigation is to design the vibrator source line such that it follows the dip of the dune surface. This approach, however, can still not prevent the risk of sliding downhill. As a general rule a terrain dip of 15° is considered the maximum angle possible for vehicle movement. Dip angles exceeding this value should be avoided. Remote sensing data can assist with such planning.

High resolution multispectral satellite imagery shows the pattern in which the dunes are arranged (d). It also provides the information necessary to interpret different dune types (c). For our study area we distinguish large, mainly stationary star dunes and highly dynamic barchan dunes, which ride on the flanks of the star dunes (red lines in (c)). The giant dune in (b) is a large star dune of about 200 m height with barchan dunes riding on the smooth star dune body.

The workflow for using remote sensing data to predict the risk of roll-over and high vibrator source signal distortion is shown in Fig. 10.18. We use the digital

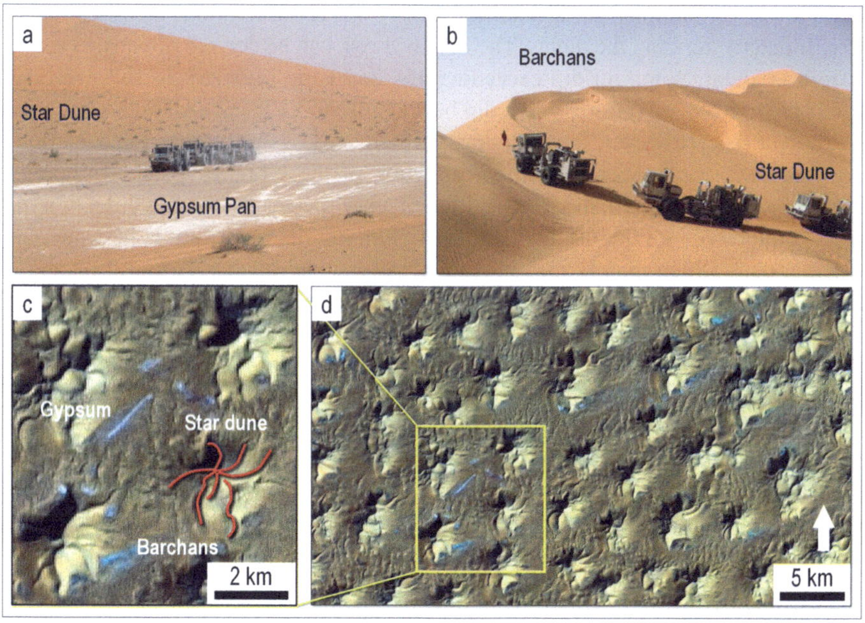

Fig. 10.17 Operational challenges in sand dune terrain. (**a**) Vibrators on a gypsum pan, (**b**) vibrators climbing steep mega-sand dunes, (**c**) dune morphology, (**d**) satellite image of dense dune field after Insley and Laake (2004). (ASTER satellite image courtesy of METI and NASA)

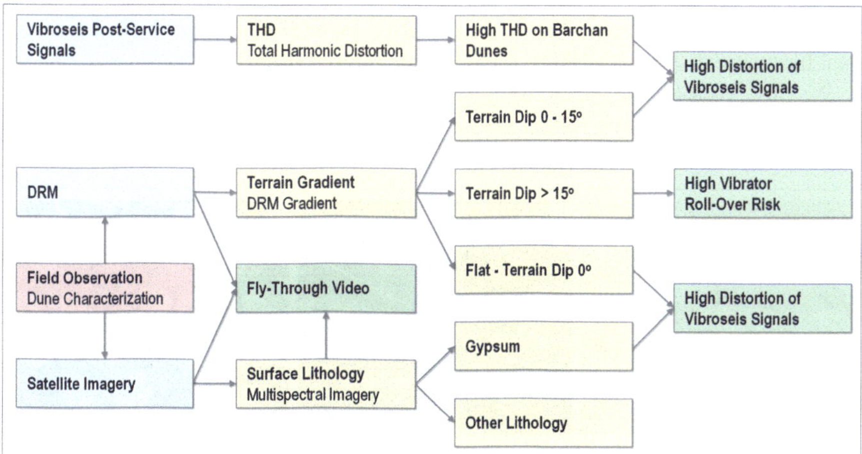

Fig. 10.18 Workflow for using remote sensing data for the detection of operational and data risk in sand dune terrain

relief model and multispectral satellite imagery. Field observations provide the criteria for characterizing the dunes. The roll-over risk can be derived directly from the terrain gradient. A terrain dip exceeding 15° bears the high risk of roll-over and needs to be avoided. Terrain with a dip between 0° and 15° normally does not pose a roll-over risk. However, Barchan dunes which are less consolidated than stationary star dunes bear the risk of high distortion of the vibrator signals. Lithology interpretation of the multispectral satellite imagery focuses on the detection of gypsum. Combined with flat terrain from the terrain gradient, gypsum is indicative of gypsum pans which mainly impacts the vibrator signal quality. A useful planning tool is the generation of a fly-through video. The digital relief model provides the topography on which the lithology interpretation of the satellite image is draped. Based on the remote sensing assessment of the terrain, the vibrator tracks are planned. The fly-through video allows the simulation of the survey prior to execution to identify and mitigate potential logistics hazards.

In Fig. 10.19 we show the process of planning the vibrator track in giant sand dune terrain using optical satellite imagery (a) and a high resolution digital elevation model (b) as well as the surface gradient derived from the elevation (c). The planned vibrator track is shown as a white line. The vibrator track should ideally be designed as straight as possible to provide the optimum processing result. However, the satellite imagery and topography indicate that several large star dunes obstruct the ideal track. The terrain dip indicates which dunes need to be avoided to prevent excessive roll-over risk, but it does not provide sufficient resolution to identify the barchan dunes. The necessary information for this task can be obtained from the optical satellite image.

In the next step we capture the vibrator signal distortion from the acquisition along the vibrator track shown in Fig. 10.19. Along the vibrator track we plot the elevation and color the area below the elevation profile according to the measured total harmonic distortion (THD) of the vibrator signal (Fig. 10.20).

We observe that star dunes, despite their height, do not pose a problem to the signal quality. The reason is that start dunes are stationary over long times which

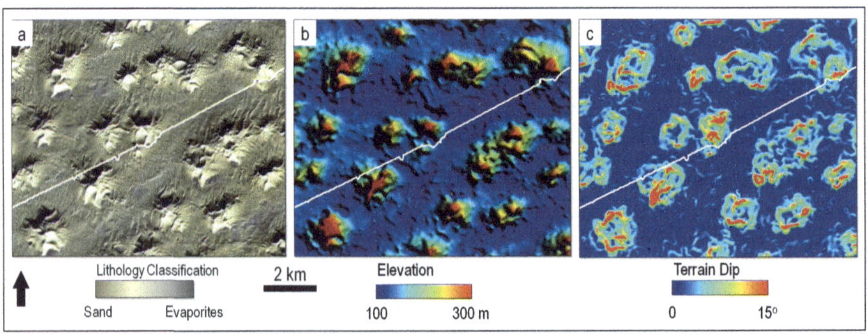

Fig. 10.19 Characterization of sand dune terrain from optical satellite imagery and DRM. (**a**) ASTER 321 RGB composite, (**b**) sun-shaded DRM, surface gradient of DRM after Laake and Insley (2004a, b, c). (ASTER DRM and multispectral imagery courtesy of METI and NASA)

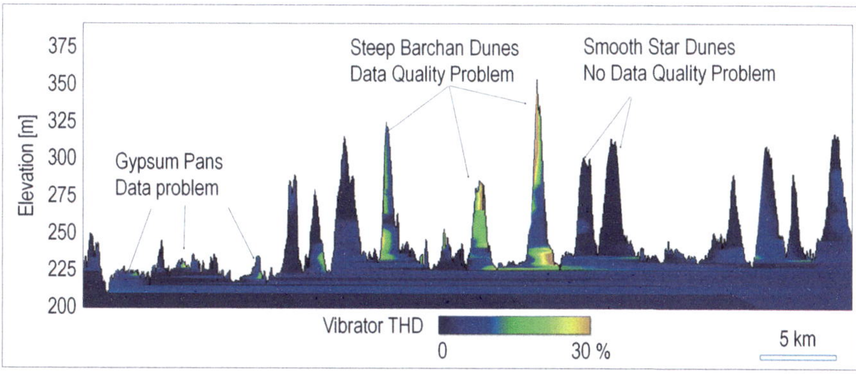

Fig. 10.20 Correlation of vibrator total harmonic distortion data with sand dune morphology after Insley and Laake (2004)

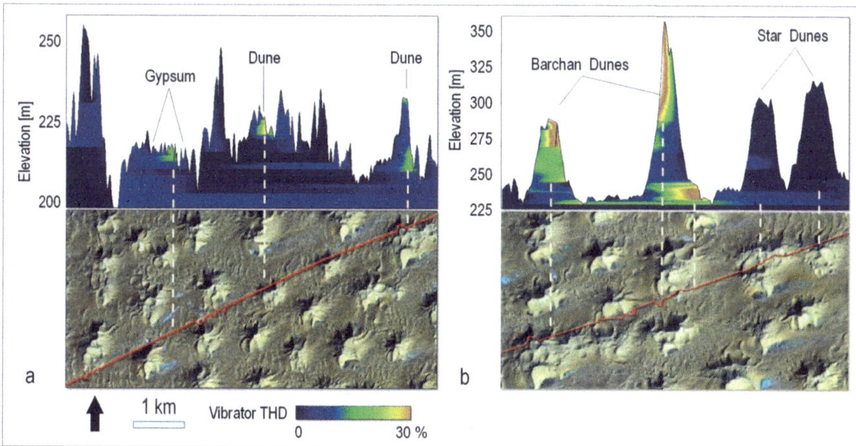

Fig. 10.21 Correlation of vibrator total harmonic distortion data with dune topography and satellite image. (**a**) Sand dunes and gypsum fields, (**b**) star dunes and barchan dunes. (ASTER satellite imagery courtesy of METI and NASA)

gives the sand time to settle and compact, which in turn results in sufficient ground support for the vibrator and good coupling of the baseplate to the consolidated sand without major sinking in of the baseplate during the sweep. Barchan dunes, however, are highly mobile which leaves the sand highly unconsolidated resulting in up to 30% THD, a level that is highly undesirable. We notice that unconsolidated Barchan dunes occur both near the foot of star dunes as well as on their crest. Another area where we observe high THD values are gypsum pans.

To study this in more detail, we zoom into the area where the gypsum pans occur (Fig. 10.21a) and correlate the THD with the location on the lithology satellite image. The gypsum pans occur in the space between small dunes which do not pose a THD problem. The high THD values are rather directly correlated with the bright spots on the satellite image, which indicate the presence of gypsum. On the same section of the vibrator track we notice also that, for the larger and broader dunes, high THD values only occur at their base where unconsolidated sand is blown in.

In (b) we focus on barchan and star dunes. The two star dunes on the right, where the vibrator climbed across significant sand volumes, do not show elevated THD levels. However, if we study the dune in the center of the section where the vibrator track passed the foot of a large star dune with a barchan dune riding on it, we observe high THD values at the base and at the crest, where the sand is most volatile. From the THD data we derive that crossing star dunes does not pose a major problem for logistics as well as data quality provided the terrain slope does not exceed 15°. Barchan dunes and gypsum pans, however, should be avoided as much as possible considering the processing of the seismic data.

10.3 Seismic Data Quality Estimation

10.3.1 Prediction of Source and Receiver Coupling in Central Algeria

The quality of 3D seismic subsurface image depends on the quality of the input signals, i.e. on the coupling of source and receiver to the ground (Laake, 2005). Most onshore operations employ hydraulic vibrators source and geophones as receivers the coupling of which is strongly dependent on the ground characteristics. In general, we can distinguish three types of source and receiver coupling to the ground (Laake et al., 2005, Fig. 10.22).

Hard ground for example hard limestone or basalt often shows a rough surface texture. The flat baseplate of the vibrator, therefore, can only achieve only partial contact to the ground resulting in so-called point loading of the baseplate (a). The transmission of the baseplate vibrations is limited in amplitude due to incomplete contact. Additionally, the transmitted frequency range is limited because point loading may result in frequency dependent flexure of the baseplate up to deformation and damage. On the receiver side, hard ground poses the problem of coupling the sensor to the ground at all. Typically, geophones are equipped with a spike with which the contact to the ground should be achieved. However, hard rock may prevent the spike from being inserted into the ground, leaving only the options of finding a fault into which the spike to insert (d) or replacing the spike with a plate and covering the geophone with a sandbag to improve ground coupling.

Intermediate ground such as gravel or soil has a compliant surface but is still hard enough to support the weight of the vibrator when shaking (b). The geophone is planted in the ground with the spike. In deserts geophones are typically buried to ensure optimum coupling as well as protection from wind noise (e). For both, source and receiver this type of ground represents the optimal environment to generate the

Fig. 10.22 Typical cases of source and receiver coupling.

best possible signals for the 3D subsurface image. If the ground is too soft, then it does not support the weight of the vibrator, the baseplate breaks through the ground and sinks into the ground (c). This results in a highly distorted source signal, both from amplitude and phase. It is highly recommended to avoid breaking the ground, which means limiting the drive level of the vibrator. This situation can also occur on apparently hard ground that overlays soft sand, which may not be detectable without running some test sweeps prior to routine operation. The receiver coupling to soft ground (f) suffers from the attenuation of amplitudes specifically at higher frequencies, when the sensor no longer follows the ground motion. Often deeper burial of the geophones is the only option to recover some quality of the receiver signal.

We will study now how satellite remote sensing data can help delineating these types of ground to assist the prediction of seismic data quality related to ground coupling (Laake, 2005; Laake & Tewkesbury, 2005; Cutts & Laake, 2008). The study area in central Algeria is located at the Bou Hadid promontory of the Tademait limestone plateau towards the Sbaa gravel plain (Fig. 10.23).

The satellite map draped on the digital relief model (c) clearly distinguishes the Bou Hadid promontory and the Tademait Plateau (dark shades, b and d) from the gravel plain (light greenish shades, a). In the south of the study area, a gypsum outcrop (e) shows as bright white color. In this special case, lithology and hardness are coupled directly. Therefore, it would be sufficient to use multispectral optical and infrared satellite imagery for the delineation of hard, intermediate and soft ground.

However, when limestone would form the surface both as hard solid limestone on the Bou Hadid promontory and its escarpments as well as intermediate limestone

Fig. 10.23 Terrain identification from satellite remote sensing data. (**a**) Intermediate ground on the flat Sbaa gravel plain, (**b**) hard ground on steep limestone escarpments of the Tademait Plateau, soft ground at blown in sand and intermediate ground on gravel plain, (**c**) Landsat 5 321 RGB satellite image, (**d**) hard limestone plateau, steep rocky slopes, soft debris fans and intermediate ground on gravel plain, (**e**) hard limestone and soft gypsum plain. (Landsat 5 satellite imagery courtesy of U.S. Geological Survey)

gravel on the flat gravel plains, then lithological discrimination alone would not be sufficient. Radar back scattering provides additional information on the hardness of the surface, which complements the lithology mapping (Fig. 10.24).

Radar back scatter intensity (a) clearly discriminates hard and rough areas (red to greenish tones) from flat and smooth areas (bluish tones). We notice, however, that radar back scatter information alone would also not be sufficient to discriminate hard, intermediate and soft ground since the lithologic information from multispectral satellite imagery (b) is missing. We, therefore, integrate both data sets and compute risk maps for hard surface risk (c). Areas shown in red are expected to present a high risk of point loading for the vibrator source and challenges with the ground coupling of the geophones. Very soft areas on the other hand (red in d) pose the problem of the vibrator baseplate breaking through the surface and the geophone coupling becoming subjected to resonance from poor ground coupling. Interestingly, the gravel plain appears to show some heterogeneity, which can be indicative of near surface heterogeneities that may result in partial baseplate breaking through the surface preferably at high frequencies.

Finally, we merge the two risk maps into a map of the combined risk (Fig. 10.25). Areas of intermediate ground that are expected to produce the best signal from a source and receiver coupling point of view are shown in green. Hard surface risks are shown in bluish colors with increasing risk towards purple color whereas soft surface risks are indicated by yellow colors increasing towards red for higher soft surface risk.

In summary, we observe that the limestone promontory and most of its escarpments are associated with a high hard surface risk, whereas gypsum and the debris fans and blown in sand at the base of the escarpments pose a significant soft surface risk. Flat gravel terrain is mostly associated with good data quality.

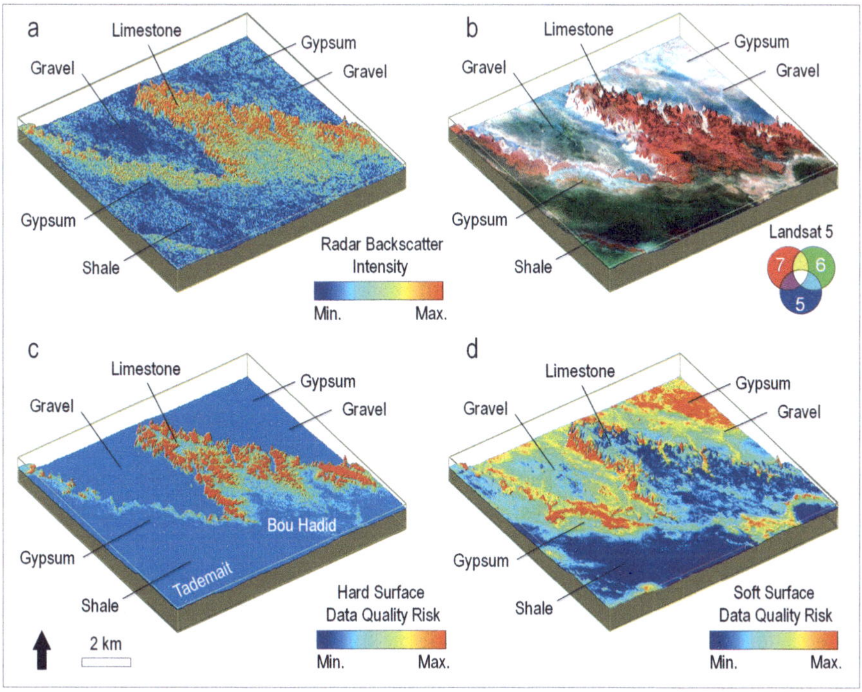

Fig. 10.24 Mapping data quality risk from satellite remote sensing data. (**a**) Radar backscatter intensity for surface roughness, (**b**) Landsat 5 infrared composite 765 RGB for lithology mapping, (**c**) hard surface data quality risk, (**d**) soft surface data quality risk. (Radarsat-1 data courtesy of CSA, DRM and satellite imagery courtesy of U.S. Geological Survey)

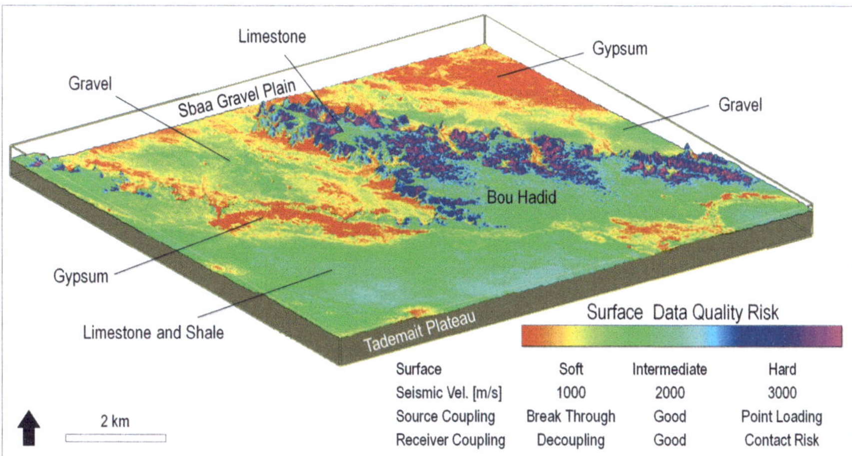

Fig. 10.25 Surface data quality risk inferred from satellite remote sensing data. (Analysis based Radarsat-1 data courtesy of CSA, DRM and satellite imagery courtesy of U.S. Geological Survey)

10.3.2 Mapping of Surface Scattering at Surface Karst, Western Desert Egypt

The near surface geologic setting impacts the way seismic waves are propagating from the source to the receiver with the surface texture influencing the propagation particularly of surface waves. Figure 10.26 shows a schematic of the complexity of such wave propagation. The seismic wavefield emitted by the seismic source is composed of body waves, that travel into the subsurface and enable seismic subsurface imaging, and surface waves, that travel at and near the surface. Body waves are only affected by features of a thickness larger than about 100 m such as the escarpment of the Marmarica limestone plateau. The effect of the escarpment can be so severe that no body waves reach the bottom of the Qattara Depression because most of the seismic wave travelling at shallow angles are reflected from the escarpment.

Surface waves, however, are impacted by surface features of a thickness as small as a few meters. Surface karst features, therefore, play a major role in complicating the propagation of surface waves, which in turn makes it very difficult to remove from the seismic record.

Surface karst features and escarpments are associated with surface texture as well as changes in lithology, which makes the digital terrain model, surface radar back scatter imagery and multispectral imagery suitable for detecting surface scatter features from remote sensing data (Fig. 10.27). The digital relief model (a) shows a rather flat but textured character on the Marmarica limestone plateau, less texture at the bottom of the Qattara Depression and the steep escarpment between the two areas. The texture of the plateau topography becomes more obvious in the gradient image (c). Smaller terrain edges of up to 10 m height are shown scales of gray depending on their size, whereas the large escarpment is highlighted in red color.

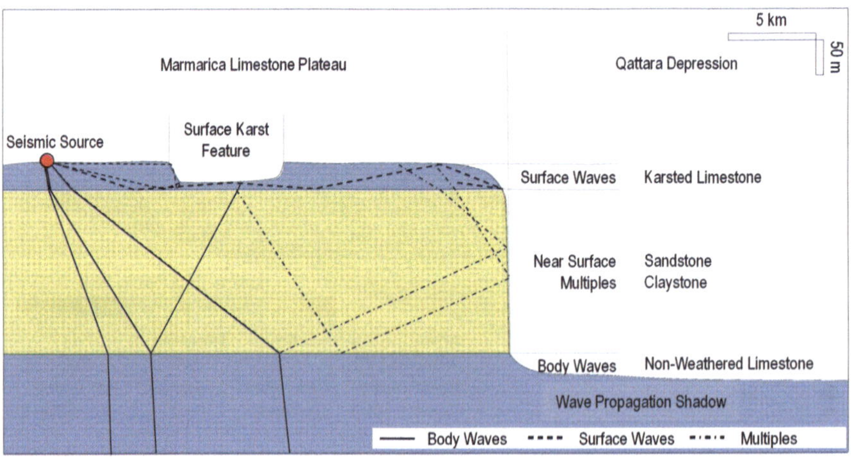

Fig. 10.26 Propagation of seismic wave an area with surface karst and a major escarpment

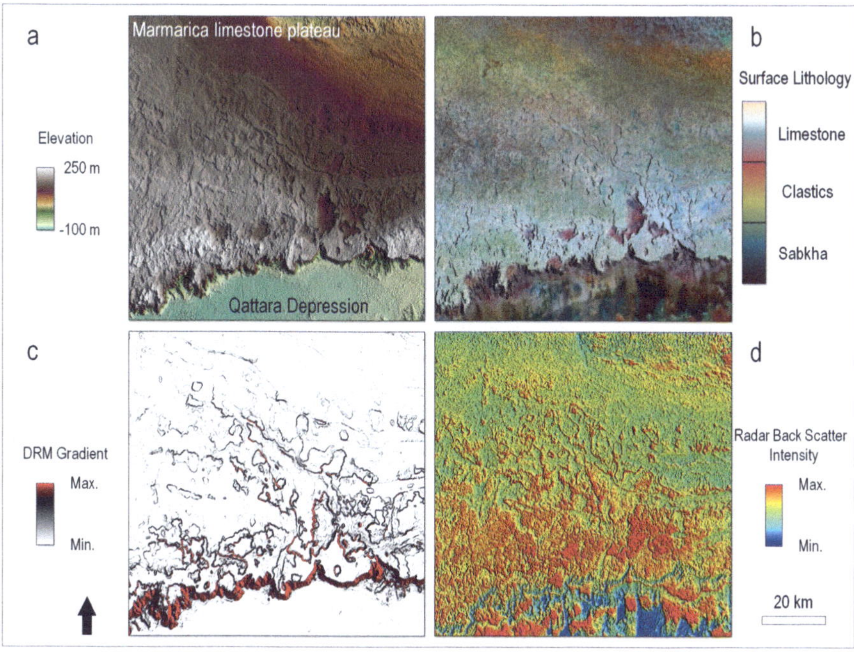

Fig. 10.27 Satellite remote sensing data used for the delineation of surface karst and related data quality risks. (**a**) DRM, (**b**) composite multispectral image from Landsat 5 742 RGB, (**c**) DRM gradient, (**d**) radar back scatter intensity. (DRM and satellite imagery courtesy of U.S. Geological Survey, Radarsat-1 data courtesy of CSA)

The multispectral composite image (b) reveals that the limestone plateau is rather homogeneous (light colors) except were surface karst has created complex shaped holes that are filled with clay (reddish tones). The sabkha at the bottom of the Qattara Depression consists of various types of evaporites and shales.

The radar surface back scatter texture (d) shows an increase of surface hardness towards the edge of the escarpment (red tones) indicating that towards the edge the thin surface cover of gravel and sand disappears.

Next, we correlate the surface DRM and back scatter texture with the lithology (Fig. 10.28a). The karsting of the limestone on the plateau results in small steps in the terrain, can hardly be seen on the lithology image (a) but are clearly visible on the overlay of the surface gradient on the lithology image (b). The radar back scatter gradient provides an even more continuous image of the edges but with less sharp expression (c). In addition to the lithology and surface texture image, radar back scatter texture also shows areas of rough surface where back scatter amplitudes are high.

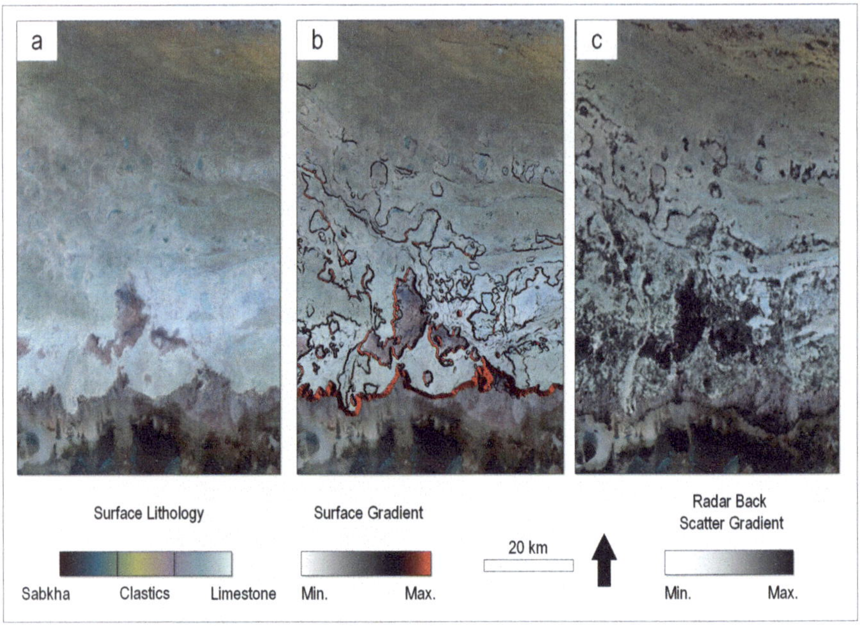

Fig. 10.28 Mapping of surface texture indicative of potential karst-related surface scatter locations. (**a**) Composite Landsat 5 742 RGB image, (**b**) DRM gradient overlaid, (**c**) radar back scatter intensity gradient overlaid. (DRM and satellite imagery courtesy of U.S. Geological Survey, Radarsat-1 data courtesy of CSA)

Finally, we render the results in 3D (Fig. 10.29). With this approach we can use vertical exaggeration and slant illumination to highlight the expression of the karst features. The lithology image (a) shows indications of those karst features that are associated with lithology changes such as clay fill (red areas).

Overlaying the surface gradient (b) emphasizes the outlines of the karst features and clearly marks the escarpment towards the Qattara Depression as a major obstacle for seismic wave propagation. The overlay with the radar surface back scatter gradient (c) further emphasizes the karst features whilst also showing rough areas. We can summarize the findings from surface lithology, surface texture and radar back scatter texture in the risk analysis (b). The Marmarica limestone plateau shows edges from karst features, which increase the risk of scattering seismic surface waves. Due to the complex shape of the karst features, the back scattering is anticipated also to be very complex, for which the Western Desert of Egypt is well known (Yanchak et al., 2013). The escarpment acts as a barrier for the propagation of seismic waves due to back scattering of the complete wave comprising surface and body waves. The Qattara Depression represents a minor risk for scattered noise which is only likely at isolated evaporite ridges.

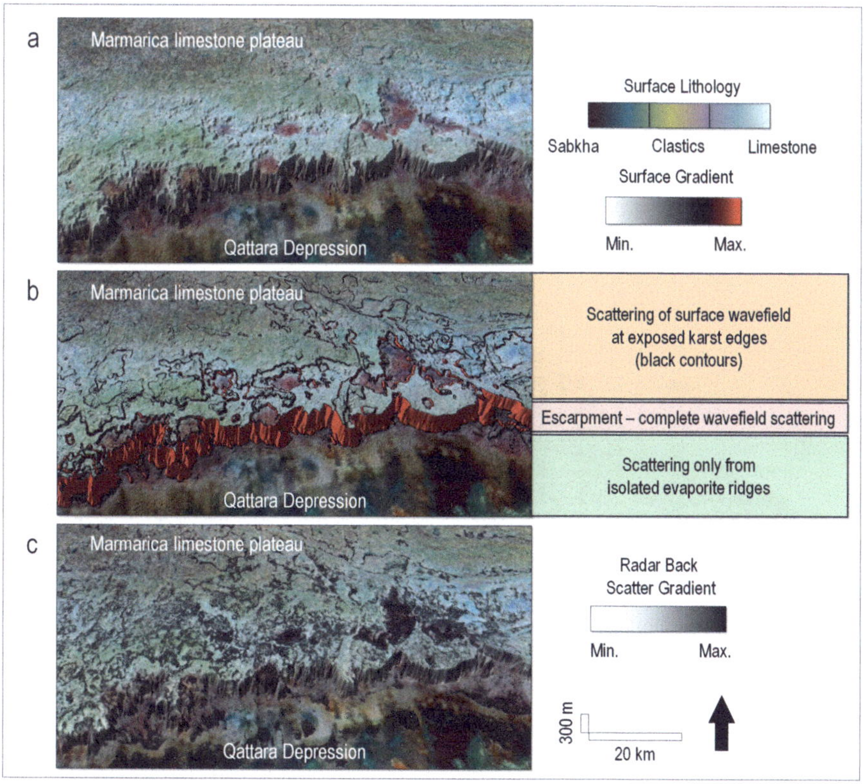

Fig. 10.29 Summary of remote sensing data interpretation for the detection of karst scatter risk in the Western Desert of Egypt. (**a**) Surface lithology, (**b**) surface texture overlaid on lithology with risk scatter interpretation, (**c**) radar back scatter gradient overlaid on surface lithology. (DRM and satellite imagery courtesy of U.S. Geological Survey, Radarsat-1 data courtesy of CSA)

10.4 Near-Surface Velocity Model from Remote Sensing Near-Surface Characterization

The heterogeneity of the weathered layers near the earth surface poses a significant challenge to data quality of seismic data (Laake & Strobbia, 2012a, b; Laake & Cutts, 2007a, b, c). After having considered the impact on the data quality in the previous section we turn now to the aspect of processing the acquired data. The processing of a seismic record requires the identification of different wave types to distinguish which waves to enhance as signal and which to attenuate because they are considered noise.

Figure 10.30 shows a simplified model of the near surface in the Western Desert of Egypt with a seismic shot record below. The shot record is the collection of all receiver signals along a line of seismic sensors that detect the seismic signal emitted from the source. The processing of this shot record ideally assumes that the direct wave between source and receiver shows as a straight linear event (DW1).

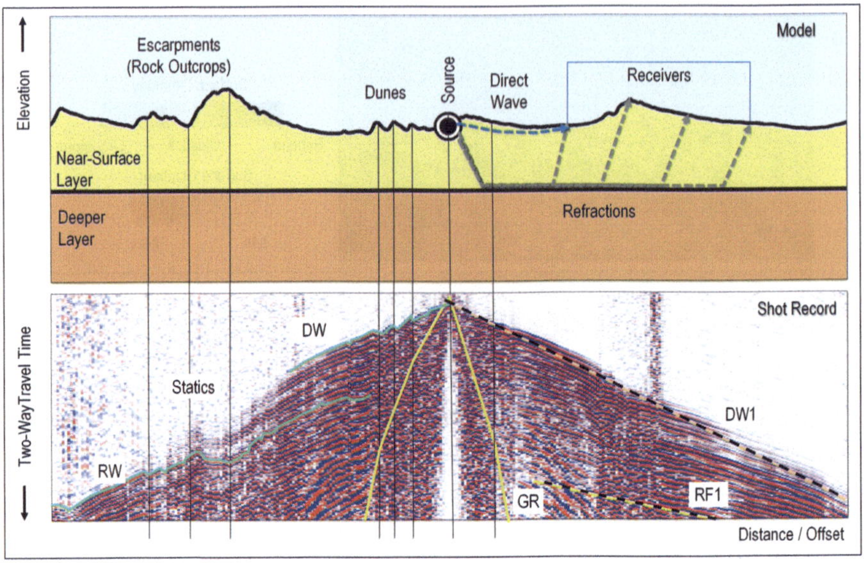

Fig. 10.30 Onshore near-surface challenges for seismic acquisition and processing. Top—simplified geologic section, bottom—split-spread seismic shot record. (Seismic data after Laake & Zaghloul (2009a, b), courtesy Apache Egypt, WesternGeco and EAGE)

With this assumption surface waves (GR) and other noise modes can be discriminated from hyperbolic reflection signals (RF1) that will deliver the subsurface image at the end of the data processing. However, in this case the rough surface topography combined with frequent changes of the near surface lithology delivers a much different picture. The recorded direct wave (DW) follows the topography and disappears at a certain offset due to attenuation in the surface. A refracted wave (RW) takes over the role of the first visible event, but again, its trace in the offset-travel time plain is strongly determined by the surface topography. Furthermore, we observe that also the ground roll component of the surface waves (GR) shows a dependency on the surface properties, in this case mainly a dependency on the surface lithology which creates the bends in the ground roll trace. These irregularities that are caused by the combined effect of surface topography and surface lithology are called static variations or short “statics”. In this chapter we will explore how remote sensing can contribute to the model building to correct for statics to enable a clean subsurface image at the end of data processing.

The conventional workflow for static corrections (Fig. 10.31) analyses each individual shot record for move out of the first breaks, i.e. direct wave and refractions aiming at a refractor velocity model to compute the subsurface travel time to the refractor and then correcting for the difference between the modeled and the recorded travel time. The goal is to achieve a straight first break signal to allow efficient enhancement of signal and suppression of noise.

A challenge associated with this approach is the sparse sampling of the surface. Seismic source and receiver points are designed along straight lines with a

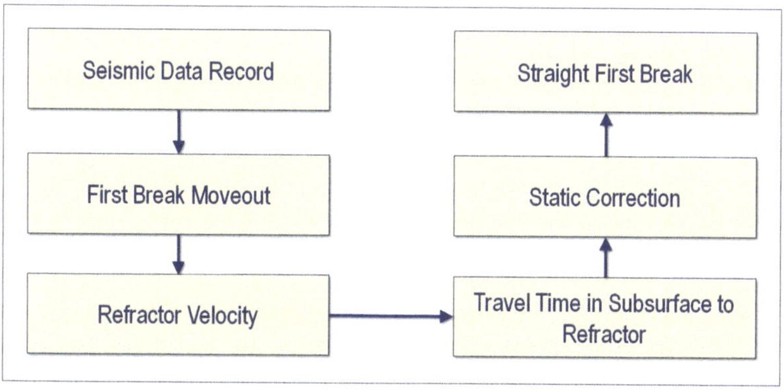

Fig. 10.31 Conventional statics workflow

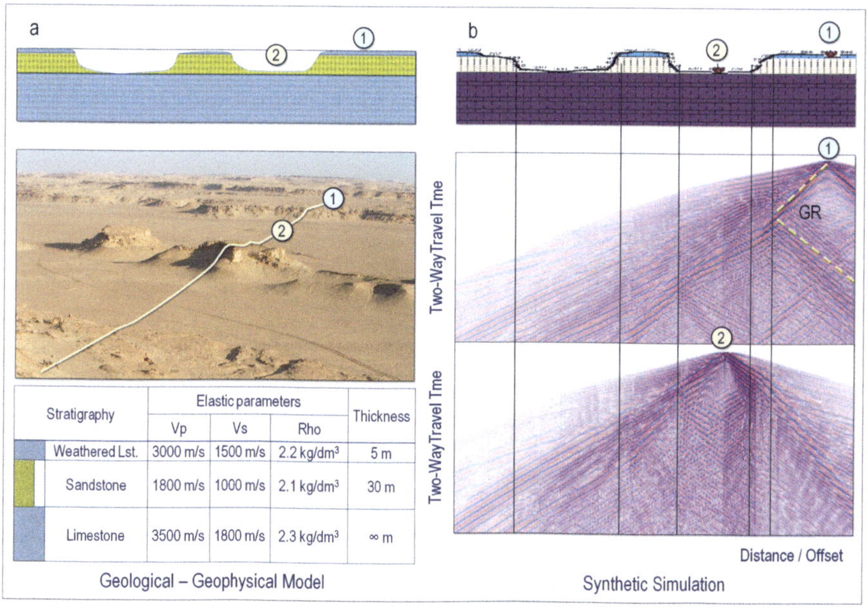

Stratigraphy	Elastic parameters			Thickness
	Vp	Vs	Rho	
Weathered Lst.	3000 m/s	1500 m/s	2.2 kg/dm³	5 m
Sandstone	1800 m/s	1000 m/s	2.1 kg/dm³	30 m
Limestone	3500 m/s	1800 m/s	2.3 kg/dm³	∞ m

Fig. 10.32 Geologic model for seismic simulation. (**a**) Geological model from remote sensing and surface observations, (**b**) synthetic seismic modeling using model (**a**)

considerable distance between adjacent lines. Consequently, the resulting velocity model is rather smooth thus not correcting for all static variations introduced by the surface topography and lithology.

One way to approach this challenge is the building of a near surface model and to generate synthetic seismic shot records to better understand the impact surface topography and lithology have on the resulting shot record (Fig. 10.32). We represent the geologic setting shown in the field photo (a) with a model the parameters of which are given below the photo. The schematic geologic profile for the section overlaid on the photo is shown above the photo. We use this model to generate synthetic seismic shot records (b).

The modeling algorithm includes direct waves, refractions and ground roll. Since only defined components of the wave field can be selected, the effect of the near surface can be studied. Sudden changes in topography such as escarpments result in backscatter of the seismic waves like ground roll (GR). Since no other noise is contaminating the shot record, these features can be studied in detail. Another aspect is that the synthetic modeling allows generating seismic receiver traces at any spatial sampling.

With this understanding, we turn to satellite remote sensing that provides densely sampled information about the surface (Fig. 10.33). We use the digital elevation model (a) and multispectral satellite imagery focusing on lithology discrimination (b) to classify both, topography and lithology for the area as a start for a 3D shallow geological model. The topography is classified into a smooth plateau (black), intermediate (green) and rough terrain in the north and at Jebel Ghazalat (light brown to white) (c).

In the multispectral composite image from Landsat 5 765 RGB, we distinguish evaporites such as salt and sabkha (light green) from sandstone and sand (blue) from limestone (orange) (d). When comparing figures (c) and (d) we notice that rough

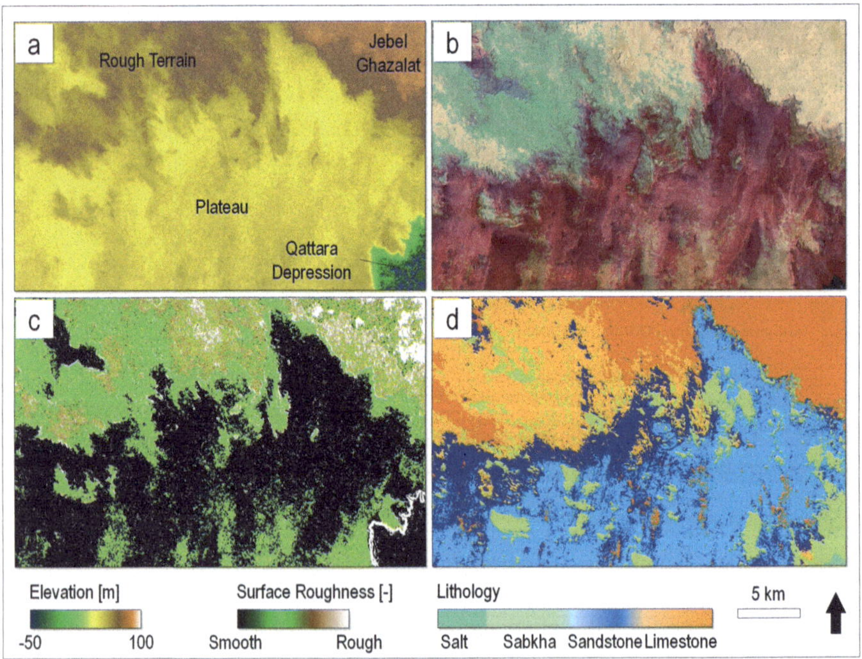

Fig. 10.33 Surface characterization from remote sensing data. (**a**) DRM, (**b**) multispectral satellite image Landsat 5 765 RGB, (**c**) terrain classification from DRM, (**d**) lithology classification from satellite image. (DRM and Landsat 5 satellite imagery courtesy of U.S. Geological Survey)

terrain coincides largely with limestone lithology for example in Jebel Ghazalat whereas smooth terrain occurs on the relatively flat plateau (yellow colors in (a)) where sandstone has weathered to soft sand. This implies that the near surface geology is relatively simple and can be approximated with a layered model.

To validate the lithology interpretation from the satellite image, we check the lithology at certain ground control points shown on the overview map from Landsat 5 742 RGB (Fig. 10.34e). For selected locations we also supply field photos with indications of the lithology and the bedding properties: a—layered limestone (top) and sandstone (base), b—gypsum outcrop, c—layered limestone (top) and sandstone (base), d—detail of limestone outcrop, f—sabkha, g—rose limestone, h—sand and sandstone plains with limestone cap at distance.

The location of the case study at the escarpment to the Qattara Depression allows the identification of all relevant zones and boundaries that are important to create the model (Fig. 10.35). We study a portion of the escarpment to the northeast of the area shown in Fig. 10.33 where limestone outcrops at the surface confirming the classification from the satellite image. Since the limestone is harder than the underlying sandstone, weathering creates a sharp escarpment, which we noticed on the digital elevation model. The heterogeneity of this first layer contributes substantially to the static variations observed in the shot record. The base of the limestone forms the lower boundary for the zone in which surface waves propagate.

Fig. 10.34 Ground truthing of remote sensing based lithology interpretation. See text for details. (Landsat 5 satellite imagery courtesy of U.S. Geological Survey)

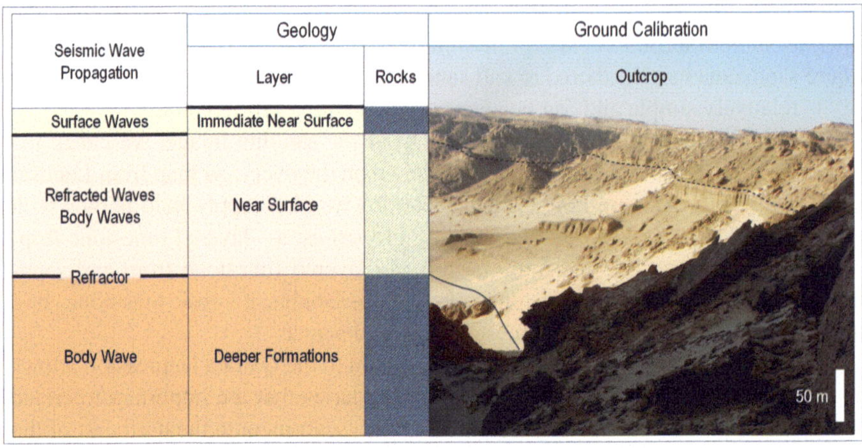

Fig. 10.35 Ground truthing at large escarpment to Qattara Depression. Geophysical and geological model (left) and interpreted outcrop photo of the escarpment (right)

Most of the escarpment comprises sandstone layers the erosion of which creates debris fans. At the plateau, this sandstone formation outcrops. The erosion of the relatively soft sandstone results in the smooth texture of the plateau. Occasionally, remnant limestone forms ridges which have been considered in the synthetic model (see Fig. 10.32).

Underlying the sandstone is a hard limestone. The lithological boundary between the soft sandstone and the hard limestone represents a major impedance boundary for the propagation of the seismic waves and acts as the interface at which refracted waves propagate.

We can now build a 3D geological model comprising two layers above the top of the hard non-weathered limestone, at the top a layer of weathered limestone and at the bottom a soft sandstone (Fig. 10.36). The multispectral composite Landsat 5 765 RGB image combines the spectral response in the thermal and shortwave infrared and allows a good distinction particularly of limestone (cyan and blue) from sandstone (mauve) as well as from clay and salt (red) in clay pans and sabkha (a). Using standardized velocities for these rock types, we can convert the lithology interpretation into a velocity model (b). In the area of the seismic survey, we distinguish faster velocities for limestone (blue) from slower for sandstone (orange). We can also include some small areas of even slower velocity for clay pans and local sabkha (cyan).

In the next step we extrapolate in 3D the outcrop information from the satellite image interpretation calibrated with the field calibration (Fig. 10.35). The outcrops show a rather flat contact surface between the limestone and the sandstone, which leads us to build the near surface velocity model shown in Fig. 10.37a. Red tones indicate Vp = 3000 m/s for limestone and yellows correspond Vp = 1800 m/s. Using this velocity model and the topography from the digital elevation model we can compute the static corrections and relate them to the datum at about 50 m below the surface of the sandstone plateau (b). The resulting static corrections amount to about

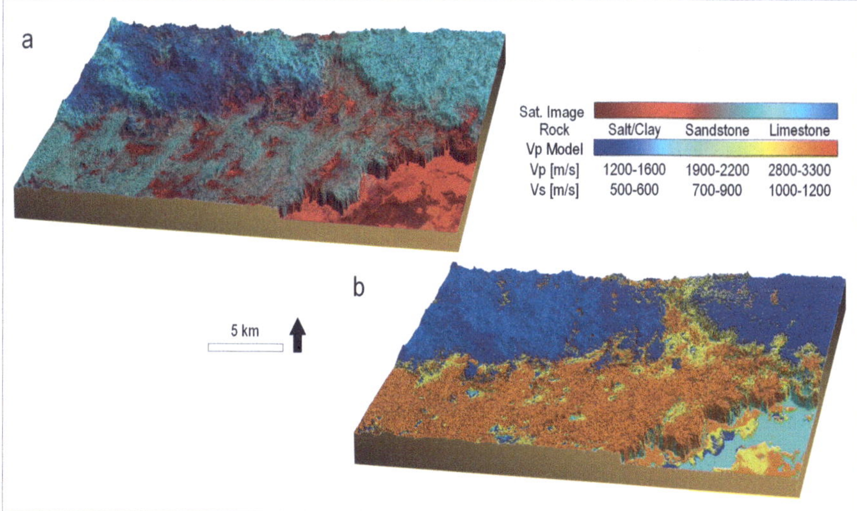

Fig. 10.36 Conversion of remote sensing-based lithology interpretation into shallow velocity model. (**a**) Landsat 5 765 RGB multispectral composite image with lithology interpretation, (**b**) surface model for seismic velocity. . (after Laake & Zaghloul 2009, DRM and Landsat 5 satellite imagery courtesy of U.S. Geological Survey)

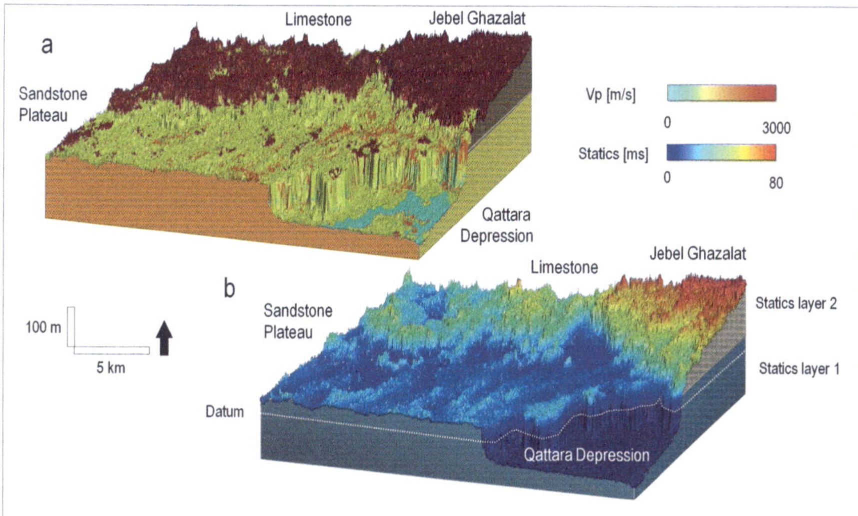

Fig. 10.37 Models for first 300 m below surface. (**a**) Velocity model based on lithology interpretation and standardized velocities, (**b**) static corrections computed from velocity model (**a**). (after Laake & Zaghloul 2009, DRM data courtesy of U.S. Geological Survey)

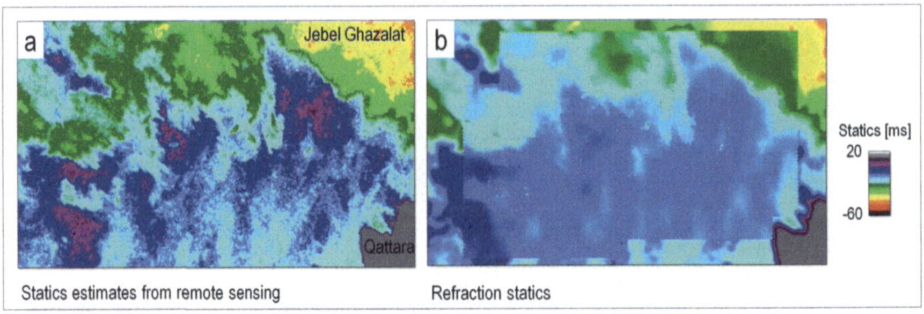

Fig. 10.38 Comparison of static corrections from remote sensing (**a**) and conventional refraction statics workflow (**b**) (after Laake & Zaghloul 2009)

15–20 ms for points on the plateau and reach up to 80 ms in Jebel Ghazalat. The statics model from remote sensing also comprises local effects from remnant limestone ridges that have been left over from the erosion of the limestone on the plateau.

Finally, we compare the static corrections computed from satellite image interpretation and 3D modeling with the statics that had been obtained purely from seismic in the conventional way (Fig. 10.38). The statics values derived from satellite image interpretation are typically within 10 ms of the refraction statics computed from 3D seismic data. However, the satellite-derived statics (a) show more detailed texture than the smooth refraction statics from seismic (b) and their texture is better correlated with the geologic features especially the remnant limestone ridges on the sandstone plateau, which are almost completely absent on the refraction statics. Providing static corrections in higher spatial resolution and with dense regular sampling allows correcting for the statics in a more surface consistent way thus enabling to better adjust surface multiples which requires a good knowledge of both, the surface topography and the spatial velocity distribution.

References

Introduction to Seismic Planning, Acquisition and Data Processing

Laake, A., & Cutts, A. (2007a). The role of remote sensing data in near-surface seismic characterization. *First Break, 25*, 61–65.

Seismic Logistics Case Studies

Crice, D., Food, P., & Walthinsen, E. (2015). Cableless seismic systems for near surface geophysics. In *SAGEEP 2015* (pp. 465–468). https://doi.org/10.4133/SAGEEP.28-066.

Laake, A., & Insley, M. (2004a). Applications of satellite imagery to seismic survey design. *The Leading Edge, 23*(10), 1062–1064.
Laake, A., & Insley, M. (2004b). Using satellite imagery aids seismic surveys planning, acquisition and processing. *World Oil, 225*(5), 27–33.

Remote Sensing Case Study Volcanic Terrain Payún, Argentina

Hernando, I. R., Llambias, E. J., Gonzalez, P. D., & Sato, K. (2012). Volcanic stratigraphy and evidence of magma mixing in the Quaternary Payún Matrú volcano, Andean Backarc in western Argentina. *Andean Geology, 39*(1), 158–179.
Laake, A. (2005). Remote sensing application for vibroseis data quality estimation in the Neuqen Basin, Argentina. In *VI Congreso de Exploracion y Desarollo de Hidrocarburos, Mar del Plata, 2005.*

Remote Sensing Case Study Sabkha Matti, UAE

Evans, G., Kirkham, A., & Carter, R. A. (2002). Quaternary development of the United Arab Emirates Coast: Ne evidence from Marawah Island, Abu Dhabi. *GeoArabia, 7*(3), 448–458.
Goodall, T. M., North, C. P., & Glennie, K. W. (2006). Surface and subsurface sedimentary structures produced by salt crusts. *Sedimentology, 47*(1), 99–118. https://doi.org/10.1046/j.1365-3091.2000.00279.x
Kendall, A. C. (1978). Continental and supratidal (Sabkha) evaporites. In G. Roger (Ed.), *Facies model* (Geoscience Canada) (Vol. 5, pp. 66–78). Geological Association of Canada Publications. https://doi.org/10.12789/gs.v5i2.3081
Sanford, W. E., & Wood, W. W. (2001). Hydrology of the coastal sabkhas of Abu Dhabi. *United Arab Emirates, Hydrogeology Journal, 9*, 358–366. https://doi.org/10.1007/s100400100137

Remote Sensing Case Study Berkine Sand Dunes, Algeria

Laake, A., & Insley, M. (2004c). Using satellite imagery aids seismic surveys in planning, acquisition and processing. *World Oil, 225*(9), 27–33.

Remote Sensing Case Study Tademait Plateau, Algeria

Cutts, A., & Laake, A. (2008). Vibrator data quality prediction from remote sensing-based near-surface analysis. In *70th EAGE, Expanded Abstracts, paper B005.*
Laake, A., & Tewkesbury, A. (2005). Vibroseis data quality estimation from re-mote sensing data. In *67th EAGE Conference & Exhibition, Expanded Abstracts, cp-1-00222, paper G017.* https://doi.org/10.3997/2214-4609-pdb.1

Remote Sensing Case Study Qattara Depression and Western Desert, Egypt

Laake, A., & Cutts, A. (2007b). The role of remote sensing data in near-surface seismic characterization. *First Break, 25*(2007), 51–55.
Laake, A., & Strobbia, C. (2012a). Multi-measurement integration for the near-surface geological characterization. *Near-Surface Geophysics, 10*(6), 591–600.
Laake, A., & Zaghloul, A. (2009a). Estimation of static corrections from geologic and remote-sensing data. *The Leading Edge, 28*, 192–196.

Remote Sensing Case Study Western Desert, Egypt

Laake, A., & Cutts, A. (2007c). The role of remote sensing data in near-surface seismic characterization. *First Break, 25*(2007), 51–55.
Laake, A., & Strobbia, C. (2012b). Multi-measurement integration for the near-surface geological characterization. *Near-Surface Geophysics, 10*(6), 591–600.
Laake, A., & Zaghloul, A. (2009b). Estimation of static corrections from geologic and remote-sensing data. *The Leading Edge, 28*(2), 192–196.

Chapter 11
Development and Oilfield Infrastructure

Abstract The development of an oilfield and the planning and building of the related infrastructure requires major investment, the planning of which needs to mitigate as many uncertainties as possible. Challenges for the planning comprise site challenges related to the geotechnical characteristics of the site. These are, however, not stationary but may vary both seasonally as well as long-term. In addition to the planning for the construction site itself, logistics and supplies need to be considered, which requires regional considerations.

Remote sensing data are often available at different resolution and with different spectral characteristics. Increasingly, country-wide high-resolution digital elevation models from LiDAR become available which are very useful for site planning. Coarser data need to be considered for regional assessment to avoid data overload.

Satellite and airborne multi-spectral imagery provide different spectral bands for the analysis of surface lithology, vegetation, water features and soil moisture. Multi-temporal imagery allows the tracking of historic changes and gives clues for predictions of future developments.

Often 3D rendering of the results on the digital elevation model delivers the most intuitive results. When processing multi-temporal images, the co-rendering of multi-temporal grayscale images is a method to give fast results about the size and direction of temporal changes at the surface.

We study five sites in different climatic settings comprising a permafrost setting in the arctic, a floodplain subject to annual flooding and coastal and deltaic environments with challenges related to river flooding and storm surge from hurricanes. Sub-tropic challenges related to monsoon flooding and the monitoring of sand dune encroachment are studied.

Keywords Climatic setting · Site characterization · Regional characterization · Flooding · Monsoon · Flooding simulation · Assessment of levees · Coastal erosion · Subsidence · Permafrost thawing · Thermokarst · Sand dune encroachment

A. Laake, *Remote Sensing for Hydrocarbon Exploration*, Springer Remote Sensing/Photogrammetry, https://doi.org/10.1007/978-3-030-73319-3_11

11.1 Introduction

The development of oilfields and the planning of the necessary infrastructure requires planning on multiple levels at different spatial and temporal scales. Starting at the choice of the site for facility development the design needs to consider the geologic and geomechanics properties of the site to ensure sufficient support for foundations. This aspect, however, is not static. Ground conditions may change on a seasonal basis, for example in the Arctic when freezing in winter solidifies the ground, whereas thawing in summer results in soft surface which makes any off-road ground transportation impossible. External influence needs to be included such as inundation as a result of river flooding or coastal storm surge and sand dune encroachment. Temporal changes at a longer scale are related among others to climate change. Examples are sea level rise and subsidence in coastal areas, permafrost thawing and thermokarsting in arctic environment. In addition to the planning at the site itself, regional aspects of logistics and supply management need to be considered. For these, the above-mentioned aspects apply as well, albeit at a larger scale.

To illustrate the different aspects of site and regional planning, we have chosen five locations in different climatic settings (Fig. 11.1). Three settings have been chosen in the U.S.A. because of the public availability of high-resolution digital elevation models as well as satellite and airborne imagery. When working in areas with lower resolution data, alternative data sets need to be considered to perform the planning task. For each of the climatic settings, we have collected in Table 11.1 the main challenges as well as preferred remote sensing data to generate solutions and recommendations for supporting non-remote sensing data. Indications for alternative data sets are given with focus on public data sets with global coverage.

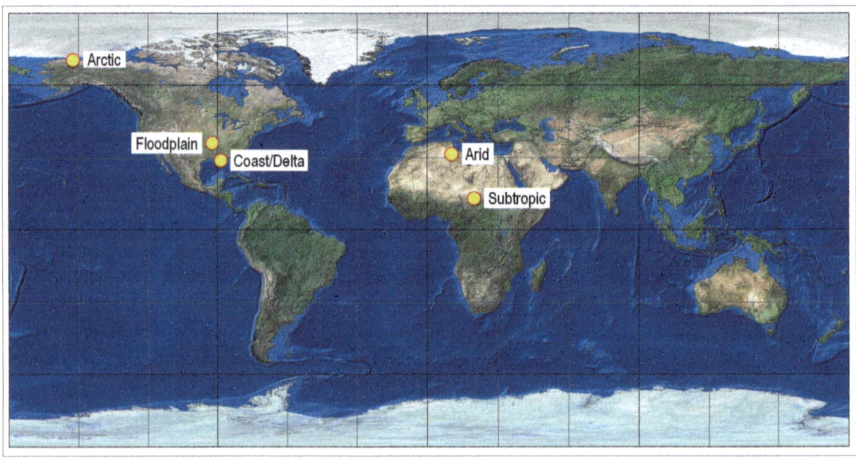

Fig. 11.1 Map showing the locations of the case histories used in this section. (Satellite imagery courtesy of U.S. Geological Survey)

Table 11.1 Challenges for development and oilfield infrastructure planning and remote sensing solutions, [a]SRTM DEM does not cover area

Setting	Scale	Time Scale	Challenge	Preferred Remote Sensing Data	Supporting Data	Solutions	Example Band Combination
Arctic	Local and Regional	Seasonal	Surface stability at sites and at transportation infrastructure	LiDAR DEM GMTED 2010*	Annual average climatic record	Multi-temporal satellite RGB from visible and VNIR bands	LS7 742 RGB, LS8 752 RGB draped on DRM
		Multi-decade	Coastal / river erosion, surface collapse following thermokarsting	LiDAR DEM (GMTED 2010*) Multi-annual VIS satellite imagery	Permafrost maps and reports	Multi-year RGB combinations from gray-scale single year satellite images	(LS1, LS2, LS3) RGB, with individual images as annual gray-scale image
Flood plain, coastal	Local and Regional	Seasonal	Flooding from rivers and hurricanes, storm surge affecting local infrastructure and regional transportation	LiDAR topo-bathymetry merge Seasonal VIS/VNIR satellite imagery	Flooding maps and reports and statists to support selection of satellite imagery	Flooding simulation	LS8 2-7-11 inverted RGB draped on DRM
	Regional	Multi-decade	Coastal erosion	LiDAR topo-bathymetry merge Multi-annual VIS/VNIR satellite imagery	Historic topographic maps and planning water engineering reports	Flooding simulation	LS7 742 RGB, LS8 752 RGB draped on DRM, flat flood horizon
						Co-rendering of satellite imagery and historic maps	Blend of topo map and LS8 2-7-11 inverted RGB draped on DRM
Subtropical	Local and Regional	Seasonal	Flooding	LiDAR, satellite DEM otherwise MODIS NDVI	Seasonal climatic records	Seasonal vegetation and moisture maps draped over DRM	NDVI from MODIS or Landsat draped on DRM
		Multi-decade	Draught leading to shortages in supply	Satellite DEM Multi-annual VIS satellite imagery	Historic topographic maps, climatic and geologic records and models, well logs	Prediction maps for draught	Multi-annual LS7 432 RGB / LS8 432 RGB draped over DRM
Arid desert	Local and Regional	Multi-decade	Sand dune encroachment	Satellite DEM Multi-annual VIS/SWIR satellite imagery	Historic topographic maps, climatic and geologic records and models, well logs	Multi-year RGB combinations from gray-scale single year satellite images	(LS1, LS2, LS3) RGB, with individual images as annual gray-scale image, LiDAR can shorten prediction interval

11.2 Arctic Climate Alaska, USA

The arctic environment presents specific challenges associated with the climatic setting where permafrost is a key aspect. When approaching arctic challenges, temporal changes need to be considered as the mechanical properties of the surface change significantly with time. We need to consider two scales of temporal changes:

- Seasonal changes: during the arctic winter the ground is frozen and can support vehicles for logistics. However, the sea is frozen as well, thus disabling logistics. During arctic summer, the upper layer of the ground thaws resulting in muddy and swampy surface that makes any logistics impossible.
- Multi-annual changes: often ignored in the past, global climate change shows an increasingly visible effect in arctic environment. Particularly, thawing of the permafrost leads to dramatically changed mechanical properties which we will study in more detail.

A critical aspect of arctic operations is the very slow pace of environmental processes. Due to the low temperatures, all biological processes are slowed down significantly, such that any environmental impact from human operation takes much longer to be compensated by natural processes than in moderate humid climate, if it is compensated at all. This applies also to the human impact on ecosystems that had adapted to the above-mentioned seasonal changes that are increasingly altered as a result of global climate change (Audubon Alaska, 2017).

We have chosen the northern coast of Alaska to study the seasonal and multi-annual processes and their implications on logistics and operations in arctic environment. The observations made in the ecologically sensitive Teshekpuk area can be transferred to most other low-lying permafrost areas in the arctic. We have chosen Alaska for the wealth of publicly available data.

We start with an overview of the geologic and present-day climatic setting of northern Alaska (Fig. 11.2). A comprehensive overview of the geological setting has been published by Gryc (1988). We overlay the climatic map showing different permafrost environments (data from Ferrians, 1998) on the digital elevation model and render it with vertical exaggeration and slant illumination to obtain a virtual 3D impression of the morphology (a) showing the correlation of the different permafrost environments with topography. Our study area in northern Alaska stretches from the rocky outcrops of the Brooks Range across the North Slope to the Arctic Ocean. From south to north, the discontinuous permafrost in the Brooks Range turns into continuous permafrost on the North Slope with increasing ground ice thickness as we approach the coast of the Arctic Ocean. In the north east we observe an anomalous area with low ground ice content (cyan color). When we overlay the permafrost environment map on the geomorphologic map created from the merge of satellite imagery and topography, we can see the correlation of the different permafrost environments with the topography.

Our focus is on the high ground ice and thick overburden area (HGTO) on the coastal plain of the North Slope in the area around Teshekpuk Lake near the Arctic

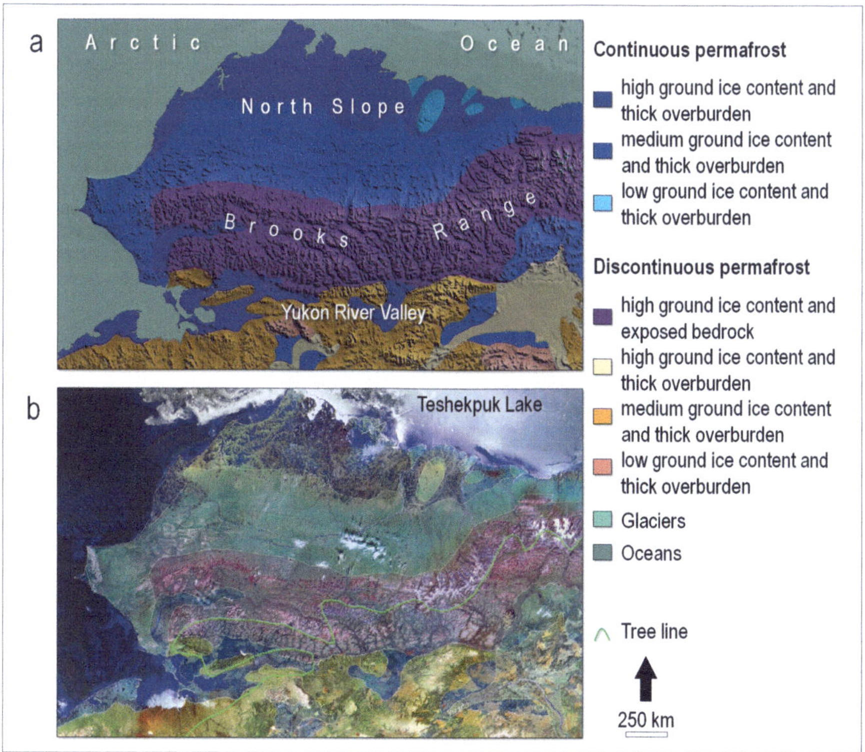

Fig. 11.2 Overview of geologic and climatic setting of northern Alaska. (**a**) permafrost distribution, (**b**) overlay of permafrost map on geomorphology. (Landsat 5 and DRM data courtesy of U.S. Geological Survey, permafrost data from Ferrians, 1998 NSIDC)

Ocean coast. Figure 11.3 shows an overview of the permafrost distribution across the North Slope together with an elevation profile. The area with dense coverage by coastal lakes is virtually flat. The lakes are directly linked to permafrost processes (Jones et al., 2011). Teshekpuk Lake is believed to be the largest thermokarst lake in the world being the result of the merge of a cluster of very shallow freshwater lakes on the coastal plain. The bottom of these lakes is formed by permafrost. Even in winter, most of these lakes do not freeze down to the ground leaving a layer of high heat capacity water to balance the temperature changes at the surface. The lake shores are formed by permafrost edges that remain stable as long as the lake does not overflow. Only in years with unusual warm summer temperatures and high precipitation, the lake shores will thaw and the lake overflows thus breaching the shores and merging adjacent lakes. This process is believed to have created Teshekpuk Lake. We will show the individual steps of this process later.

For the analysis of the arctic setting to assess logistics and operational aspects for development, we use a workflow that employs digital elevation models and multispectral satellite imagery (Fig. 11.4). For the regional overview we use the ETOPO1 global topography and bathymetry model where the bathymetry is required and the

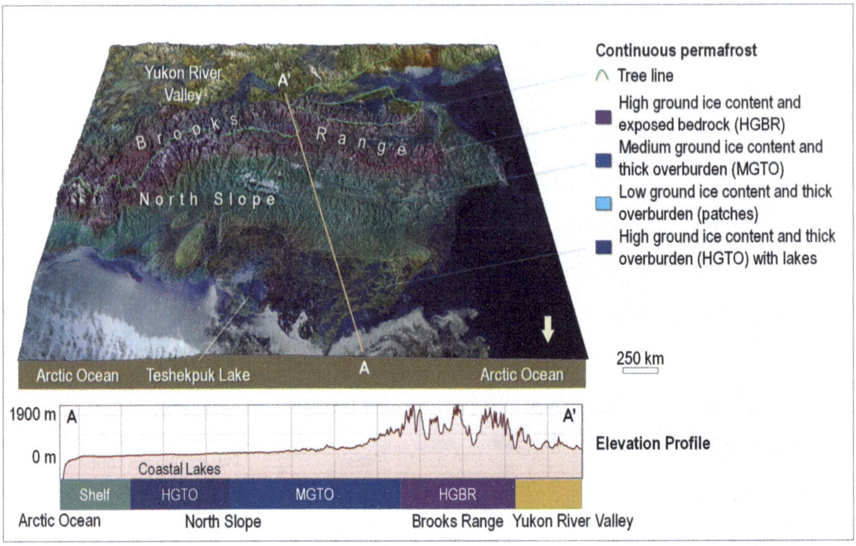

Fig. 11.3 Permafrost distribution and elevation profile for North Slope, Alaska. (Landsat 5 and DRM data courtesy of U.S. Geological Survey, permafrost data from Ferrians, 1998 NSIDC, elevation profile from GoogleEarth)

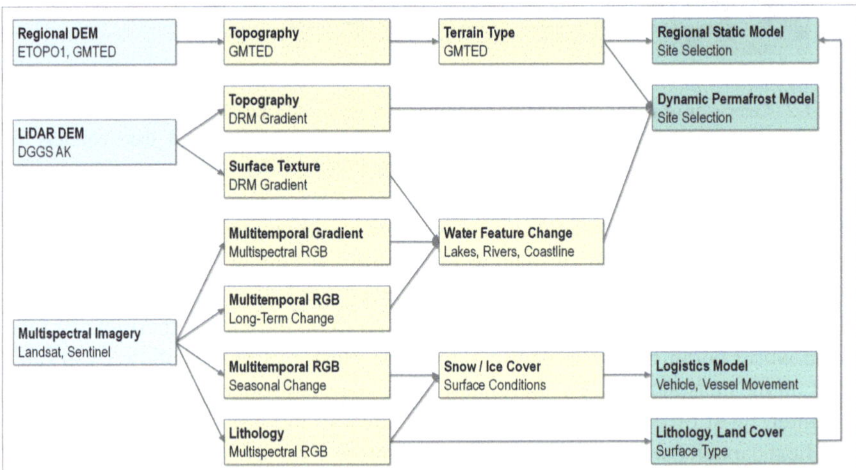

Fig. 11.4 Workflow for arctic land cover and logistics assessment based on remote sensing data

GMTED elevation model for onshore. The multispectral satellite imagery originates from Landsat 5 for regional work using regional mosaics. For detailed work, Sentinel imagery provided imagery at 10 m resolution using pan-sharpening. Seasonal aspects have best been covered by a combination of Landsat 5 and Landsat 8 imagery employing the Landsat Data Continuity Mission (USGS, 2020). From the digital elevation models the topography has been obtained. The gradient provided the surface texture. For the permafrost lakes, the LiDAR DEM offered a vertical resolution in the range of decimeters which allowed using the gradient for the delineation of the lake shores and the detection of secondary thermokarst features such as collapse faults. Multispectral imagery was processed for land cover to discriminate water features from soil. We used the same approach on seasonal imagery for the delineation of lake and sea ice as well as for the detection of frozen ground. The combination of the LiDAR DEM with multispectral imagery allowed the mapping of long-term changes in lake cover due to permafrost degradation which resulted in increased erosion of river and lake banks as well as coastal erosion (Chambers et al., 2006; Nitze et al., 2018; Jones et al., 2011; Gatto, 1995; Mars & Houseknecht, 2007). The interpretation of all results yields a static model that is modified by seasonal aspects and a dynamic model that incorporates longer term changes of permafrost.

We zoom into the area of Teshekpuk Lake and assess the combination of satellite imagery, digital elevation model and surface gradient in the context of permafrost (Fig. 11.5). The multispectral satellite image (a), the gradient map (c) and the elevation profile BB′ (d) reveal an area of elongated lakes north of Teshekpuk Lake (TL) in a very flat area with coastal permafrost with the lake surfaces at only few meters above sea level. The lake bank height ranges from 1 m to a few meters. South of the lake, a dense cover by rounded lakes indicates inland permafrost. The digital elevation model (b) and the elevation profile AA′ (d) show a topography between 60 and

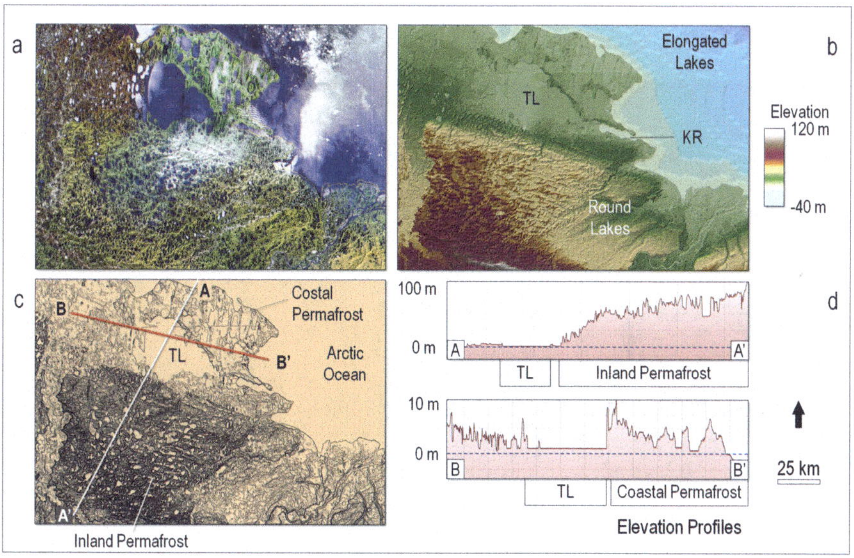

Fig. 11.5 Permafrost and geomorphological setting of the Teshekpuk Lake area. (**a**) multispectral satellite image, (**b**) 5 m resolution DRM onshore and ETOPO1 DRM offshore, arrows indicate errors in the DRM, (**c**) gradient of DRM with locations of elevation profiles, (**d**) elevation profiles. *TL* Teshekpuk Lake, *KR* Kogru River. (NED 5 m IFSAR DRM and Landsat 5 satellite imagery courtesy of U.S. Geological Survey, elevation profile from GoogleEarth)

120 m with the lake banks at 5–10 m height. When studying the long-term change of the permafrost we will find that the shallow lakes located in the area of coastal permafrost are most vulnerable to thermal karsting and coastal erosion.

Next, we focus on lakes in the area of inland permafrost southwest of Teshekpuk Lake (Fig. 11.6) and compare two Landsat images from 1984 (a) and 2020 (c). On the 2020 image, some bright areas on the lake banks indicate an obvious decrease in lake size over 36 years. However, when we plot the differences between the two images in red for lake size decrease and in cyan for increase image (c), we notice that in the southern part of the study area some lakes increased the size slightly. Overall, the lake size decrease amounted to approximately 5% over the period of 36 years.

The effect of permafrost thawing on elongated lakes in the coastal permafrost area can be illustrated at a former lake south of Kogru River (see Fig. 11.7b for location). The lake surface prior to lake collapse had only been about 2 m above sea level as can be obtained from the digital elevation model (a) and the elevation profile (b). Permafrost had formed the banks indicated by the sharp outline of the lake in the elevation model. Thawing of the permafrost weakened the banks and increased precipitation lead to an overflow of the lake resulting in an erosional channel (c) cutting into the lake bank and draining the lake. The permafrost below the lake bottom was no longer thermally protected by the water body and started thawing leaving behind thermokarst in the shape of collapse fault features.

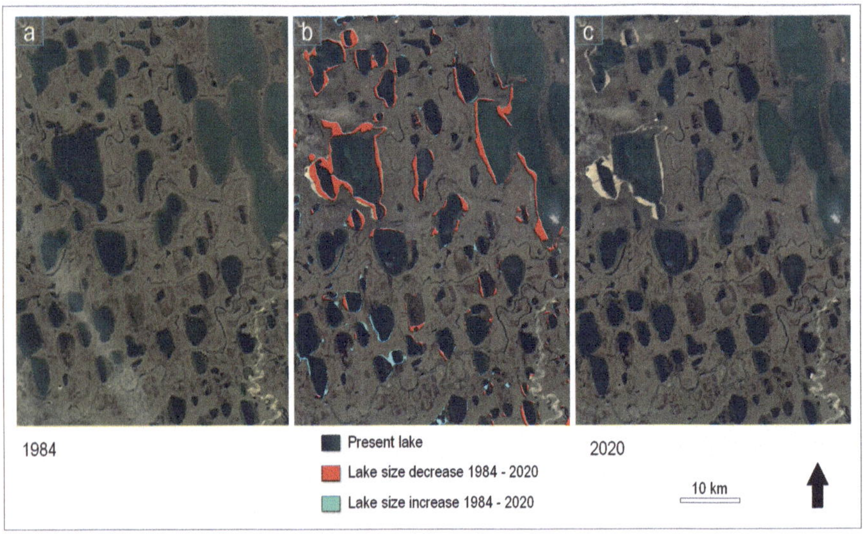

Fig. 11.6 Change of permafrost in the rounded lake area southwest of Teshekpuk Lake represented by change in size of lakes between 1984 and 2020. (**a**) satellite image 1984, (**b**) difference between 2020 and 1984 satellite images, (**c**) satellite image 2020. Land 5 and 8 multispectral satellite imagery courtesy of U.S. Geological Survey

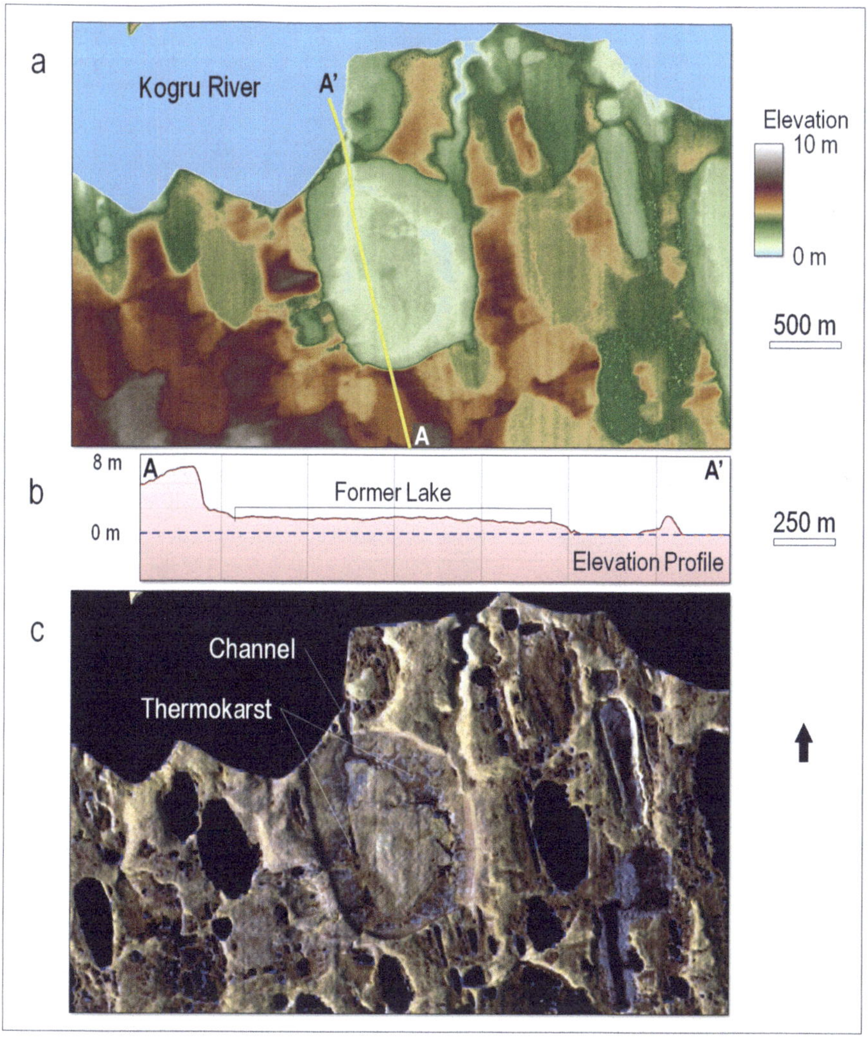

Fig. 11.7 Lake drainage as a result of permafrost thawing. (**a**) DRM, (**b**) elevation profile, (**c**) satellite image draped over DRM. (NED 5 m IFSAR DRM courtesy of U.S. Geological Survey, satellite imagery from Copernicus Sentinel data 2017, elevation profile from GoogleEarth)

These are also indicative of increasingly unstable ground that changes its mechanical properties to support infrastructure with the seasonal change of surface temperatures.

The complete history of such a lake drainage as a result of permafrost thawing can be observed at the north shore of the Barrow Peninsula (Fig. 11.8). Along a profile the elevation of which is shown in (b) series of erosional stages from original lake (1), overflowed and drained lake (2) to coastal erosion of the lake bank (3) to complete marine flooding of the former lake (4) can be studied. The original

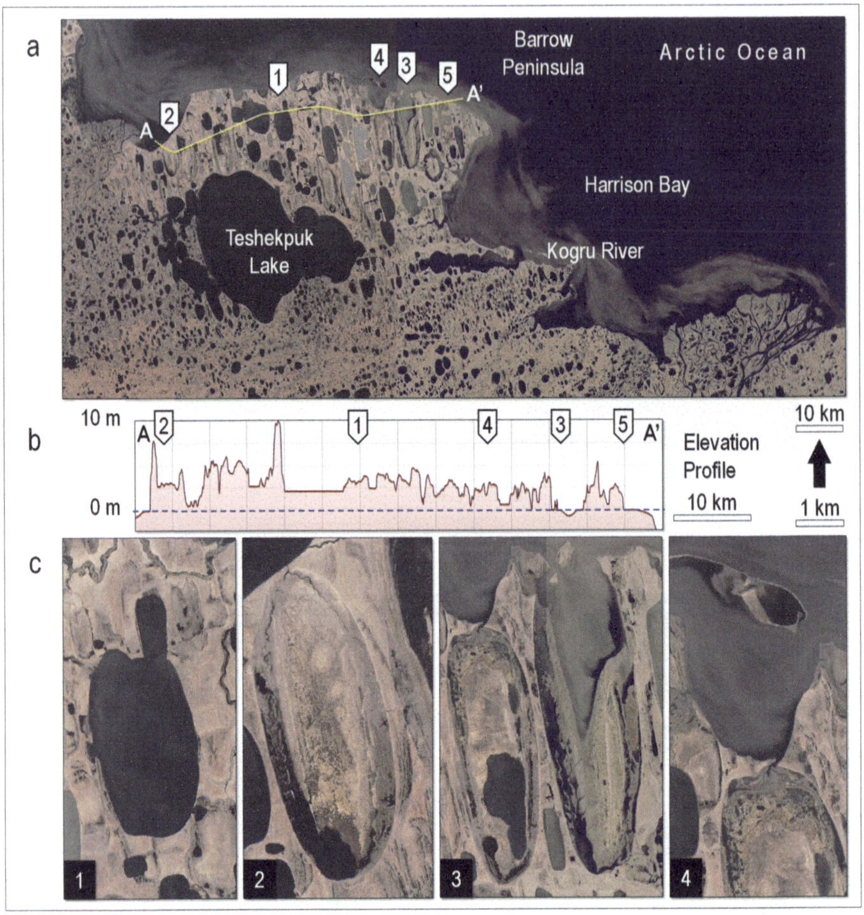

Fig. 11.8 Stages of impact of permafrost thawing on lake drainage and coastal erosion. (**a**) Overview map with locations of the cases of lake drainage in (**c**), (**b**) arbitrary elevation profile, (**c**) selection of stages of lake drainage. (NED 5 m IFSAR DRM courtesy of U.S. Geological Survey, satellite imagery from Copernicus Sentinel data 2017, elevation profile from GoogleEarth)

elongated lake (c1) shows the well-defined bank that is formed by permanently frozen soil.

When permafrost thawing and high precipitation cause the lake to overflow, channel erosion into the bank results in lake drainage (c2) leaving behind thermokarst that is indicative of soft unconsolidated soil exposed to high weathering during the summer. Permafrost thawing affects the stability of the coastal cliff (5) that also consists of frozen soil. Thawing of the permafrost at the coast removes the protective cliff and allows the seawater to erode the seaward bank of the former lake (c3) thus changing the habitat from freshwater to saltwater with significant impact on the ecology (Audubon Alaska, 2017). Tidal currents finally remove the former lake banks completely and turn the former lake into a bay (c4). Increased thawing of the permafrost is expected to accelerate this process (Mars & Houseknecht, 2007).

When combining the study of permafrost thawing with seasonal imagery from July and September (Fig. 11.9), an additional erosional component becomes evident. Whilst the shallow waters offshore Barrow Peninsula appear not to show any motion even when the sea ice has thawed in July (a), currents are setting in during the arctic summer. Longshore coastal currents (b) may add to the coastal erosion resulting from wave action and tidal currents. The LiDAR DRM (c) at 200× vertical exaggeration reveals that this effect is most likely strongest in the north where the elevation of the onshore lakes is almost at sea level. This implies that the development of infrastructure in coastal areas should include a multi-temporal study of the area integrating time-lapse satellite imagery with climatic records to assess the level of historic coastal erosion and to make predictions about its future evolution.

The seasonal study of using multi-spectral satellite images can reveal key parameters that are important for logistic planning. In Fig. 11.10 we have correlated five seasonal satellite images with climatic data collected by Urban and Clow (2014). The first cloud-free image is available for April, when the entire peninsula as well as the adjacent sea is frozen and covered in ice (cyan). In May, the land is free of snow, but still frozen and the sea ice starting to break up. In June the sea is almost

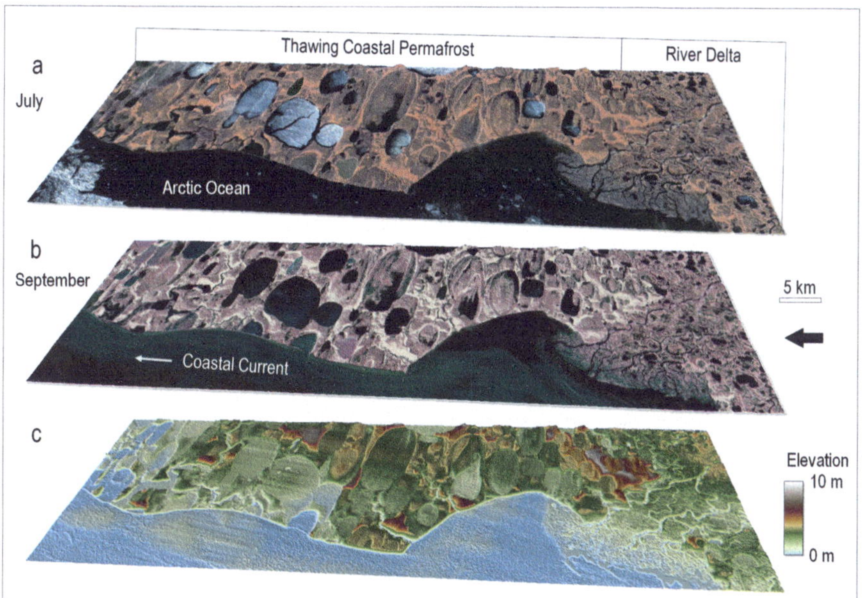

Fig. 11.9 Seasonal imagery of the northern coast of the Barrow Peninsula revealing the effects of thawing permafrost on coastal erosion in 200× vertical exaggeration. (**a**) July image, (**b**) September image showing a coastal current, (**c**) LiDAR digital elevation model. (NED 5 m IFSAR DRM courtesy of U.S. Geological Survey, satellite imagery from Copernicus Sentinel data 2017 and 2020)

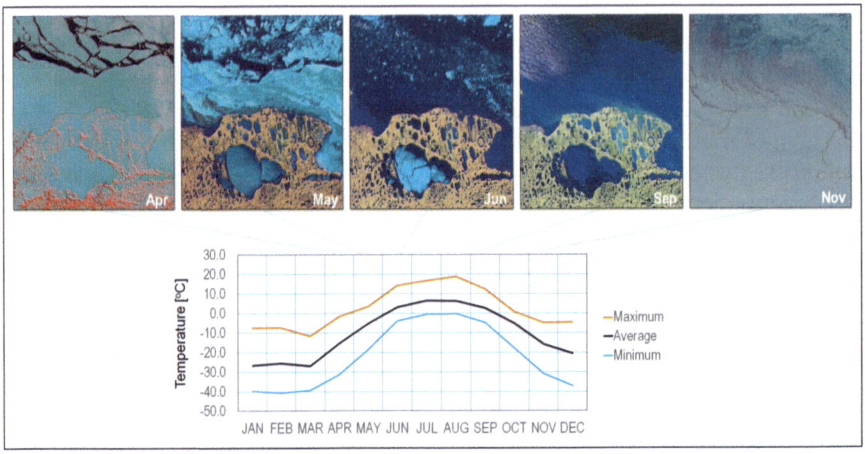

Fig. 11.10 Correlation of regional ice and snow distribution with temperature curves for Teshekpuk area, Alaska. (Climatic data from Urban and Clow, U.S. Geological Survey 2011, Landsat 8 imagery courtesy of U.S. Geological Survey)

free of ice, whereas the lakes still show partial ice cover. The remnant ice thaws completely in July. Until September, land, lakes and sea are completely ice free. Around end October the ice returns and by November the entire area is frozen and covered in ice again. Images from the winter are not available due to high cloud cover. The correlation of the satellite images with the temperature curves shows that around end of May the average air temperature exceeds the freezing point at which time one should expect the ground to thaw, whereas floating ice on water still persists. By end September, the average air temperature dip back below freezing and the ground is expected to freeze again. Due to the high heat capacity of the water, lake and sea ice may occur with a delay to the freezing of the soil.

For a more detailed study of the freezing and ice cover of land, lakes and sea, we have collected multi-temporal satellite images on a monthly basis (Fig. 11.11). Until end of May land, lakes and sea are completely covered in ice and snow. The frequent dense cover by clouds indicates that snow storms and bad weather will have a significant impact on logistics. Transportation, if possible at all, will be limited to onshore using ice roads (Arp et al., 2019). In June, when the air temperature starts to exceed freezing, the ground starts to thaw, making logistics in the thawed areas impossible. Lake and sea ice still persist but may no longer be suitable to support logistics. From July to the mid of October the sea appears mostly free of ice which would allow for offshore logistics. Towards end October, winter returns with snow and ice cover and frequent storms.

In Table 11.2 we have summarized the seasonal logistics options in arctic climate related to the air temperature and obtained from interpretation of seasonal satellite imagery. The typical period for arctic operations is during the period of complete ground freezing from November to beginning of May. This period would also not interfere with bird migration, nesting and breeding. The impact of arctic operation on land mammals during winter still needs to be considered (Audubon Alaska,

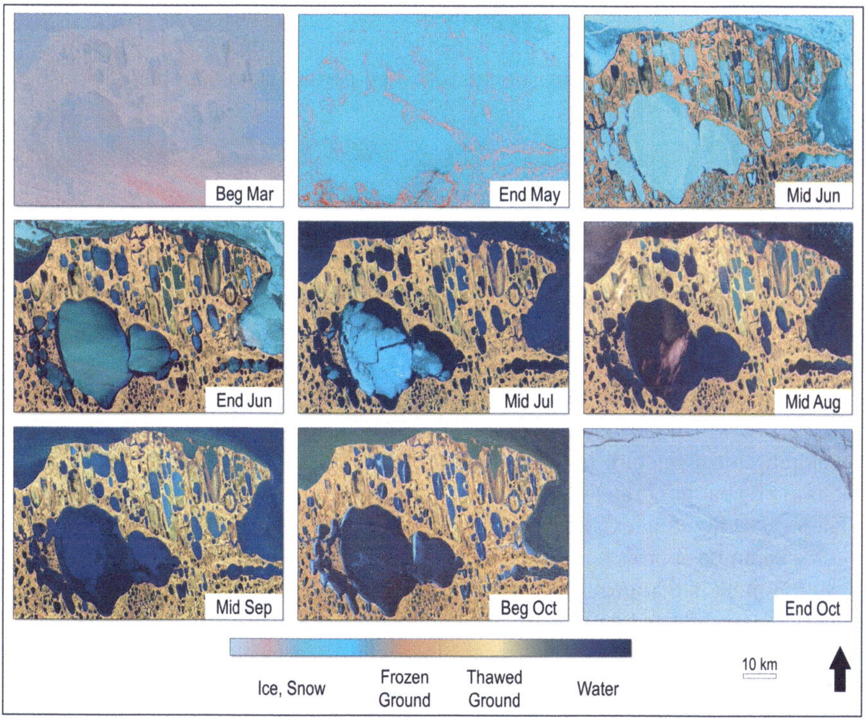

Fig. 11.11 Regional seasonal changes of snow and ice cover for the Teshekpuk area, Alaska showing stages of land, lake and sea freezing. (Landsat 7 and 8 imagery courtesy of U.S. Geological Survey)

Table 11.2 Summary of seasonal logistics for arctic climate, temperatures after Urban and Clow (2014)

Month	Air temperature	State of freezing	Logistics
Jan to May	−45 to 0 °C	Complete ice cover	Land: ice roads
			Sea: no logistics
June	0 to +20 °C	Land thawed, Lake and sea ice	Land: no logistics
			Sea: no logistics
July to Aug	+5 to +25 °C	Land thawed, Completely ice-free	Land: no logistics
			Sea: possible
Sep	+0 to 18 °C	Land thawed, Lake ice forming, Sea ice-free	Land: no logistics
			Sea: no logistics
Oct	−15 to +8 °C	Land freezing, Lake and sea ice	Land: ice roads
			Sea: no logistics
Nov to Dec	−40 to 0 °C	Complete ice cover	Land: ice roads
			Sea: no logistics

2017). The long-term change of the air temperature as reported by Arp et al. (2019) will certainly change the above-mentioned observations with the prospect of a shift of the thawing period into May, a longer ice-free period in summer and the push of the re-freezing into November.

11.3 Humid Floodplain Setting Arkansas, U.S.A.

Floodplains are a preferring setting for infrastructure since these areas are often developed due to the ease of planning and building infrastructure and logistics. However, floodplains are the result of active river channels depositing sediments that create the floodplain. Large floodplains are often a sign of substantial deposition, part of which occurs through flooding. The key challenges for planning and developing infrastructure in a floodplain are the measures that have already been taken to prevent flooding above a critical threshold. It is also important to estimate what level of flooding existing flood protection measures can sustain.

One common approach for flood inundation modelling are high resolution digital elevation models that provided for example from LiDAR (NOAA, 2012). For our study we have selected an area in the White River–Black River floodplain between Batesville and Jonesboro in Arkansas (Fig. 11.12) for which the 1 m statewide LiDAR is available (Arkansas GIS Office, 2018). At this location the White River leaves the Ozark Mountains and joins the Black River and its floodplain near

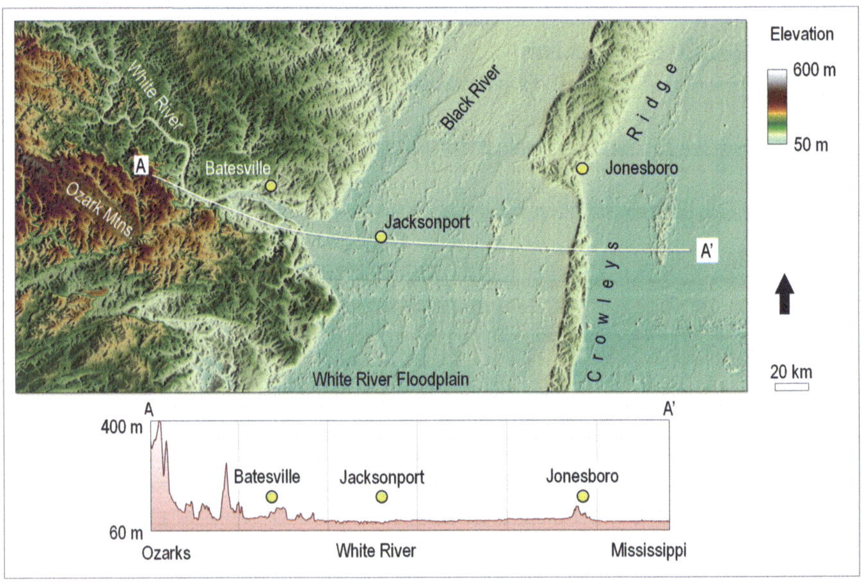

Fig. 11.12 LiDAR DEM of the White River–Black River confluence in Arkansas. (Data courtesy Arkansas GIS Office, elevation profile courtesy GoogleEarth)

Jacksonport. The elevation profile stretches from Batesville across the floodplain to the ridge at Jonesboro and to the east into the Mississippi floodplain. The White River floodplain is very flat with typical elevation changes of less than 20 m across the entire floodplain, which makes it susceptible to widespread flooding in the case of heavy rain in the large catchment area of the combined White River and Black River watersheds.

We can simulate the effect of flooding taking Jacksonport as a reference (Fig. 11.13). The LiDAR DEM (b) and its gradient, the surface dip (a), reveal the substantial meandering of both river courses, a feature that is indicative of very shallow slope of the river course.

We simulate the flooding in steps of 1 m and overlay semi-transparent layers of blue color onto the LiDAR DEM (c). With this approach we can delineate the potentially inundated areas as a result of the flooding.

When we render the different flooding stages in 3D (Fig. 11.14) we notice that the inundated part of the floodplain increases non-linear with the rise of flooding level. We have simulated four stages of flooding where the levels from 4 to 6 m occur on an annual basis. Flooding levels of 7 and 8 m are considered exceptional. We have added the worst flooding event with a head of 11 m that occurred in 1945, when almost the entire floodplain was inundated.

Next we convert the flooding simulation to flood risk maps (Fig. 11.15) that provide an information about which areas in the floodplain are susceptible to flooding of a specific head.

This information can guide the design of flood mitigation such as elevated platforms for critical equipment. We drape the flood risk (b) onto the LiDAR DEM (a) and include the extreme case of the 1945 flood (c) to provide the worst-case scenario.

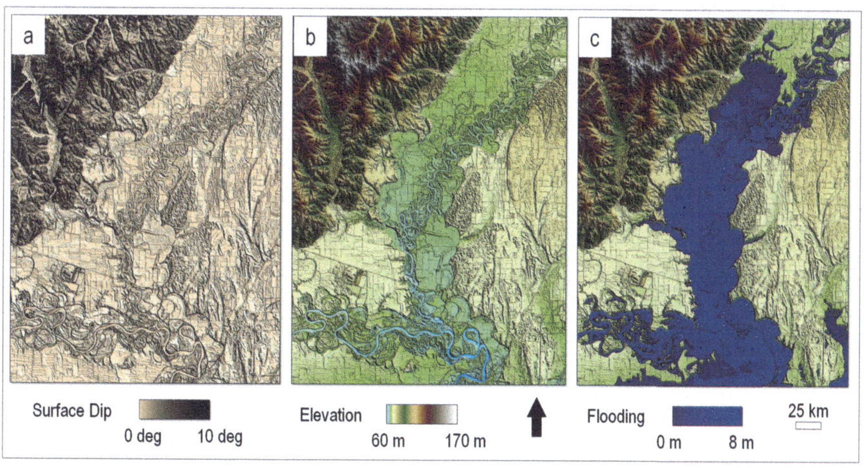

Fig. 11.13 Flooding model based on LiDAR DEM. (**a**) surface dip from gradient of DEM, (**b**) LiDAR DEM, (**c**) flooding simulation using LiDAR DEM. (LiDAR DEM data courtesy Arkansas GIS Office)

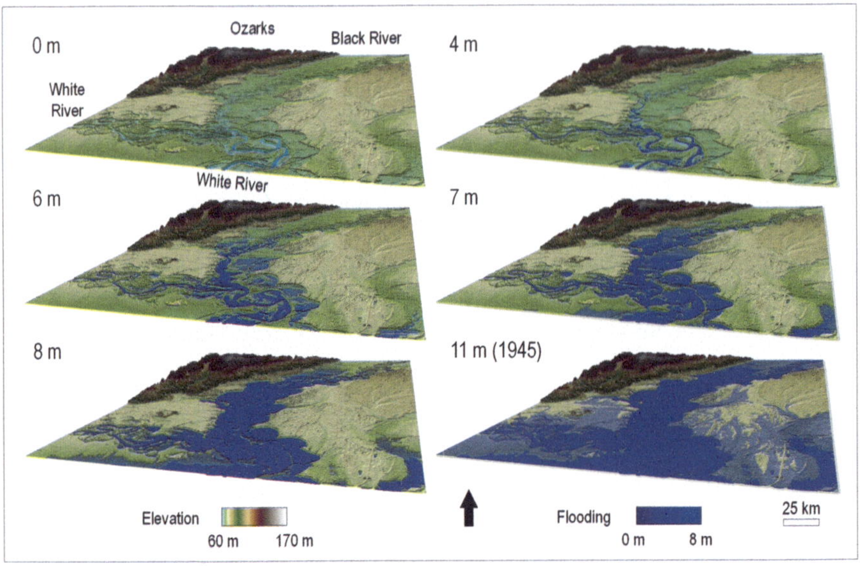

Fig. 11.14 Flooding simulation based on LiDAR DEM for 0, 4, 6, 7, 8, and 11 m flooding levels. (LiDAR DEM data courtesy Arkansas GIS Office)

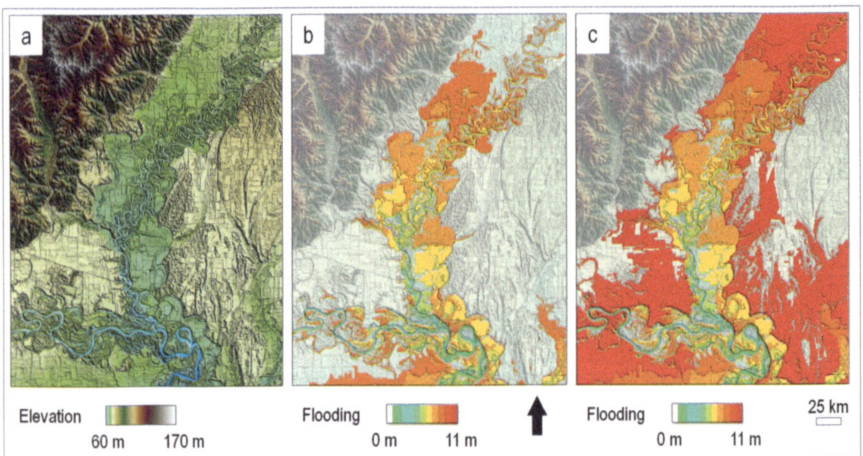

Fig. 11.15 Flood risk maps using the LiDAR DEM. (**a**) LiDAR DEM, (**b**) flooding levels from 1 to 8 m, (**c**) flooding including the 1945 flood. (LiDAR DEM data courtesy Arkansas GIS Office)

For the validation of these simulations we use satellite imagery of actual flooding (Fig. 11.16). Due to persistent cloud cover, optical imagery is not always available from the peak flood situation. For the peak flood event on May 17, 2017 Landsat 8 imagery not obstructed by cloud cover is available. We create the multiband 753 RGB composite that is indicative of landuse.

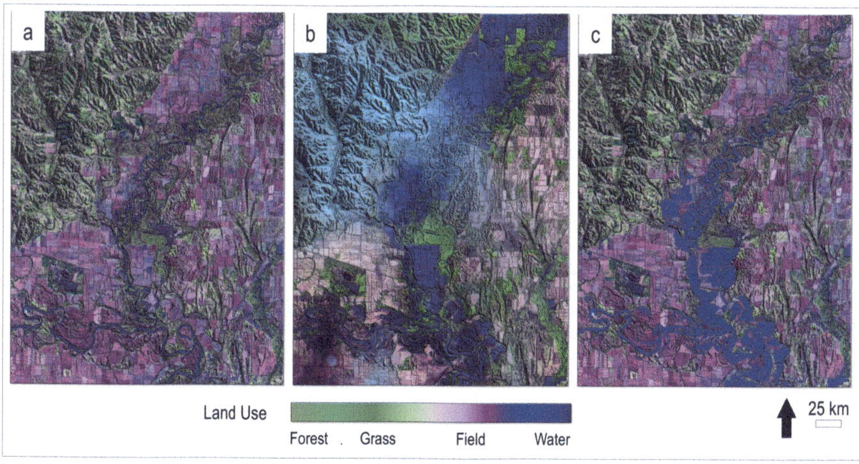

Fig. 11.16 Validation of flooding simulation using Landsat 8 optical imagery. (**a**) non-flood situation on Oct 11, 2018, (**b**) peak flood situation on May 17, 2017, (**c**) overlay of 6 m flood simulation on non-flood satellite image. (Landsat 8 satellite imagery courtesy of U.S. Geological Survey)

The non-flood situation from October 11, 2018 (a) shows widespread agricultural use of the floodplain (purple patches). The color indicates that the crops on most of the fields have already been harvested thus exposing the bare earth. The fields contrast sharply with the green tints of the forest covering the Ozark Mountains and some meadows in the Black River floodplain. The image of the peak flood situation on May 17, 2017 shows the fields earlier in the year in a beige tone indicating that the vegetation is still growing.

The forest in the Ozark mountains shows a vivid green from fresh vegetation. The Black River floodplain is largely inundated as indicated by the intense blue color. Please note that the light blue tint in the center of the image and above the Ozark Mountains originates from thin cloud cover that has not been removed.

The flooding simulation uses a 6 m flood head overlaid on the Landsat image from the non-flooded situation (c). The comparison with the flooding image (b) shows that in the lower half of the image the simulation matches the actual flooding well. In the upper half the simulation predicts significantly less flooding, which implies that a new model has to be used to capture possible obstacles for the flow that have not been considered in the flooding model using Jacksonport as the reference.

Using the 7 m flooding model we can now estimate the total inundated area (Fig. 11.17). We render the satellite image from the peak flood (a) and the simulated 7 m flooding event (c) in 3D to obtain a better impression of the inundated portions of the floodplain. In the center we extract the water volume associated with the inundation at a flood head of 6 m (b).

The area shown in Fig. 11.17 comprises a total of 920 km^2, of which 87 km^2 or almost 10% are inundated. Since most of the flooded surface is farmland, the damage to the economy is considerably higher than given by the percentage of the area inundated by the flood.

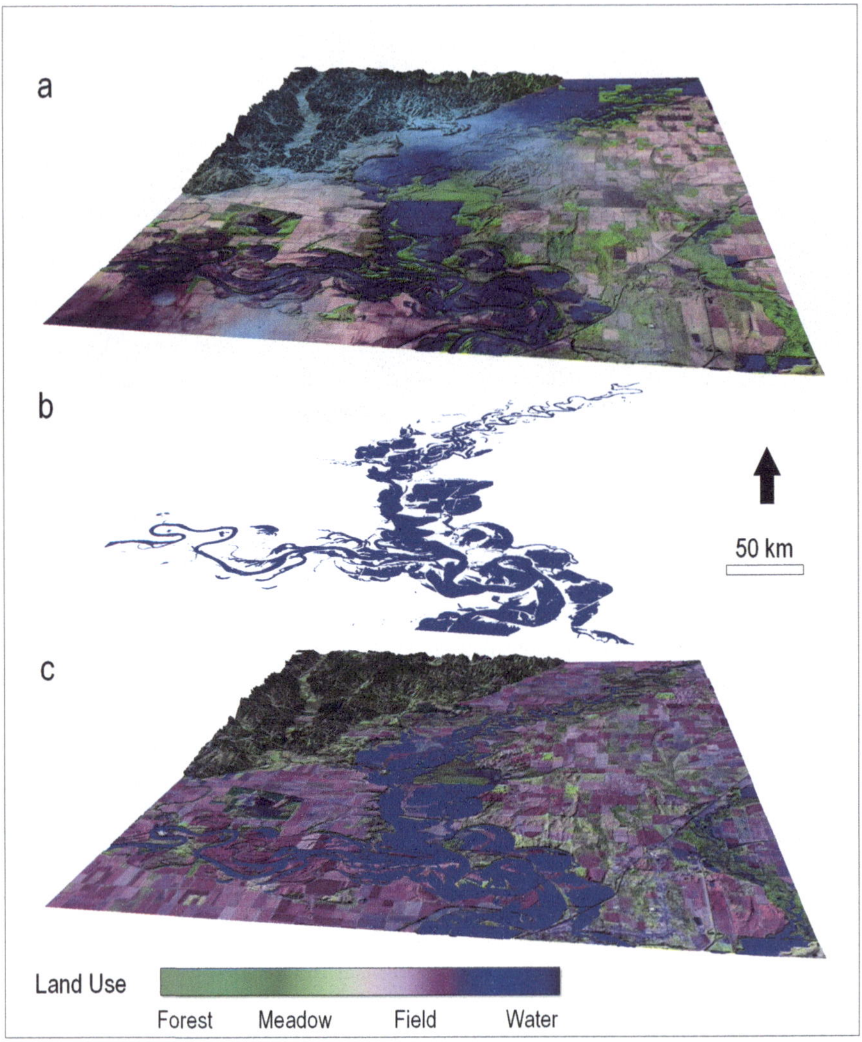

Fig. 11.17 3D representation of flooding. (**a**) peak flooding situation on May 17, 2017 on Landsat 8 image, (**b**) flooded areas for 6 m flooding model, (**c**) overlay of 6 m flooding model on non-flood situation on October 11, 2018. (Landsat 8 satellite imagery courtesy of U.S. Geological Survey, LiDAR DEM data courtesy of Arkansas GIS Office)

In addition to the flooding of the farmland, the logistics is also affected by the flooding. All logistics routes, i.e. roads and rails that are not elevated will also be affected by the flooding. Therefore, the flood risk map (Fig. 11.15) covers both, the flooding risk for infrastructure as well as logistics routes.

Starting at the beginning of the twentieth century, levees to mitigate the perennial flooding were erected. The devastating flood of 1927 led to increased efforts to expand the levees. Under the coordination of the US Army Corps of Engineers the

Mississippi River and Tributary project provided a flood plan for the entire region. Levee grade (height) and section (width) were standardized and calculated to withstand the floods higher than the 1927 flood (USACE, 2014). For infrastructure planning it is essential to assess the flood level the levees can withstand.

We have chosen the town of Jacksonport AR to assess the levees for their maximum flood protection (Fig. 11.18). The LiDAR DEM provides a high-resolution basis for the assessment (Schumann & Bates, 2018; Courty et al., 2019). The assessment is done first in 2D with all data draped on the LiDAR DEM, rendered with 50× vertical exaggeration and slant illumination to enhance the surface texture.

For the LiDAR DEM (a) we choose a colormap that highlights the non-flooded river surfaces in light blue, shows the banks and low-lying terrain in shades of green and high ground outside the floodplain brown. For the assessment of the levees we compute a relative elevation map that highlights the levee height above the terrain (b). For high-resolution imagery we chose the 1 m resolution airborne imagery from the National Agriculture Imagery Program available through the U.S. Geological Survey (c). Based on the LiDAR DEM we run a flood simulation for flood heads from 4 to 8 m and color code the flooding risk related to the flood head. The map shows that the town site of Jacksonport will be safe from flooding for flood heads of less than 8 m.

We complete the study with the simulation of a 7 m flood to confirm that the levees are designed to protect the site safely (Fig. 11.19). We render the LiDAR DEM (a) and the NAID airborne (b) in 3D and mark the location of Jacksonport.

Fig. 11.18 Flooding simulation at Jacksonport AR. (**a**) LiDAR DEM, (**b**) relative elevation draped on LiDAR DEM, (**c**) high-resolution aerial photography, (**d**) flooding risk for flood levels from 4 to 8 m. *JP* Jacksonport, *WR* White River, *BR* Black River. (LiDAR DEM courtesy of Arkansas GIS Office, airborne imagery courtesy NAID program of U.S. Geological Survey)

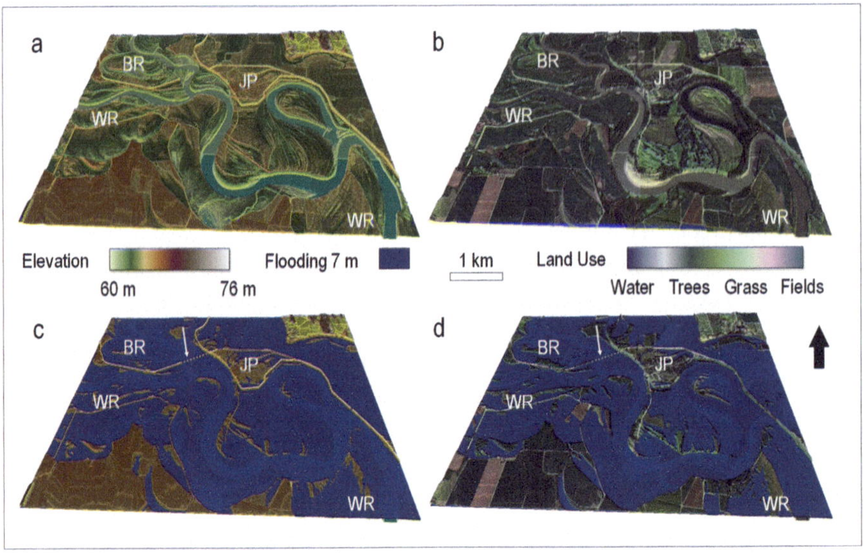

Fig. 11.19 Flood risk simulation for protective levees at Jacksonport AR. (**a**) LiDAR DEM, (**b**) high-resolution aerial photography, (**c**) simulation of 7 m flooding on LiDAR DEM, (**d**) simulation of 7 m flooding on high-resolution aerial photography. *JP* Jacksonport, *WR* White River, *BR* Black River. (LiDAR DEM courtesy of Arkansas GIS Office, airborne imagery courtesy NAID program of U.S. Geological Survey)

Then we plot the 7 m flooding surface on top of both 3D rendered surfaces (c and d). We notice that the road connecting Jacksonport with the higher ground in the east as well as across the Black River with the west is laid out on a levee that withstands a 7 m flood.

The bridge across the Black River has been added manually (white arrow) because the LiDAR DEM is corrected to bare ground and, therefore, does not comprise the bridge feature. The levees around the townsite of Jacksonport also protect efficiently against a flood of 7 m head. Infrastructure planning should confirm with the U.S. Geological Survey the actual river level measurements to confirm the protection the levees will provide.

11.4 Coastal Setting Louisiana, U.S.A.

Coastal setting and delta floodplains challenge infrastructure planning through subsidence, coastal erosion and flooding from hurricanes as well as from large rivers. Water engineering is used to mitigate some of these threats, but often are the natural events outperforming human efforts.

We chose the western Mississippi Delta and the adjacent Louisiana coast for a study how remote sensing can support the identification, assessment and mitigation of coastal challenges. The Mississippi watershed of 3.2 million km^2 is the fourth largest in the world and consequently is susceptive to substantial flooding following snow melt in spring. On the other hand, the Mississippi is the most important transportation highway between the Gulf of Mexico and the Great Lakes. To provide optimum conditions for shipping whilst minimizing the flood impact on settlements and infrastructure along the river the flood control planning for the Mississippi and its tributaries has been established (USACE, 2014). The US Army Corps of Engineers has designed extensive flood control measures for the entire Mississippi delta plain which comprises the Atchafalaya Basin. The main goal is to allow up to 50% of the water volume of the Mississippi to be diverted into the Atchafalaya Basin thus promoting the generation of two Atchafalaya River deltas.

Figure 11.20 shows a false color satellite image of the western Mississippi delta and the Atchafalaya Basin. We used a Landsat 8 multiband 2-7-11 inverted RGB composite of the blue visible band, a shortwave infrared band and a thermal infrared band which enhances the distinction of water and wet terrain from dry terrain.

The blue band is sensitive to water, whereas the shortwave infrared allows the discrimination of vegetation from soil. The thermal infrared band is sensitive to both, open water and water saturated soil. The composite image is draped over the SRTM digital elevation model to enhance the coastline and the river banks. We have chosen an image taken on February 28, 2020 when heavy rains and snow melt caused flooding of the Mississippi. The flood control upstream of Baton Rouge appears to be effective. The river is confined to its bed, whereas most of the Atchafalaya Basin is flooded as indicated by the cyan to blue colors. The Mississippi flood control prevents the large inhabited area west of the Atchafalaya Basin from

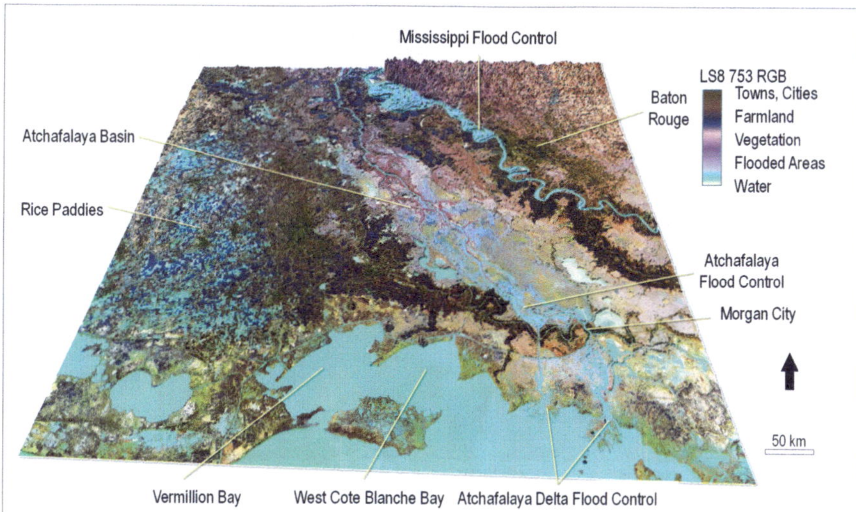

Fig. 11.20 Overview of challenges to infrastructure planning in the Mississippi Delta and Louisiana coast. (Landsat 8 2-7-11 inverted RGB color composite during high flood stage on 28 February 2020. DRM and satellite imagery courtesy of U.S. Geological Survey)

becoming flooded. The relatively dry farmland, therefore, shows dark colors. In the far west of the image, rectangular cyan patches dot the landscape. These are rice paddies that are flooded in early spring during the growth of the rice plants.

In our study we will examine the effectiveness of the Mississippi flood control in the area of Baton Rouge and the impact water engineering has on the Lower Atchafalaya and its deltas and its adjacent areas.

In Fig. 11.21 we compare different remote sensing approaches to characterize the Mississippi flooding immediately upstream of Baton Rouge. The digital elevation model (a) reveals that the Mississippi floodplain is very flat with only about 20 m elevation profile changes across the floodplain, the eastern border of which is clearly defined by the wooded hills north of Baton Rouge.

The dry season image (b) shows a Landsat 8 band combination 753 RGB typically used for landuse mapping. It is comprised of the shortwave infrared band 7 for lithology detection, which we use to delineate farmland, the very near infrared band 5 for the detection of vegetation and the visible red band 3 for mapping infrastructure. The band combination distinguishes bare earth in farmed areas (bright purple) from wooded areas (dark green), meadows (bright green) and water surfaces of the Mississippi River (dark blue). The city of Baton Rouge with buildings and road surfaces shows as dark purple. We draped the satellite image on the DRM and exaggerated it vertically by a factor of 10 to enhance the texture of the floodplain. The flood image (c) combines the visible blue band 2 with the short-wave infrared band

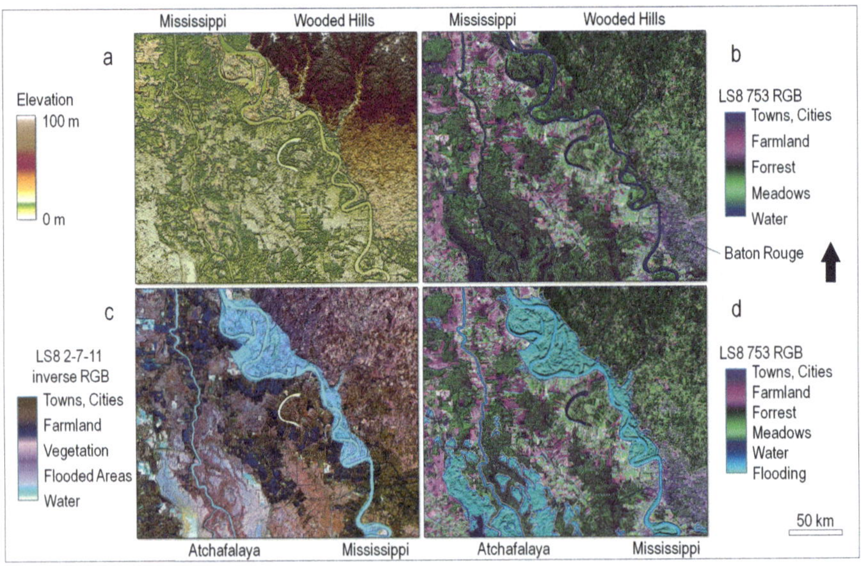

Fig. 11.21 Assessing the impact of the Mississippi flooding. (**a**) DRM, (**b**) Landsat 8 dry season landuse image 753 RGB, (**c**) Landsat 8 flooding image 2-7-11 inverse RGB, (**d**) flooded area extracted from image (**c**) superimposed on image (**b**). (SRTM DEM V3 and Landsat 8 imagery courtesy of the U.S. Geological Survey)

7 and the thermal infrared band 11. We then invert the RGB band combination such that the resulting image shows water features in cyan tones and dry land in red to brown tones. Since the muddy flood waters attenuates the optical bands completely, the raw RGB image would not show the correlation with the dry season river bed. This correlation can be obtained when we drape the image on the digital elevation model. Image (c) allows distinguishing the dry season river course (solid cyan) from the flooded areas that are completely submerged (textured cyan areas) and areas that are just inundated or water soaked (rose to reddish pink tones). Since the band combination used in the flooding image does not distinguish very well wet areas from forest, we need the digital elevation model for support to extract the entire area affected by flooding. In image (d) we have overlaid the flooded area on the dry season satellite image to assess the impact of the flooding on the area around Baton Rouge. We notice that the Mississippi flooding in February 2020 was successfully contained to the river flood plain, which is only covered in meadows with isolated trees. The city of Baton Rouge and the farmland west of the Atchafalaya Basin have been kept dry, whereas the Atchafalaya Basin itself was flooded as per USACE plan.

Figure 11.22 summarizes our findings. We overlay the area affected by flooding as extracted from the Landsat 8 2-7-11 inverted RGB image on the landuse Landsat 8 753 RGB dry season image and use the same blue tone, in which the 753 RGB band combination shows water. The resulting image is draped on the SRTM DEM V3, 10 times vertically exaggerated and illuminated with slant lights from the southwest. With this approach, even subtle elevation changes are highlighted so that we can correlate the flood impact with the topography. For orientation we have drawn in the I-10 motorway (dashed line) and the city of Baton Rouge. The motorway crosses the Atchafalaya Basin on a bridge. This ensures that Baton Rouge is reliably connected to the west even during significant flooding. The connection to the north and into the Mississippi Delta is secured through roadways that are placed on the

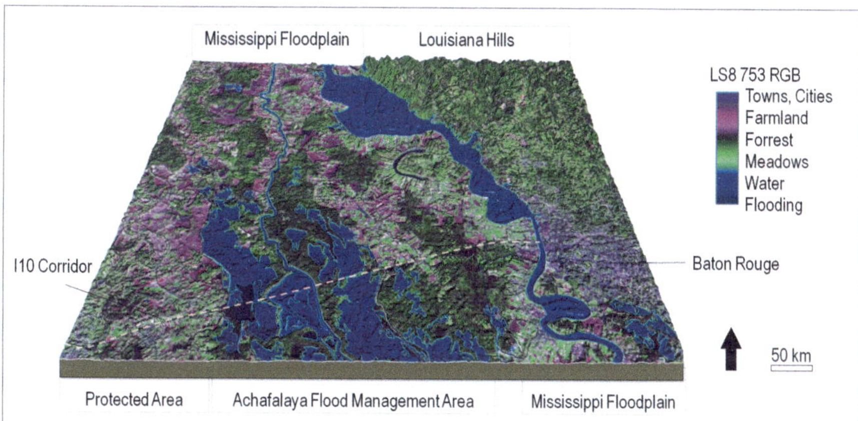

Fig. 11.22 Assessment of flood control efficiency in the Baton Rouge area using the Landsat 8 2-7-11 inverted RGB band combination draped on the SRTM DEM V3 DRM. (DRM and satellite imagery courtesy of the U.S. Geological Survey)

elevated eastern shoulder of the Mississippi floodplain. For infrastructure planning considering to avoid the impact of flooding, the awareness of the flood control design laid out in the USACE plan complemented with support from remote sensing using multi-temporal images is the best approach to assess the challenges and provide the necessary geospatial recommendations.

The flood controls partially deviating the water of the Mississippi river to the Atchafalaya river impact the lower Atchafalaya and its deltas. To assess the long-term development of this area we compare a US topographic map from 1963 with recent satellite imagery both from dry season and flooding stage (Fig. 11.23). For the digital elevation model, we use the NGOM topobathymetric elevation model that has been compiled by the USGS and is accessible via the NOAA (2014). The Atchafalaya Deltas are one of the few areas along the Louisiana coast that do not suffer from subsidence (Bernier et al., 2006; Morton et al., 2015).

The US topographic map from 1963 (a) shows no delta features for the Lower Atchafalaya River (LAR) and the man-made canal at Wax Lake Outlet (WL). We

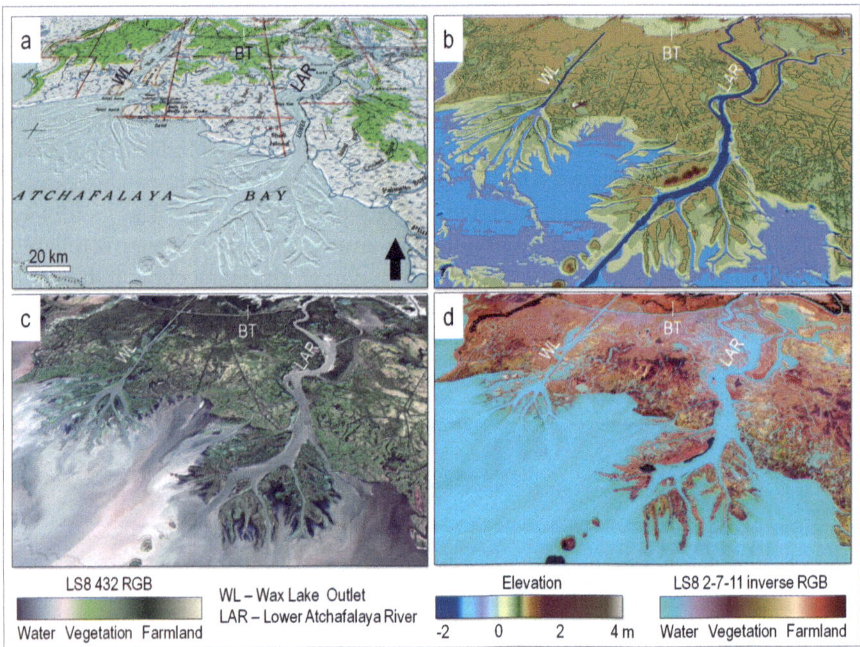

Fig. 11.23 Mapping the long-term impact of flood control engineering on the Atchafalaya Deltas south of Morgan City LA. (**a**) US topographic map New Orleans from 1963, (**b**) NGOM topobathymetric elevation model, (**c**) Landsat 8 432 RGB from 17 Oct 2017, (**d**) Landsat 8 2-7-11 inverted RGB from 28 Feb 2020, *BT* Bayou Teche, *LAR* Lower Atchafalaya River, *WL* Wax Lake Outlet. (US topographic map, NGOM DRM and satellite imagery courtesy of U.S. Geological Survey)

have draped the topographic map on the high-resolution DRM to show the difference between the coast in 1963 and today. The high-resolution DRM (b), the main component of which is a LiDAR digital elevation model at 1 m lateral resolution and a vertical resolution of a few dm. We note that most of delta area of approximately 10,000 km^2 has a maximum elevation of 2 m above sea level, which makes it prone to storm surge during hurricanes. The satellite image taken during the dry season (c) on October 17, 2017 shows the full extent of the Atchafalaya and Wax Lake deltas.

The symmetric birdfoot delta of the Wax Lake Outlet shows a morphology different from the asymmetric Atchafalaya delta because the Wax Lake channels are not dredged. Compared to the 1963 topographic map, the Wax Lake Delta has added approximately 2500 km^2 of coastal wetlands whereas the Lower Atchafalaya Delta added about 3600 km^2.

The satellite image taken during flooding season (d) on February 28, 2020 highlights the extent of inundation resulting from river flooding. Only the farmland and settlements in the Bayou Teche (BT) area are not affected because they are protected by the levees of the Atchafalaya Basin system.

The formation of the Wax Lake Outlet delta as a man-made delta is the result of extensive water engineering of the Lower Atchafalaya River (Fig. 11.24). In the US topographic map from 1963 (a) we have marked the lakes in the lower Atchafalaya Basin by a dark blue polyline and transferred this outline to the flooding satellite image from 2020 (b). We notice that Six Mile Lake (6ML) and Flat Lake (FL) have been drained to become wetlands. They appear not to be flooded even during a normal spring flood. The reason is the diversion of substantial amounts of water from the Lower Atchafalaya River to the Wax Lake Outlet to protect the infrastructure at Morgan City (MC) from flooding. The same levee system also protects the settlements, infrastructure, farms, roads and rail lines along Bayou Teche (BT).

Infrastructure planning in coastal areas should, therefore, obtain guidance from historic maps, which show the historic morphology and water features, consider the large area flood planning and use multi-temporal satellite imagery to confirm the flood risk for areas that are protected by levee systems.

We complete this section with a detailed look at the flood protection of the city of Baton Rouge (Fig. 11.25). The natural color Landsat 8 432 RGB (a) shows that the Mississippi River in Baton Rouge has significantly less width than outside the city which would increase the flooding risk for the city. However, upstream of Baton Rouge a large area has been set aside to capture the flooding successfully as can be seen on the flooding image (b). The surplus of flood waters that the flooding area cannot capture has to be deviated to the Atchafalaya Basin at the Old River flood installation. Through this mitigation and levees along the Mississippi River downstream of Baton Rouge, the city and the settlements and farms along the river are protected at least from annual flooding.

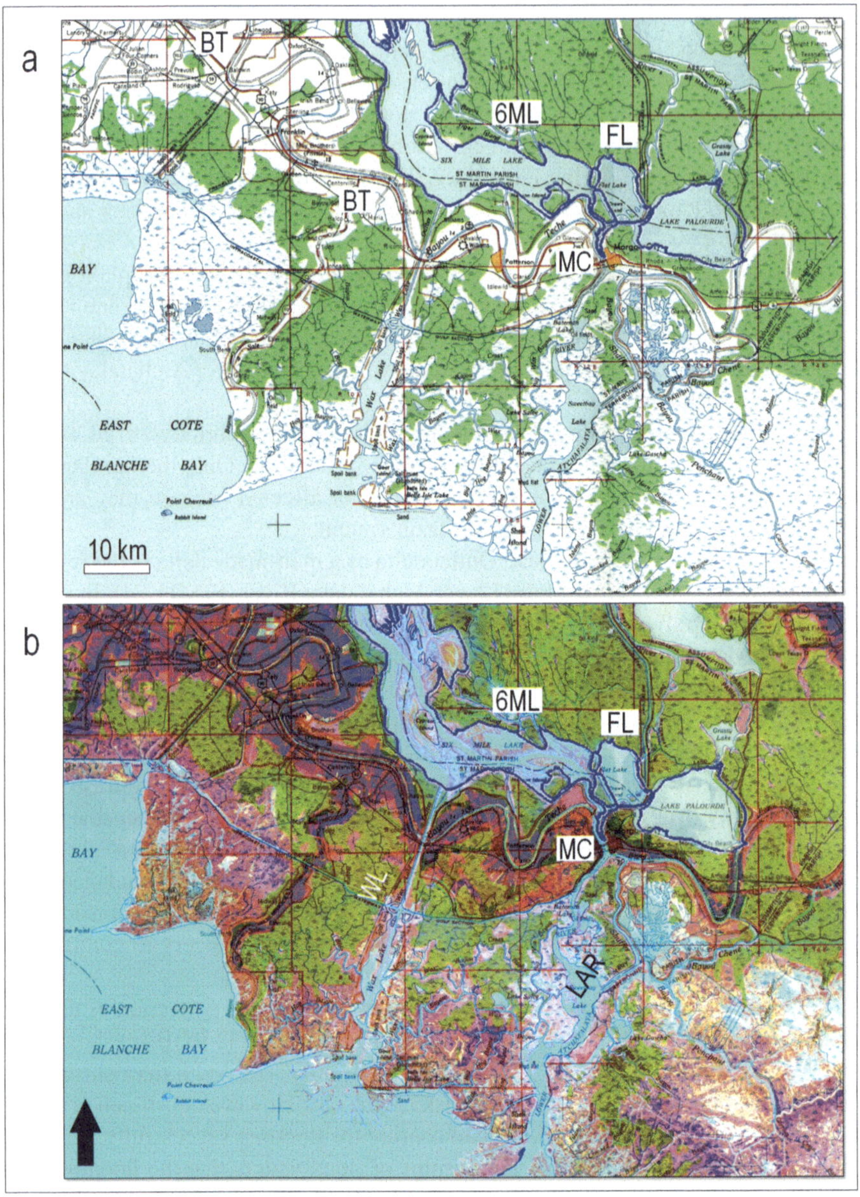

Fig. 11.24 Impact of water engineering on the morphology of the Lower Atchafalaya. (**a**) US topographic map from 1963, (**b**) Landsat 8 2-7-11 inverted RGB image taken during flooding on 28 Feb 2020. Outlines of 1963 water bodies as blue polylines, *MC* Morgan City. (Topographic map, NGOM DRM and satellite imagery courtesy of U.S. Geological Survey)

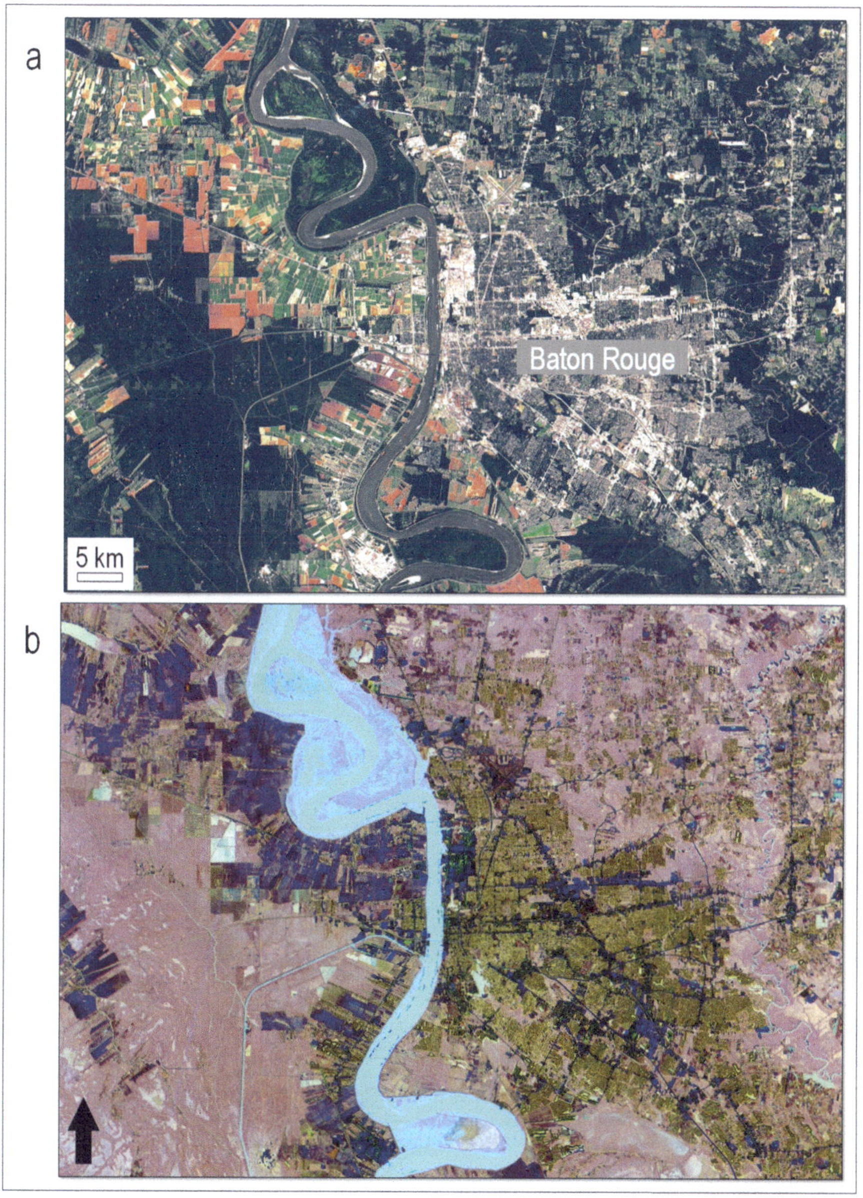

Fig. 11.25 Impact of Mississippi River flood control on the city of Baton Rouge. (**a**) dry season Landsat 8 natural color 432 RGB composite taken on 17 Oct 2017, (**b**) flooding image from Landsat 8 2-7-11 inverted RGB taken on 28 Feb 2020. (Satellite imagery courtesy of U.S. Geological Survey)

11.5 Subtropical Climate Chad

Sub-tropical climate is characterized by significant annual variations in precipitation (Armitage et al., 2015), which impacts infrastructure and logistics especially where it is undeveloped.

Typical for major sub-tropical settings is the monsoon that can provide substantial precipitation leading to the risk of flooding and consequently interruption of logistics. Understanding the seasonal climatic changes is, therefore, a key ingredient for the planning of infrastructure, the sourcing of supplies and logistics in general.

The Lake Chad basin in Central Africa is an example for an area the climate of which is dominated by the seasonal changes related to the West African monsoon. Figure 11.26 shows the geographic setting of the Lake Chad basin (a) as well as two elevation profiles in north-south (b) and west-east (c) direction. The basin is located between the Aïr, Tibesti and Ennedi mountains (from north-west to east) and

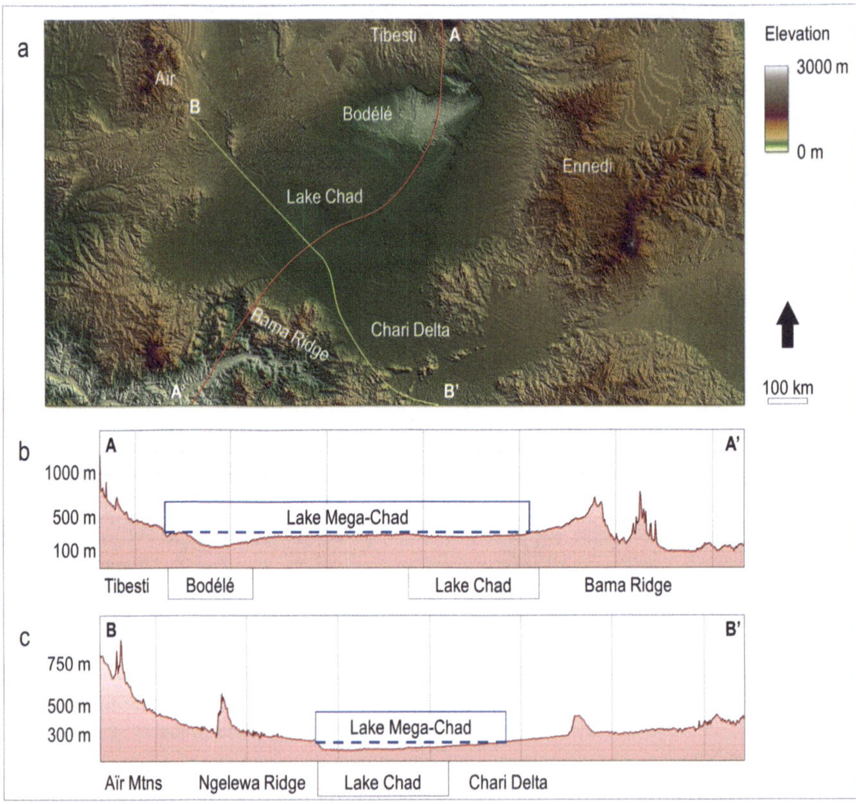

Fig. 11.26 Geographic setting of the Lake Chad Basin. (**a**) digital elevation model, (**b**) north-south elevation profile, (**c**) west-east elevation profile. (GMTED DRM data courtesy of U.S. Geological Survey, elevation profiles courtesy GoogleEarth)

separated from the Niger basin in the south-west by the Bama Ridge. The basin itself has very smooth topography with elevation changes of less than 50 m with the exception of the Bodélé Depression. This implies that Lake Chad is very shallow, minor variations in the lake level of Lake Chad have a significant impact on the size of the lake.

The Lake Chad basin is one of the prominent areas where long-term and short-term climate change have caused major desertification of large parts of the basin. For the period of 15,000–5000 YBP, Armitage et al. (2015) have shown that the climate had been much wetter than today resulting in the monsoon extending not only across the entire area of present-day Chad but also across the entire Sahel zone (Fig. 11.28). Since the Lake Chad basin is an isolated intra-cratonic basin, significantly increased precipitation from the monsoon had created Lake Mega-Chad, at that time the largest lake in Africa (blue outline in Fig. 11.27). The lake level was only controlled by a canyon cut into the Bama Ridge in the southwest. At around 5000 YBP the wet period came to a sudden end. The northern extent of the monsoon moved south by almost 1000 km, which dramatically reduced precipitation resulting in the lake level to drop by about 80 m. That drop was enough to isolate the lakes in the Bodélé Depression in the north from present-day Lake Chad in the south (light grey outline).

Since about 1000 YBP the desertification accelerated, shrinking Lake Chad to its present size (green outline) and creating one of the driest deserts on earth in the Bodélé Depression. We will look at the recent development of Lake Chad at the end of this section.

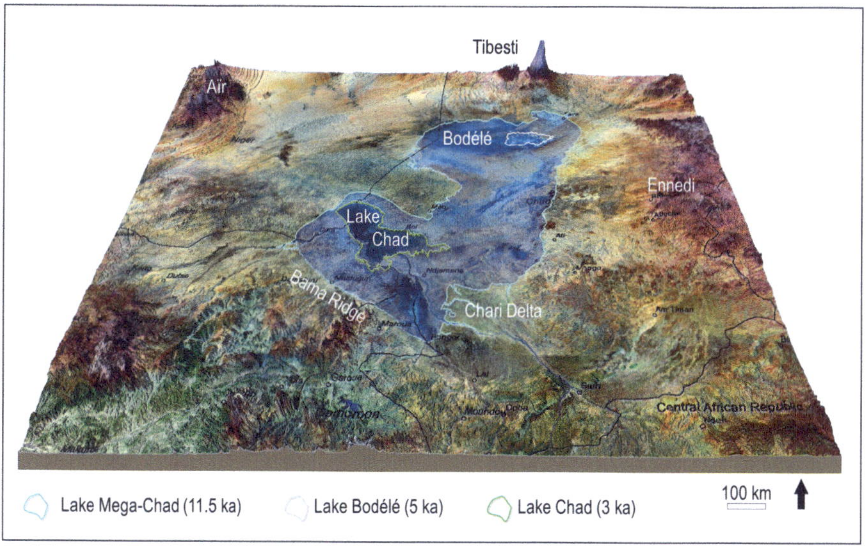

Fig. 11.27 Paleo-geography and extent of Lake Mega-Chad in the Lake Chad Basin. (GMTED2010 DRM and Landsat 5 satellite imagery courtesy of U.S. Geological Survey, outlines of paleo-lake contours after Armitage et al., 2015)

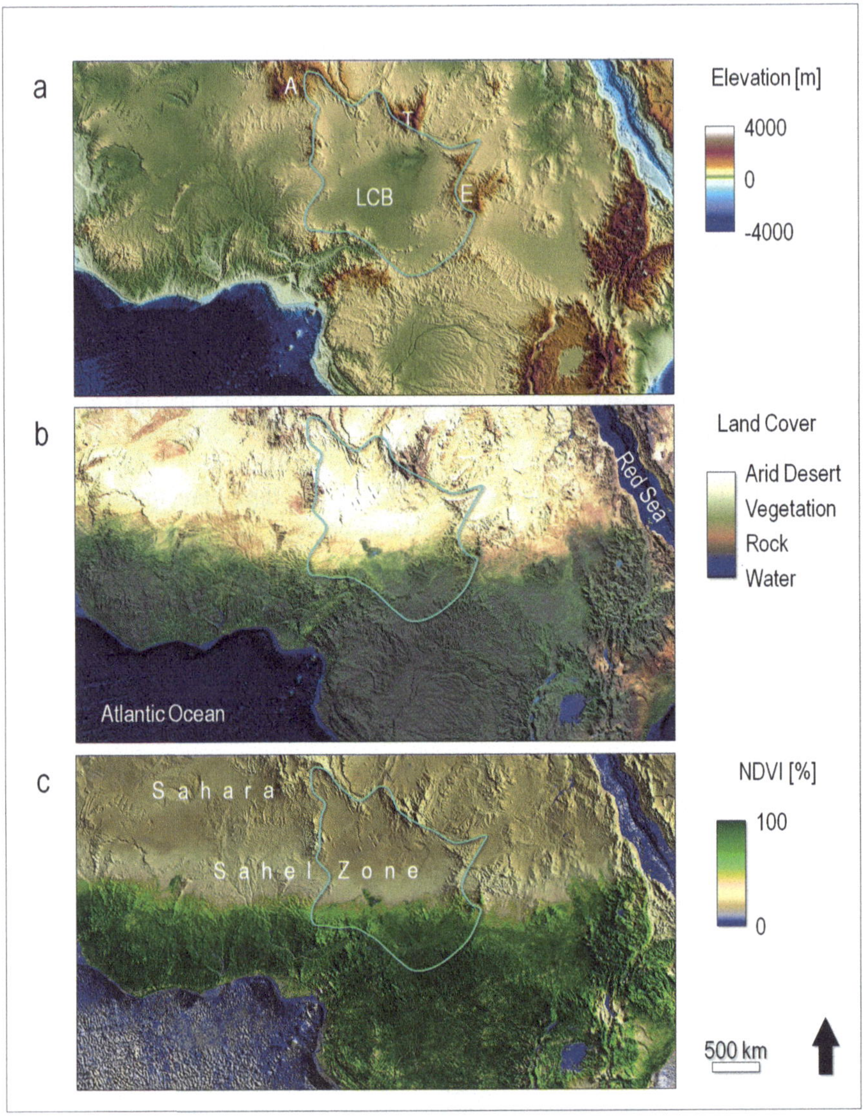

Fig. 11.28 Present-day geographic and climatic setting of the Lake Chad Basin in the Central African monsoon belt. (**a**) DRM, (**b**) Landsat 5 landcover image, (**c**) MODIS NDVI vegetation index for October 2019. *LCB* Lake Chad Basin, Mountains: *A* Aïr, *T* Tibesti, *E* Ennedi. (GMTED DRM and Landsat 5 satellite imagery courtesy of U.S. Geological Survey, MODIS NDVI data from Didan, 2015, NASA)

We will focus now on the present climatic setting of the Lake Chad basin. The location as a land-locked basin in central Africa (cyan outline in Fig. 11.28a) makes it dependent on rain that is generated by large regional climatic processes.

The catchment of Lake Chad and the rivers flowing into it is about 2.3 million km^2 and comprises parts of Sudan, Chad and parts of Niger, Nigeria and Algeria. Today 60% of this area is hot arid desert with only minimal precipitation and limited vegetation cover as shown in the land cover satellite image (b). A MODIS satellite image from the month of October, that has been processed for the normalized difference vegetation index (NDVI), shows the extent of dense vegetation at that time of year. The Sahel zone to the north of the vegetated area represents the transition from humid climate to desert climate without vegetation. From a planning point of view, the Sahel Zone coincides with the zone where water is scarce and water supply needs to be factored into the planning process.

For the assessment of the flooding risk associated with the monsoon, we map the extent of the area that is exposed to seasonal changes (Fig. 11.29). From a series of monthly MODIS NDVI images the zone affected by the monsoon can be mapped (bright green zone). It covers the southern half of Chad's territory. The northern half between Ndjamena and Faya does not receive monsoon and hence is a hot and arid desert. In the Congo Basin south of the monsoon zone the tropical climate provides precipitation around the entire year.

We complete the regional assessment with a series of bi-monthly NDVI images (Fig. 11.30) that reveal the extent and intensity of the rain and its impact on

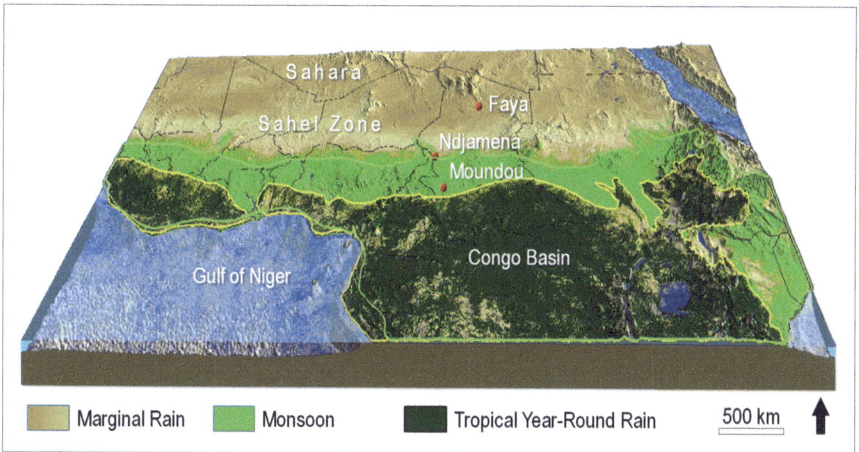

Fig. 11.29 Climatic zones of the Central Africa monsoon belt indicating temporal variation of moisture and flooding potential. (GMTED2010 DRM courtesy of U.S. Geological Survey, MODIS NDVI data from Didan, 2015, NASA)

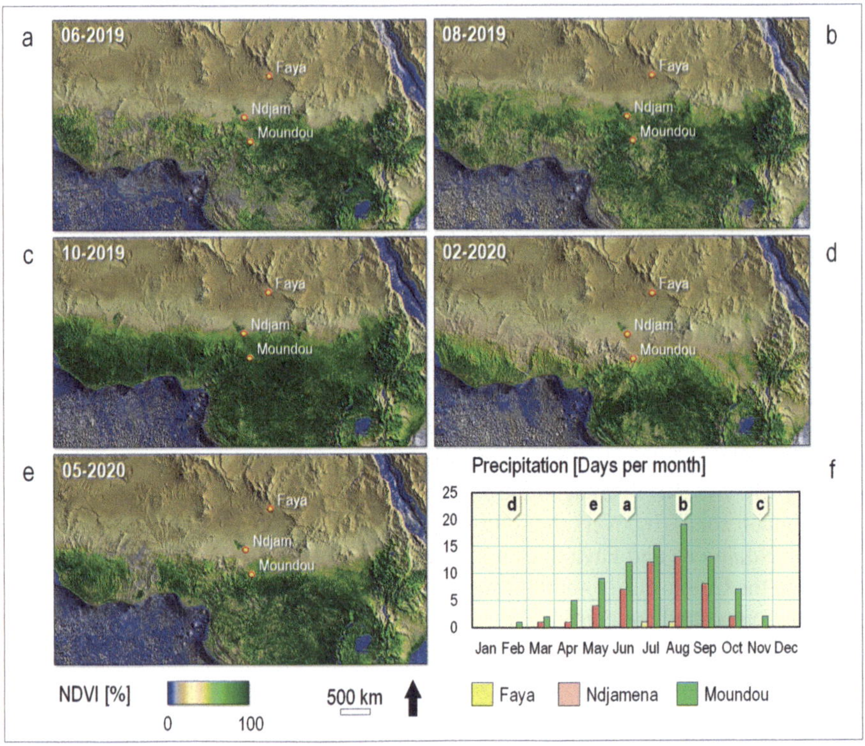

Fig. 11.30 Annual climatic variations shown from MODIS NDVI data draped on the DRM and correlated with the climatic diagram. (**a**) 06-2019, (**b**) 08-2019, (**c**) 10-2019, (**d**) 02-2020, (**e**) 05-2020. Ndjam–Njdamena. (GMTED2010 DRM courtesy of U.S. Geological Survey, climatic data from https://www.climatestotravel.com/climate/chad [accessed 23 Aug 2020])

vegetation. During the dry season from November to March the precipitation is limited to the tropical areas of the Congo Basin and the Guinea coast. With the onset of the monsoon the vegetation expands north up to about Ndjamena, with a slight time delay between the onset of the monsoon and the start of the vegetation growth. The maps in (a) to (e) show NDVI averages over 2 weeks. There relation to the precipitation curve is indicated by related arrows.

In summary of the regional observations, the region north of Ndjamena does not receive precipitation from the monsoon and is not affected by flooding. However, due to the arid and hot climate water scarcity needs to be considered during planning. The southern half of Chad is affected by the monsoon, which provides maximum precipitation from July to September. During this time the entire region should be considered subject to flooding.

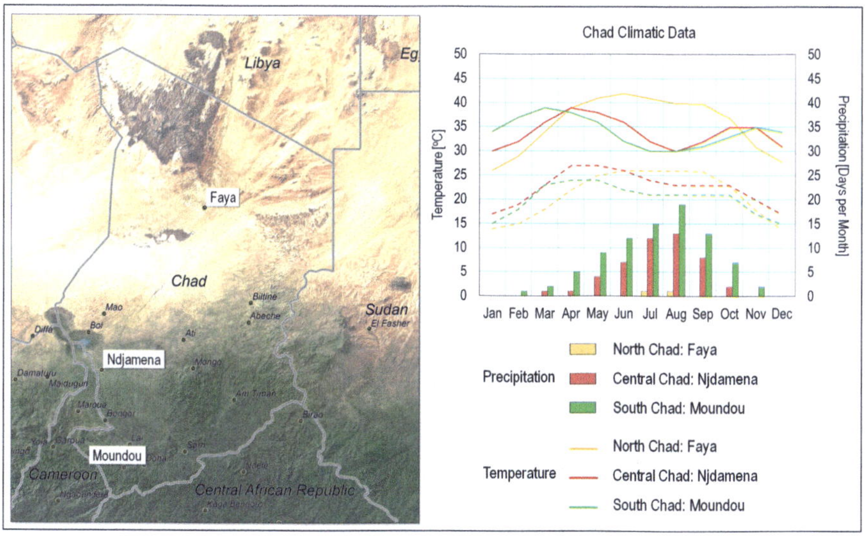

Fig. 11.31 Climatic zones of southern Chad. Landsat 5 satellite imagery blended with GMTED2010 DRM. (DRM and Landsat imagery courtesy of U.S. Geological Survey, climatic data from https://www.climatestotravel.com/climate/chad [accessed 23 Aug 2020])

Focusing now on the territory of Chad, we study the monthly changes in temperature and precipitation as number of rain days per month (Fig. 11.31). The monsoon season starts in June, reaches its peak in July and August and ends end September. Most of the precipitation falls between July and September whereas the precipitation from November to March is negligible. The high maximum summer temperatures in Faya indicate that it is not affected by the monsoon, whereas Ndjamena and Moundou clearly show the effect of cloud cover and precipitation during the peak of the monsoon season. The difference in precipitation between Ndjamena and Moundou in the far south is not significant in terms of planning. For the entire southern part of the country, the peak monsoon period in July and August needs to be considered for logistics planning.

In Fig. 11.32 we have collected bi-monthly MODIS NDVI images centered around the monsoon season 2019/2020 (c–h). For orientation we have overlaid the subsurface basin outlines on the digital elevation model (a) and on the landcover satellite image from October (b). Yellow dots indicate cities for orientation. The zone with intense vegetation grows from June to August, peaks in October and then retreats quickly to the tropical zone with the dry season peaking in March. The flooding risk prevails especially in the Chari River floodplain between Sarh, Bongor and Njdamena between June and October.

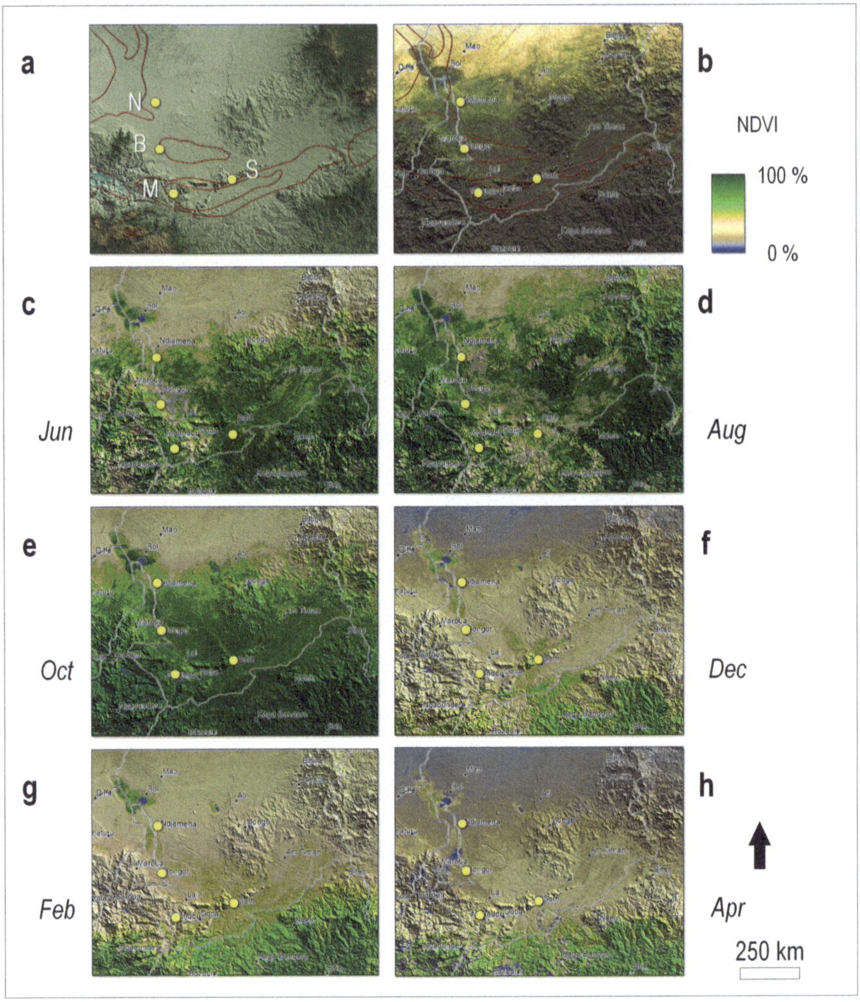

Fig. 11.32 Seasonal climatic variation across southern Chad obtained from MODIS NDVI. (**a**) DRM with outline of subsurface basins, (**b**) satellite image draped on DRM, (**c**) 06-2019, (**d**) 08-2019, (**e**) 10-2019, (**f**) 12-2019, (**g**) 02-2020, (**h**) 04-2020. *N* Njdamena, *B* Bongor, *M* Moundou, *S* Sarh. (GMTED2010 DRM and satellite imagery courtesy of U.S. Geological Survey, climatic data from https://www.climatestotravel.com/climate/chad [accessed 23 Aug 2020])

11.6 Desert Environment, Grand Erg Oriental Tunisia

Sand dune migration and encroachment on settlements and infrastructure is a major risk in desert environment across the globe. Multi-temporal remote sensing using optical bands over a period of minimum 10 years has been established as the preferred method for the monitoring sand dune migration and prediction of their future movements (Baird et al., 2019; Els et al., 2015; Engel et al., 2018; Hereher, 2010; Hermas et al., 2012; Maghsoudi et al., 2017; Yao et al., 2007). Image processing methods for the extraction have been proposed by Ali et al. (2020), Sparavigna (2013), and Varma et al. (2014). Remote sensing observations may be compared against numerical modeling (Zhang et al., 2012). Using this method, Vermeesh and Drake (2008) detected and monitored the fastest migrating barchan dunes in the Bodélé Depression. Recently, the availability of multi-temporal LiDAR opened a new avenue for high-resolution sand dune tracking over shorter time spans (Xia & Dong, 2016). Sand dune encroachment on human settlements and infrastructure has been studied country-wide in Kuwait (Al-Helal & Al-Awadhi, 2006), Egypt (Ghadiry & Koch, 2010), Libya (Pradhan et al., 2018) and Tunisia (Afrasinei et al., 2018).

We start the remote sensing study with a morphologic and lithologic overview of the study area, collect multi-temporal imagery and co-render it as multi-temporal composite RGB images. These enable distinguishing the areas of surface changes from migrating sand dunes from those that remain stationary. Then we correlate migrating sand dunes with the geomorphology and create a sand dune encroachment map. We complete the study with the measurement of direction of migration and celerity using the multi-temporal RGB images.

For a study we have chosen the south east Grand Erg in southern Tunisia (Fig. 11.33). The digital elevation model (a) shows sand dune patterns of the Grand Erg Oriental in the west and rocky outcrops of Al Hamra in the east. This is confirmed by the satellite image (b) processed for classification of the surface cover using the optical multiband composite of Landsat 8 432 RGB. The elevation profile (c) indicates that the study area is located on the slope from the Al Hamra ranges in the east to the dune-covered plains of the Grand Erg Oriental in the west.

The Grand Erg shows regular patterns composed mainly of linear seif dunes. A 50 m high dune belt (arrow in c) marks the edge of the dune field towards the gravel plain at the foot of the Al Hamra ridge which may indicate that perimeter of the dune field is stationary. The limestone outcrops of the Al Hamra ridges (dark areas in b) are partially covered by linear sand patches (light areas) running in East-West direction. We will study these patches for mobile sand.

The oilfield infrastructure such as drilling rigs, well pads and regional gathering centers (facilities) as well as airstrips for access and accommodation for service personnel are located on the gravel plains (Fig. 11.34). The main road and a pipeline run almost parallel to the boundary of the dune field and cross several of the sand patches and meandering wadis.

The focus of the study is to detect mobile sand and to predict which parts of the infrastructure are potentially affected by sand encroachment. To assess this challenge, we use Landsat multi-temporal imagery rendered as optical multispectral 432 RGB images taken in December of the years 1984 (Landsat 5), 1995 (Landsat 5), 2005 (Landsat 7) and 2016 (Landsat 8, Fig. 11.35). This analysis is possible because of the Landsat data continuity mission (USGS, 2020) which ensured the continued acquisition of similar spectral bands over decades.

Since the visual comparison of the imagery spanning 31 years does not reveal obvious differences even over periods of 10 years, we use the technique of multi-temporal RGB rendering (Fig. 11.36). RGB satellite images from three different dates are used as input. We need to convert the individual images to grayscale to

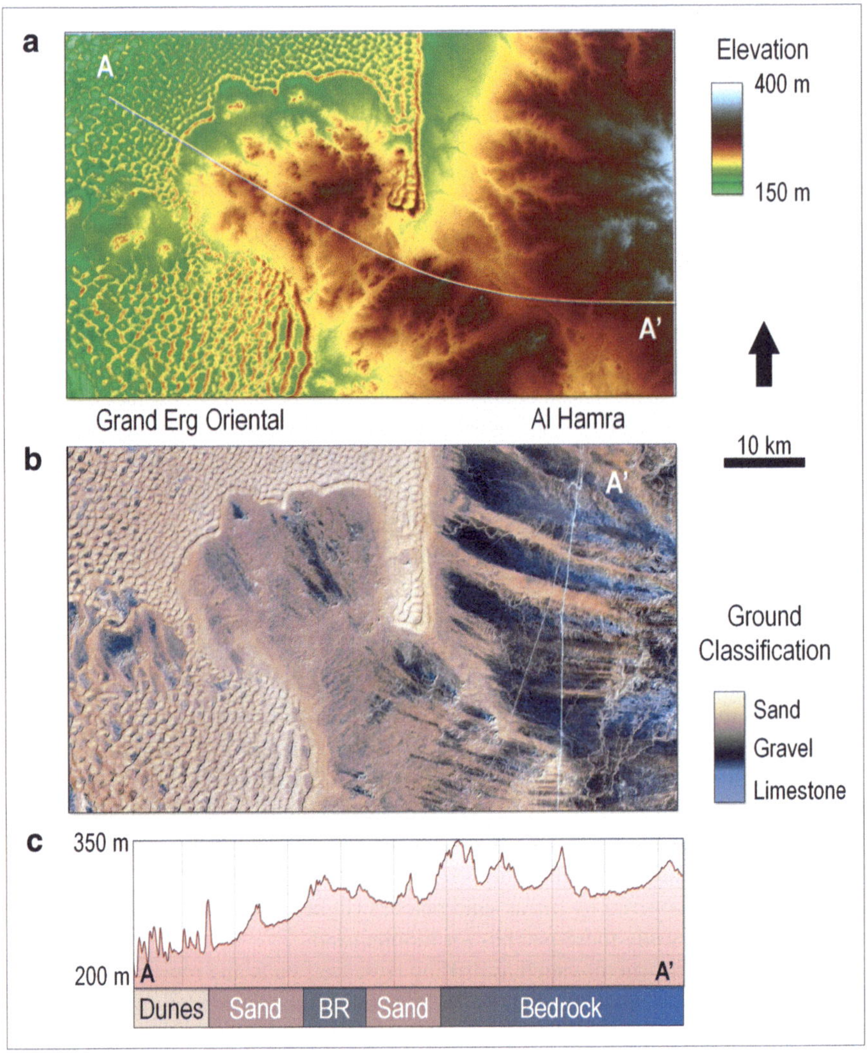

Fig. 11.33 Geographic setting of mobile sand dune belt in Grand Erg Oriental, Tunisia. (**a**) DRM, (**b**) Landsat 8 composite 432 RGB image, (**c**) west-east elevation profile. (DRM and satellite imagery courtesy of U.S. Geological Survey, elevation profile courtesy of Google Earth)

enable creating a balanced multi-temporal composite. To achieve this, we can either select the band most sensitive to temporal changes or convert the RGB image to a grayscale image.

Prior to loading them into the color channels of the multi-temporal RGB image, the grayscale images should be balanced to a similar RMS value across the entire image to achieve the best detection of temporal changes. The interpretation of the individual colors is shown in the matrix on the right-hand side.

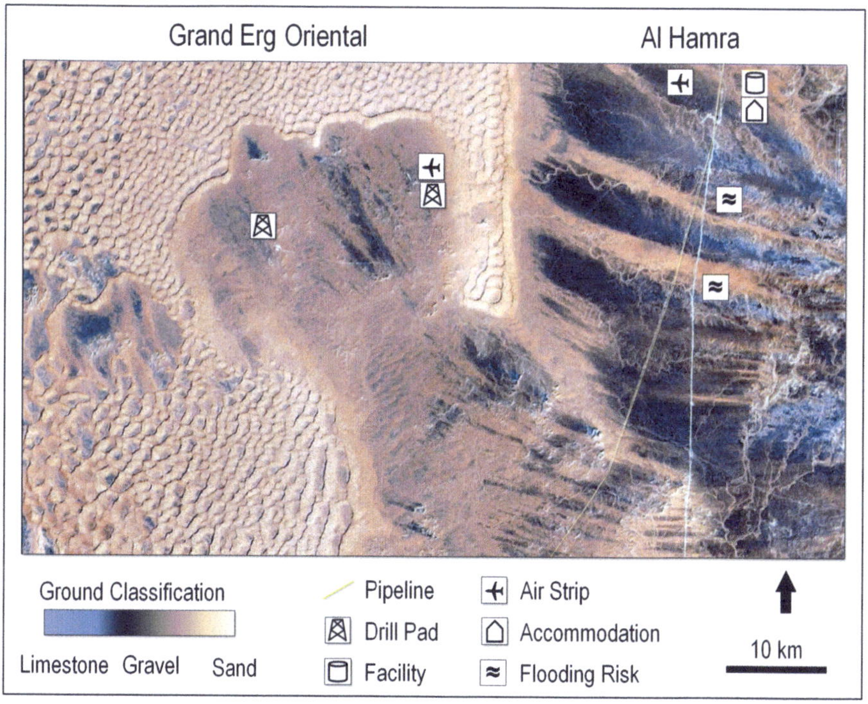

Fig. 11.34 Grand Erg—Oilfield infrastructure in study area shown on Landsat 8 432 RGB composite image. (Landsat 8 imagery courtesy of US Geologic Survey)

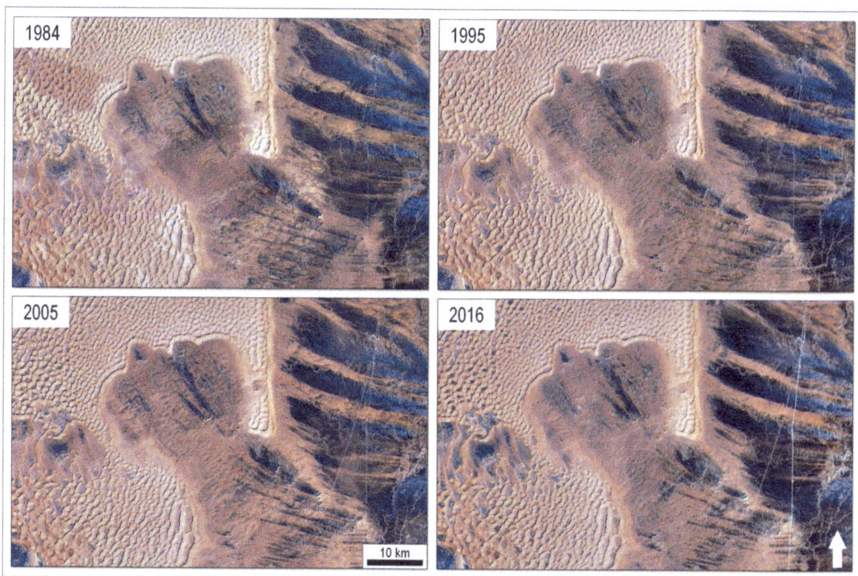

Fig. 11.35 Multi-temporal satellite images of the study area. Landsat 321 RGB and Landsat 432 RGB composite images, respectively, for 1984, 1995, 2005, and 2016. (Landsat imagery courtesy of U.S. Geological Survey)

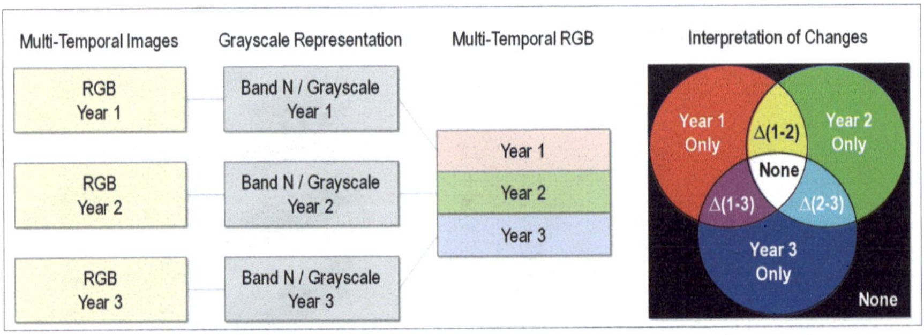

Fig. 11.36 Technique of multi-temporal image RGB composite

For the multi-temporal composite in our study we have chosen the images from 1984, 1995 and 2016 (Fig. 11.37). The RGB composite (a) reveals the exposed rock outcrops of the Al Hamra ridges, select areas of the gravel plains and most of the seif dunes in the Grand Erg Oriental have been stable over more than 30 years as indicated by either very bright or very dark colors. Particularly, the front of the dune field (bright areas) has been remarkably stable which indicates a very stationary wind direction in this area (Vermeesh & Drake, 2008; Zhang et al., 2012). The stable areas of the gravel plain (dark areas) are mostly correlated with exposed elevated ridges, where a higher wind velocity does not support the deposition of sand.

Colored areas in the multi-temporal composite image indicate changes that are interpreted as sand movement. These occurred across large areas of the gravel plain. In the center of the image, areas of sand movement follow mainly the orientation of the dark ridges of exposed rock which appear to channel the wind direction. We observe the same trend in the north east, where mobile east-west trending sand patches follow the topography.

We can consolidate these observations in a sand migration risk map (b) where reddish tones mark mobile sand areas and green tones indicate mostly stationary areas. We have draped the risk map on the elevation model to highlight the correlation of mobile sands with relative lows in the topography.

We can assess the correlation between mobile sand and topography (Fig. 11.38). The overview (a) indicates the direct correlation of mobile sand with lows in the topography, whereas exposed ridges do support sand cover (dark areas). The present-day optical composite (b) indicates exposed limestone on the Al Hamra ridges and sand in the valleys. The multi-temporal composite RGB highlights that these sand patches are highly mobile.

The infrastructure is located outside the main mobile sand zones. Airport and accommodation in the north (area 1) use a limestone ridge, but the colors in the multi-temporal composite indicate a sand migration risk. The main road in north-south direction crosses several mobile sand patches and is most likely subject to significant maintenance to mitigate the drifting sand.

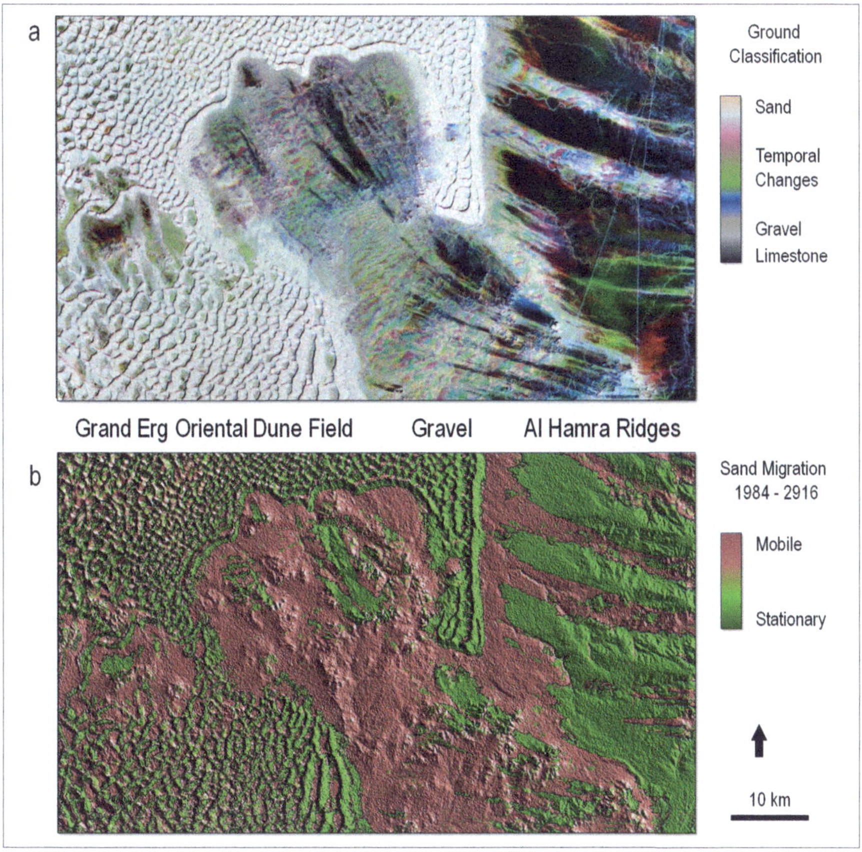

Fig. 11.37 Multi-temporal RGB composite and sand dune encroachment risk. (**a**) RGB composite of individual grayscale images acquired 1985, 2005, and 2016, (**b**) sand migration risk for the time between 1985 and 2016 draped on DRM. (DRM and Landsat imagery courtesy of U.S. Geological Survey)

The airstrip and drill pad in area 2 are exposed to sand migration, especially related to sand dune encroachment as shown by the colored sand dune patches (c). To mitigate this challenge, the access road has been designed to follow the gravel ridge, the exposure of which prevents the accumulation of sand. The area of the drill pad south west of the airstrip shows exposure to drifting sand. The access road, which crosses several patches of significant sand migration, may need to be relocated from time to time to accommodate the sand migration.

We summarize our findings on the 3D sand migration risk map (Fig. 11.39), which is composed of the risk map itself draped on the digital elevation model. Ridges and limestone outcrops yield the lowest sand migration risk, whereas lows between the ridges expose a high risk. In addition to the sand migration risk, these lows most likely exhibit flash flood risk should occasional rain fall in the area.

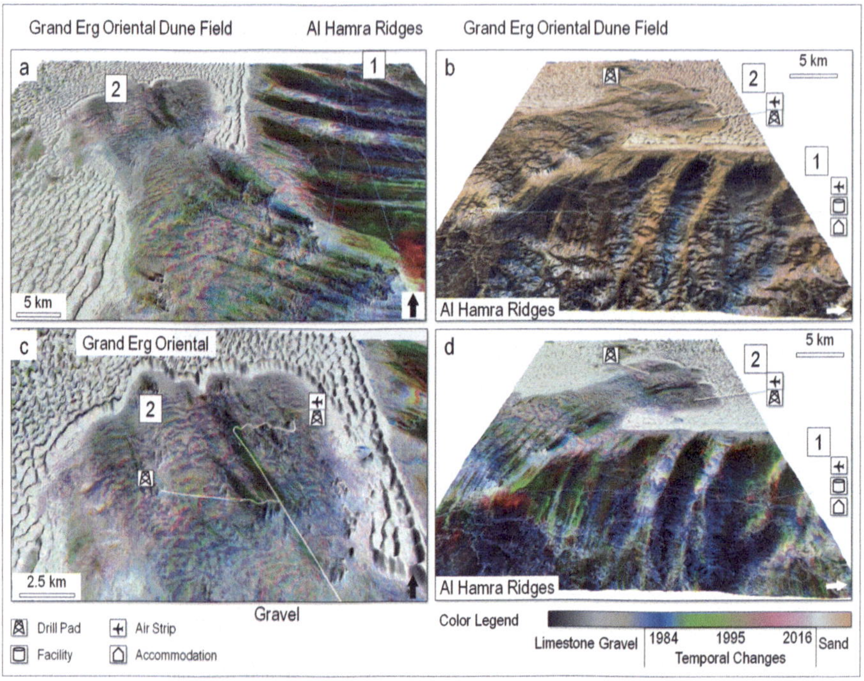

Fig. 11.38 Zoom into selected area of mobile sand using 1985, 1996 and 2016 imagery. (**a**) overview of study area, (**b**) 2016 optical 432 RGB composite image of northern area with infrastructure, (**c**) mobile sand area in the center of the study area, (**d**) multi-temporal RGB image corresponding to (**b**), multi-temporal RGB unless specified otherwise, all images are draped on the DRM. (RDM and Landsat imagery courtesy of U.S. Geological Survey)

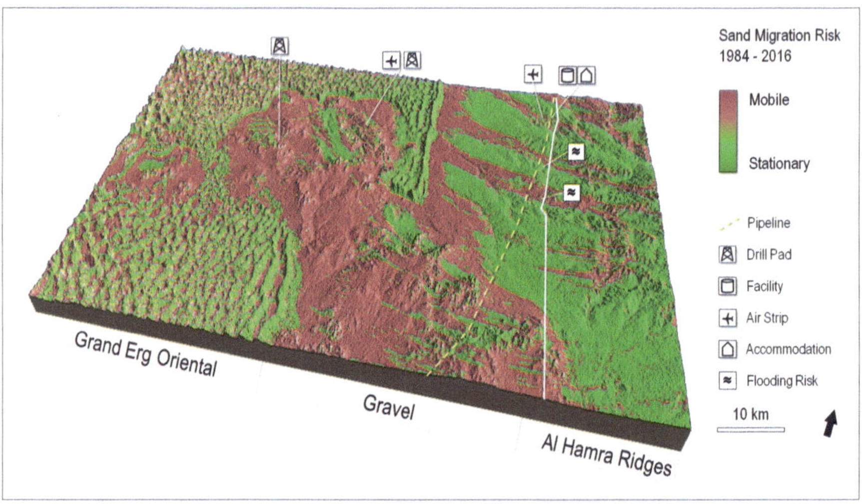

Fig. 11.39 Sand migration risk for oilfield installations and infrastructure

We conclude the study with a quantitative assessment of the sand encroachment risk (Fig. 11.40). For the multi-temporal composite image for area 2 (a) we have chosen the images from 1984, 2005 and 2016 to study both, the long-term and the

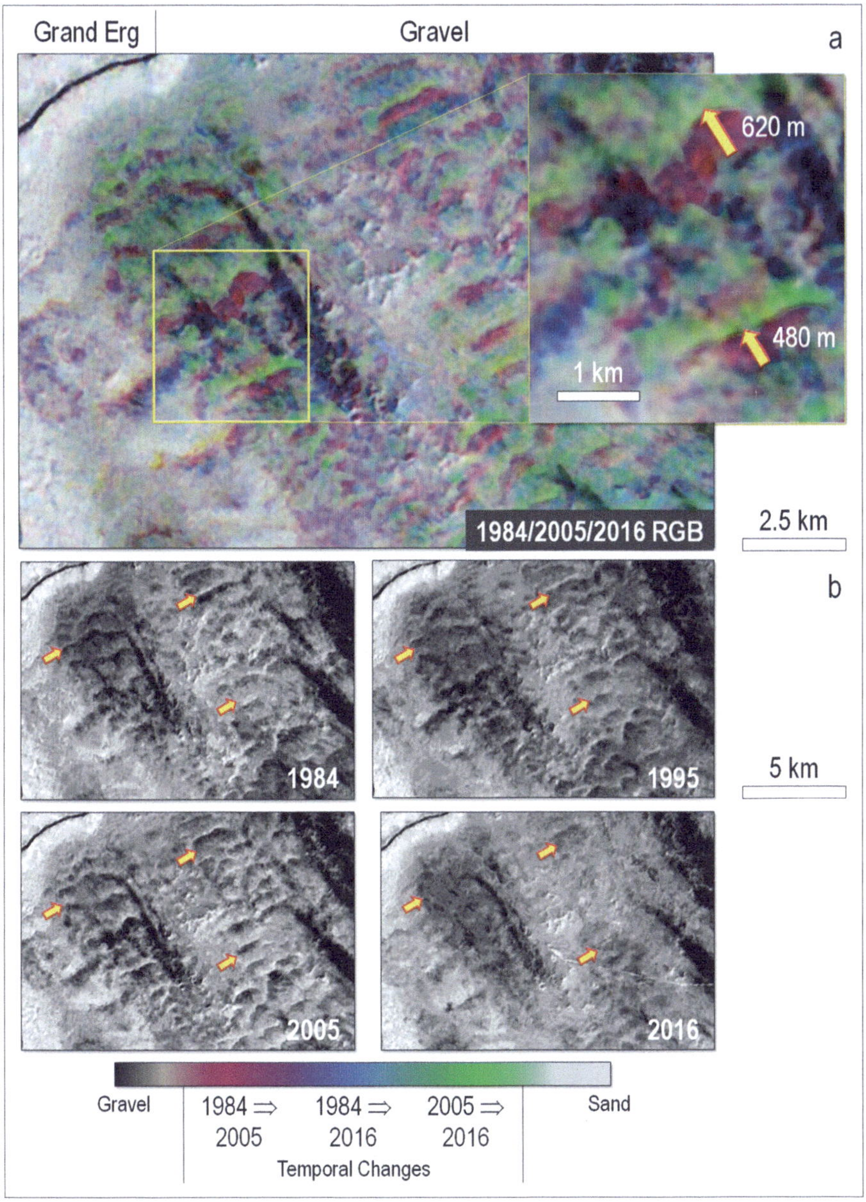

Fig. 11.40 Measuring the direction and amount of sand encroachment over 32 years for area 2. (**a**) 1984–2005–2016 Landsat multi-temporal RGB composite image, insert shows the sand dune migration in 32 years, (**b**) single time Landsat grayscale image used to generate the multi-temporal RGB image in (**a**). (Satellite imagery courtesy of U.S. Geological Survey)

intermediate changes. The areas of long-term change between 1984 and 2016 are represented in purple. The width of these patches gives clues for the amount and direction of the dune migration. We observe migration intervals from 480 to 620 m over 32 years which corresponds to a dune celerity of 15–20 m/year. At these values of celerity, the planning of permanent oilfield infrastructure, accommodation and airstrips should consider the risk map shown in Fig. 11.37. The 3D risk map (Fig. 11.39) and the celerity information obtained from Fig. 11.40 provides input from long-term prediction of mobile sands and sand dune migration. The risk maps also provide information for the planning of roads and pipelines, especially for the prediction of the amount of maintenance that is necessary to protect pipelines from sand encroachment and to keep roads free of sand. To minimize the maintenance related to sand dune encroachment, priority during planning should be given to areas showing dark colors in the multi-temporal RGB image, since, in addition to low sand migration risk, dark color indicates also indicate a gravel surface cover which is advantageous for road construction.

The satellite images for the individual years (Fig. 11.40b) show certain characteristic dune features that can assist the selection for the estimation of the dune celerity and direction (yellow arrows).

In summary, the risk from sand dune migration and sand encroachment can be detected and quantified from multi-temporal satellite imagery. The co-rendering of multi-temporal satellite images in a composite RGB image provides insights into the long-term celerity and direction of sand dune migration and allows creating a sand dune migration risk map. Critical to the significance of the colors in multi-temporal RGB map is the preprocessing of the input satellite images. Seasonal and atmospheric anomalies should be removed. For optimum co-rendering in RGB, the input images should either be converted to grayscale or the band with the highest sensitivity to the change should be chosen. Rendering the results in 3D provides the added benefit of correlating the sand migration features to the topography, which assists long-term forecasting.

References

LiDAR DEM

Audubon Alaska (2017). assessment of wildlife values in the Teshekpuk Lake wetlands complex within the National Petroleum Reserve – Alaska (NPRA), https://ak.audubon.org/sites/default/files/special_report_on_teshekpuk_lake_wildlife_values_2017_dec2017-final.pdf [accessed 1 Feb 2021].

NOAA. (2012). *Lidar 101: An introduction to Lidar technology, data and applications, revised.* National Oceanic and Atmospheric Administration (NOAA), Coastal Service Center.

Remote Sensing Case Study Teshekpuk, Alaska, U.S.A.

Chambers, M., Lilly, M., White, D., Hilton, K., & Prokein, P. (2006). *Lake survey data for the Kuparuk Foothills Region: Spring 2006, Kuparuk Foothills Hydrology Project, Report No. INE/WERC 06.05.*

Ferrians, O. (1998). *Permafrost map of Alaska, USA, Version 1*. NSIDC: National Snow and Ice Data Center.

Gatto, L. W. (1995). *Soil freeze-thaw effects on bank erodibility and stability*. US Army Corps of Engineers, Cold Regions Research & Engineering Laboratory, Special Report 95-24.

Gryc, G. (Ed.). (1988). *Geology and exploration of the National Petroleum Reserver in Alaska, 1974–1982*. U.S. Geological Survey Professional Paper 1399.

Jones, B. M., Grosse, G., Arp, C. D., Jones, M. C., Anthony, K. M. W., & Romanovsky, V. E. (2011). Modern thermokarst lake dynamics in the continuous permafrost zone, northern Seward Peninsula, Alaska. *Journal of Geophysical Research, 116*, G00M03. https://doi.org/10.1029/2011JG001666

Mars, J. C., & Houseknecht, D. W. (2007). Quantitative remote sensing study indicates doubling of coastal erosion rate in past 50 yr along a segment of the Arctic coast of Alaska. *Geology, 35*(7), 583–586. https://doi.org/10.1130/G23672A.1

Urban, F. E., & Clow, G. D. (2014). *DOI/GTN-P climate and active-layer data acquired in the National Petroleum Reserve—Alaska and the Arctic National Wildlife Refuge, 1998–2011: U.S. Geological Survey data series* (Vol. 812). USGS. https://doi.org/10.3133/ds812. ISSN 2327-638X.

Operations Case Studies in Arctic Climate

Arp, C. D., Whitman, M. S., Jones, B. M., Nigro, D. A., Alexeev, V. A., Gaedeke, A., Fritz, S., Daanen, R., Liljedahl, A. K., Adams, F. J., Gaglioti, B. V., Grosse, G., Heim, K. C., Beaver, J. R., Cai, L., Engram, M., & Uher-Koch, H. R. (2019). Ice roads through lake-rich Arctic watersheds: Integrating climate uncertainty and freshwater habitat responses into adaptive management. *Arctic, Antarctic, and Alpine Research, 51*(1), 9–23. https://doi.org/10.1080/15230430.2018.1560839

Nitze, I., Grosse, G., Jones, B. M., Romanovsky, V. E., & Boike, J. (2018). Remote sensing quantifies widespread abundance of permafrost region disturbances across the Arctic and Subarctic. *Nature Communications, 9*, 5423. https://doi.org/10.1038/s41467-018-07663-3

Remote Sensing Case Study White River Floodplain, Arkansas, U.S.A.

Arkansas GIS Office. (2018). *DEM 1 m*. http://gis.arkansas.gov/product/dem-1m-2018/

USACE. (2014). *Mississippi River and tributaries project, Levee system evaluation report for the Nation Flood Insurance Program*. U.S. Army Corps of Engineers, Mississippi Valley Division. https://www.mvd.usace.army.mil/Portals/52/docs/MRC/LeveeSystem/MRT_levee_system_eval_report_for_NFIP.pdf

Operations Case Studies in Floodplains

Courty, L. G., Soriano-Monzalvo, J. C., & Pedrozo-Acuna, A. (2019). Evaluation of open-access global digital elevation models (AW3D30, SRTM, and ASTER) for flood modelling purposes. *Journal of Flood Risk Management, 12*(Supp. 1), e12550. https://doi.org/10.1111/jfr3.12550

Schumann, G. J.-P., & Bates, P. D. (2018). The need for a high-accuracy, open-access global DEM. *Frontiers in Earth Science, 6*, 225. https://doi.org/10.3389/feart.2018.00225

Remote Sensing Case Study Louisiana Coast, U.S.A.

Bernier, J. C., Morton, R. A., & Barras, J. A. (2006). Constraining rates and trends of historical wetland loss, Mississippi River Delta Plain, south-central Louisiana. In Y. J. Xu & V. P. Singh (Eds.), *Coastal environment and water quality* (pp. 371–382). Water Resources Publications.

Morton, R. A., Bernier, J. C., Barras, J. A., & Ferina, N. F. (2015). *Rapid subsidence and historical wetland loss in the Mississippi Delta plain: Likely causes and future implications, U.S. Geological Survey Open-File Report 2005-1216*, 124 p.

NOAA. (2014). *2014 USGS CoNED Topobathymetric DEM (1888 to 2013): Northern Gulf of Mexico.* https://coast.noaa.gov/htdata/raster2/elevation/Louisiana_Coned_Topobathy_DEM_2014_4958/

Remote Sensing Case Study Chad

Armitage, S. J., Bristow, C. S., & Drake, N. A. (2015). West African monsoon dynamics inferred from abrupt fluctuations of Lake Mega-Chad. *Proceedings of the National Academy of Sciences, 112*, 8543. https://doi.org/10.1073/pnas.1417655112

Operations and Subsurface Case Studies in Subtropical Setting

Didan, K. (2015). *MOD13Q1 MODIS/Terra Vegetation Indices 16-Day L3 Global 250m SIN Grid V006 [Data set]*. NASA EOSDIS Land Processes DAAC. https://doi.org/10.5067/MODIS/MOD13Q1.006

Remote Sensing Case Study Grand Erg Oriental, Tunisia

Afrasinei, G. M., Melis, M. T., Arras, C., Pistis, M., Buttau, C., & Ghiglieri, G. (2018). Spatiotemporal and spectral analysis of sand encroachment dynamics in southern Tunisia. *European Journal of Remote Sensing, 51*(1), 352–374. https://doi.org/10.1080/22797254.2018.1439343

Operations Case Studies in Desert Setting

Al-Helal, A., & Al-Awadhi, J. (2006). Assessment of sand encroachment in Kuwait using GIS. *Environmental Geology, 49*, 960–967. https://doi.org/10.1007/s00254-005-0134-8

Ali, E., Xu, W., & Ding, X. (2020). Improved optical image matching time series inversion approach for monitoring dune migration in North Sinai Sand Sea: Algorithm procedure, application, and validation. *ISPRS Journal of Photogrammetry and Remote Sensing, 164*, 106–124. https://doi.org/10.1016/j.isprsjprs.2020.04.004

Baird, T., Bristow, C. S., & Vermeesch, P. (2019). Measuring sand dune migration rates with COSI-Corr and Landsat: Opportunities and challenges. *Remote Sensing, 11*, 2423. https://doi.org/10.3390/rs11202423

Els, A., Merlo, S., & Knight, J. (2015). Comparison of two satellite imaging platforms for evaluating sand dune migration in the Ubari Sand Sea (Libyan Fazzan). In *The International Archives of the Photogrammetry, Remote Sensing and Spatial Information Sciences, Volume XL-7/W3, 2015, 36th International Symposium on Remote Sensing of Environment, 11–15 May 2015, Berlin, Germany* (pp. 1375–1380). https://doi.org/10.5194/isprsarchives-XL-7-W3-1375-2015

Engel, M., Boesl, F., & Brueckner, H. (2018). Migration of barchan dunes in Qatar – Controls of the Shamal, teleconnections, sea-level changes and human impact. *Geosciences, 8*, 240. https://doi.org/10.3390/geosciences8070240

Ghadiry, M., & Koch, B. (2010). Developing a monitoring system for sand dunes migration on Dakhla Oasis, Western Desert, Egypt. In *30th EARSel Symposium: Remote Sensing for Science, Education and Culture, Paris 31 May - 3 June 2010* (pp. 313–320).

Hereher, M. E. (2010). Sand movement patterns in the Western Desert of Egypt: An environmental concern. *Environmental Earth Sciences, 59*, 1119–1127. https://doi.org/10.1007/s12665-009-0102-9

Hermas, E. S., Leprince, S., & El-Magd, I. (2012). Retrieving sand dune movements using sub-pixel correlation of multi-temporal optical remote sensing imagery, northwest Sinai Peninsula, Egypt. *Remote Sensing of Environment, 121*, 51–60. https://doi.org/10.1016/j.rse.2012.01.002

Maghsoudi, M., Navidfar, A., & Mohammadi, A. (2017). The sand dunes migration patterns in Mesr Erg region using satellite imagery analysis and wind data. *Natural Environment Change, 3*(1), 33–43. https://doi.org/10.22059/jnec.2017.225011.62

Pradhan, B., Moneir, A. A. A., & Jena, R. (2018). Sand dune risk assessment in Sabha region, Libya using Landsat 8, MODIS, and Google Earth Engine images. *Geomatics, Natural Hazards and Risk, 9*(1), 1280–1305. https://doi.org/10.1080/19475705.2018.1518880

Sparavigna, A. C. (2013). A case study of moving sand dunes: The barchans of the Kharga Oasis. *International Journal of Sciences, 2*, 95–97. https://doi.org/10.18483/ijSci.241

USGS. (2020). *Landsat data continuity mission history*. https://www.usgs.gov/land-resources/nli/landsat/landsat-data-continuity-mission-history?qt-science_support_page_related_con=0#qt-science_support_page_related_con

Varma, S., Shah, V., Banerjee, B., & Buddhiraju, K. M. (2014). Change detection of desert sand dunes: A remote sensing approach. *Advances in Remote Sensing, 3*, 10–22. https://doi.org/10.4236/ars.2014.31002

Vermeesh, P., & Drake, N. (2008). Remotely sensed dune celerity and sand flux measurements of the world's fastest barchans (Bodélé, Chad). *Geophysical Research Letters, 35*, L24404. https://doi.org/10.1029/2008GL035921

Xia, J., & Dong, P. (2016). A GIS add-in for automated measurement of sand dune migration using LiDAR-derived multitemporal and high-resolution digital elevation models. *Geosphere, 12*(4), 1316–1322. https://doi.org/10.1130/GES01329.1

Yao, Z. Y., Wang, T., Han, Z. W., Zhang, W. M., & Zhao, A. G. (2007). Migration of sand dunes on the northern Alxa Plateau, Inner Mongolia, China. *Journal of Arid Environments, 70*, 80–93. https://doi.org/10.1016/j.jaridenv.2006.12.012

Zhang, D., Narteau, C., Rozier, O., & Courrech du Pont, S. (2012). Morphology and dynamics of star dunes from numerical modelling. *Nature Geoscience, 5*, 463. https://doi.org/10.1038/ngeo1503

Chapter 12
Geologic Analogs

Abstract The interpretation of subsurface structures for hydrocarbon reservoirs from 3D seismic data is a challenging task since seismic reflections are not uniquely correlated with rock and fluid types of depositional environments. Geologic analogs assist building subsurface models and correlating them with present-day geological settings thus enabling the explorer to assess the risk of success when planning to drill certain structures. Following the exploration strategy of delineating the structural framework first and then interpreting the available data for lithology and depositional environments, we have structured the section on geologic analogs accordingly:

1. **Structural analogs**. These comprise analogs assisting the interpretation for regional structural styles such as rifting and pull-apart settings and delineation of regional faults through the alignment of karst features. Structural trap risking analogs comprise relay ramps and extensional fault settings. Salt tectonic analogs can be used for both, potential trap delineation as well as trap risking.
2. **Depositional analogs**. We sub-divide these into source analogs for example from algal mats and marshes, reservoir and seal analogs. For reservoir analogs we distinguish carbonate settings with reefs, carbonate platforms, shoals and tidal channels, whereas analogs for clastic settings contribute fluvio-marine settings from floodplains, deltas, tidal channels and fan deltas.
3. **Erosional analogs**. We distinguish between carbonate settings comprising different types of karsting such as dolines and fluvial karsting. Clastic settings are represented by various glacial settings from sub-glacial tunnel valleys, glacial valleys, sub-glacial features and moraines. Glacial analogs can also indicate the direction of the ice advance through analysis of the orientation of sub-glacial drumlins.

For each analog we assess the remote sensing method best suited to extract the geologic characteristics needed for subsurface basin and reservoir modeling. Often, we combine data sets and render them in specific ways to highlight the wanted characteristics.

Keywords Structural: relay ramp, extensional faulting, karst alignment to faults, salt tectonics, salt glacier, slat diapir, rifting, pull-apart basin, fold belt ·

A. Laake, *Remote Sensing for Hydrocarbon Exploration*, Springer Remote Sensing/Photogrammetry, https://doi.org/10.1007/978-3-030-73319-3_12

Depositional: carbonate reefs, sabkha, shoal, tidal channel, salina, algal marsh, lagoon, marine delta, intracratonic delta, fan delta, river floodplain, tidal floodplain, birdfoot delta · Erosional: karsting, doline, carbonate reservoir, karst cave alignment, glacial valleys, tunnel valleys, moraines, finger lakes, drumlins, direction of ice advance, glacial reservoirs

12.1 Introduction

Geologic analogs assist the understanding of subsurface structures using present-day structural and depositional settings that are mappable at the surface. We use digital elevation models and their derivatives and combine them with satellite imagery in various band combinations to generate geomorphologic surface models that highlight the basin and reservoir features we expect to see in the subsurface. We need to be aware that this approach has two weaknesses:

1. Remote sensing can only detect features, texture and lithology at the earth's surface. Only microwave radar allows limited penetration into the subsurface (see Sect. 2.4).
2. Erosional processes may remove portions of depositional environments. This limitation is identical to geologic outcrop analogs (de Wickens & Bouma, 1995).

The three categories of geologic analogs are associated with the petroleum system elements of potential hydrocarbon reservoirs. Structural analogs contribute to models of structural traps, depositional analogs contribute to reservoir features. Carbonate analogs are widely used as source, reservoir, seal and trap analogs whereas erosional analogs can be applied to both, reservoir and stratigraphic trap modeling. Table 12.1 gives the complete overview of the analogs discussed in this section.

Table 12.1 Overview of geologic analogs discussed in this section

Section	Geological analogs	Present-day analog	SRC	RES	Seal	Trap
12.2	**Structural analogs**					
12.2.1	Relay Ramp	Arches NP, UT USA				■
12.2.2	Faults and karsting	Hassi R'Mel, Algeria				■
12.2.3	Salt glaciers	Zagros, Iran				■
12.2.4	Salt diapirs	Louisiana coast, USA				■
12.2.5	Rift tectonics	Afar Triangle - ETH				■
12.2.6	Pull-apart basins	Wadi Gharib, Sinai Egypt				■
12.2.7	Fold and fault belt	Appalachians, PA USA				■
12.3	**Depositional analogs**					
12.3.1	Analogs for carbonate and evaporite deposition	UAE coast, Lake MacLeod Australia		■	■	
12.3.2	Carbonate depositional processes	Caicos Islands UK (OT)	■	■	■	■
12.3.3	Marine deltas	Irrawaddi and Sittang, Myanmar		■		■
12.3.4	Tidal floodplain	North Frisia, Germany		■		
12.3.5	Flood plain channels	Black River AR USA		■		
12.3.6	Intra-cratonic deltas	Okavango, Botswana and Kufra, Libya		■		
12.4	**Erosional analogs**					
12.4.1	Limestone karsting	Karst Plateau, Slovenia/Italy		■		■
12.4.2	Glacial valleys	Jutland, Denmark		■		■
12.4.3	Regional glacial mapping of moraines	Great Lakes, Canada/USA		■		■
12.4.4	Glacial valleys and finger lakes	Lake Ontario, USA		■		■

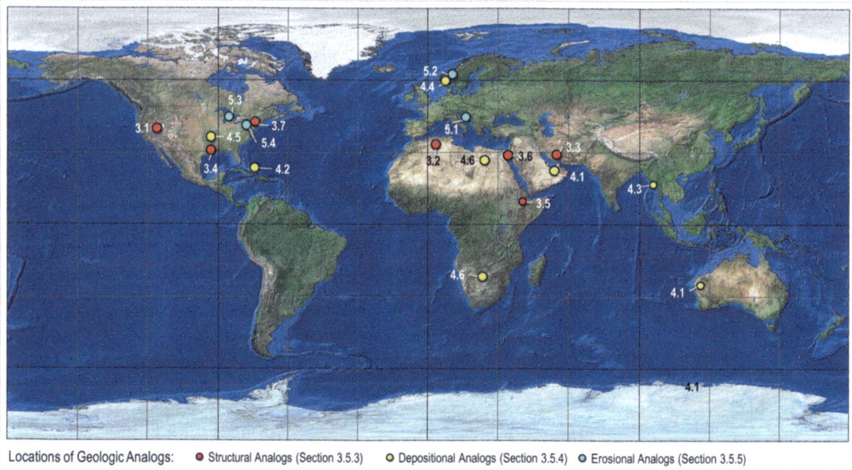

Fig. 12.1 Map of geologic analogs discussed in this section. *SA* structural analogs, *DA* depositional analogs, *EA* erosional analogs. (Satellite imagery courtesy of U.S. Geological Survey)

Figure 12.1 shows on a map the locations of the geologic analogs for each category. The label refers to the abbreviated section number.

12.2 Structural Analogs

12.2.1 Relay Ramps, Arches NP UT, U.S.A.

Relay ramps are a structural feature that is important for the fault sealing assessment of structural traps (Rotevatn et al., 2007; Botter et al., 2017; Dimmen et al., 2020). One of the most frequently used geological outcrop analogs is the relay ramp near Delicate Arch in the Arches National Park in Utah (Doelling, 2000; Rotevatn et al., 2007; Torabi et al., 2019; van Gent & Urai, 2020). It exposes extensional faults of the Paradox Fault and Fold Belt in a clastic setting (Fig. 12.2a). The faults are related to eroded salt walls and collapsed overburden (Doelling, 2000).

The sequence of normal faults (Fig. 12.2b) reveals a classic relay ramp (RR) connecting the upthrown blocks in the north to the downthrown blocks in the south. The location of Delicate Arch (DA) serves for spatial orientation. The relay ramp has a width of 500–1000 m. In a reservoir setting, a ramp of such dimensions could provide a pathway for fluid migration, which is the reason for its extensive study.

For the overview image we have used Sentinel imagery that is available through the USGS EarthExplorer portal. We used the visible band combination 432 RGB. Usually we would choose a combination of SWIR and visible bands. However, the Salt Valley and Paradox Fault and Fold Belt reveal the different lithology also in the visible bands-sandstone is bright red, carbonates and evaporites are bluish white-so that we can use the higher spatial resolution of visible bands.

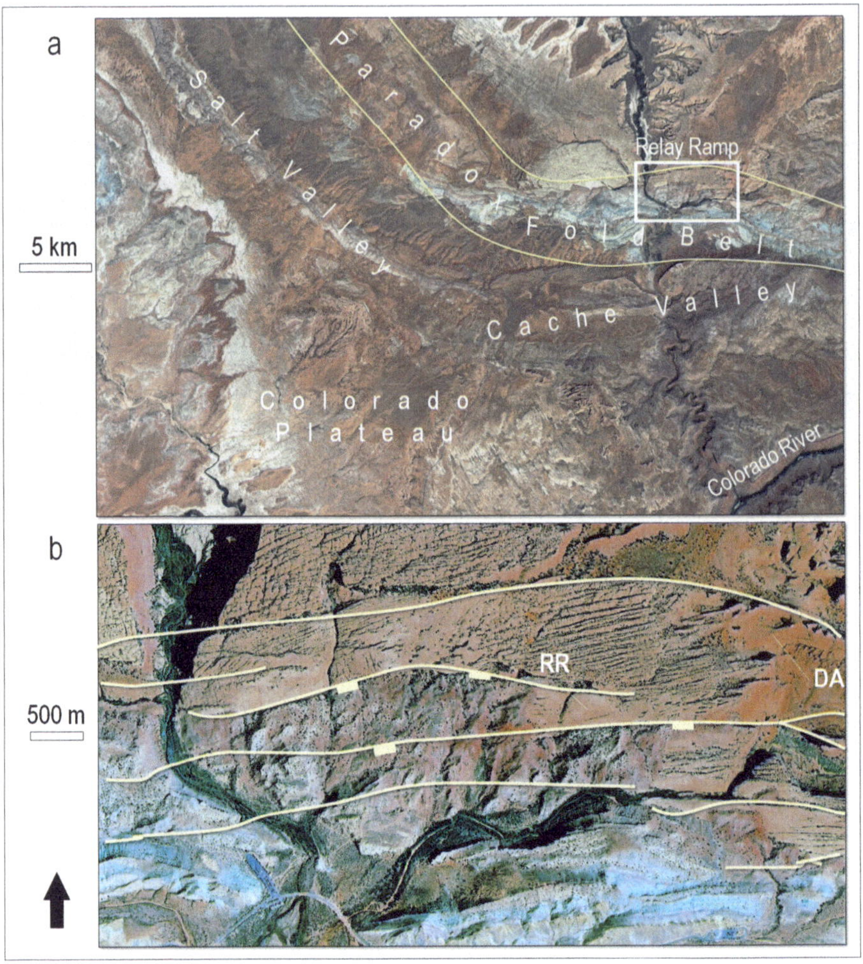

Fig. 12.2 Relay ramp at Arches National Park, UT. (**a**) Sentinel 432 RGB satellite image of structural overview of the location of the relay ramp in the Paradox Fault and Fold Belt, (**b**) detailed structural interpretation after Doelling (2000) on Sentinel 432 RGB image sharpened with NAIP airborne imagery, *RR* relay ramp, *DA* delicate arch. (Copernicus Sentinel data 2020 courtesy of ESA, NAIP airborne imagery courtesy of U.S. Geological Survey)

The detail image of the relay ramp uses the same Sentinel satellite image, but we sharpened it with a high-resolution airborne photo provided by the NAIP program. With this technique we retain the spectral characteristics of the Sentinel satellite image, which provides 10 m spatial resolution, and benefit from the 60 cm spatial resolution of the NAIP aerial photo.

We have used the Arches relay ramp as an analogue in a seismic study on the northwest shelf of Australia (Laake, 2018). Figure 12.3 compares the interpreted satellite image from the Arches relay ramp with a Tertiary subsurface structural

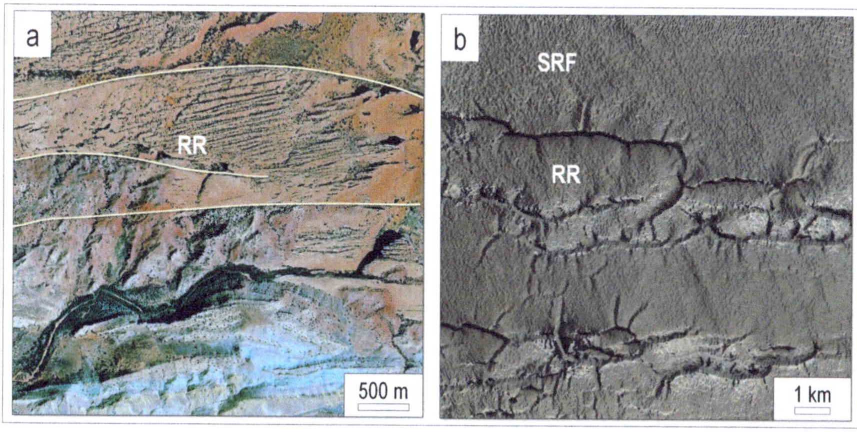

Fig. 12.3 Application of Arches Relay Ramp as a structural geologic analog. (**a**) structural interpretation after Doelling (2000) on Sentinel 432 RGB image sharpened with NAIP airborne imagery, (**b**) subsurface structural image from seismic data from the northwest shelf offshore Australia (after Laake, 2018), *RR* relay ramp, *SRF* stress release fault. (Copernicus Sentinel data 2020 courtesy of ESA, NAIP airborne imagery courtesy of U.S. Geological Survey, seismic image courtesy WesternGeco)

horizon interpreted from 3D seismic data offshore Australia. The relay ramps in both cases are roughly of the same dimensions. The structural architecture follows the observations by Khalil and McClay (2017). When comparing the two images of relay ramps we notice that the Arches relay ramp is affected by erosional process which appear to smooth the fault contours. The seismic image shows a buried structural horizon that appears not to have been exposed to erosion. Consequently, there are more detailed traces of the tectonic evolution comprising for example stress release faults (SRF) that had been observed for example by Destro (1995).

12.2.2 Faults and Karsting, Hassi R'Mel, Algeria

When faulting introduces weak zones into rocks, surface water has a pathway into the rock that may lead to chemical erosion along the faults. Rock types that are susceptible to karsting comprise carbonates and evaporites. In this section we show how remote sensing can delineate surface faults in carbonates through the detection of surface karst features. The detection of karst erosion features in general is included in the section onshore erosional analogs.

In this section we study how remote sensing can assist the detection of karsting at the surface, i.e. epigenic karsting (Klimchouk, 2017a). For this study we chose an area in Algeria near the Hassi R'Mel gas field. The area is located on the slope from the Atlas Mountains eastward to the Grand Erg basin. Its surface is mainly composed of relatively pure carbonates, i.e. marble and limestone. Surface faulting is mainly oriented perpendicular to the Atlas ranges, i.e. in a northwest-southeast direction. Figure 12.4a shows a Landsat 8 natural color 432 RGB composite, which has been enhanced with the digital elevation model to highlight the correlation of surface lithology and topography.

In this rendering, carbonate outcrops are represented by reddish to brownish tones, whereas surface sand accumulations show as greenish tones. Water and stream features are represented as bluish tones.

Karst features show as dark greenish spots because the areas, where near surface karst voids collapsed, are filled with soil. The fault system linking the karst features acts as a pathway for water which vegetation uses to thrive in the depressions, that are shielded from the wind. Using these karst features, we have marked the regional fault trends by dashed lines. For a more detailed analysis we focus on the area marked by the yellow rectangle in the south east corner of the overview image.

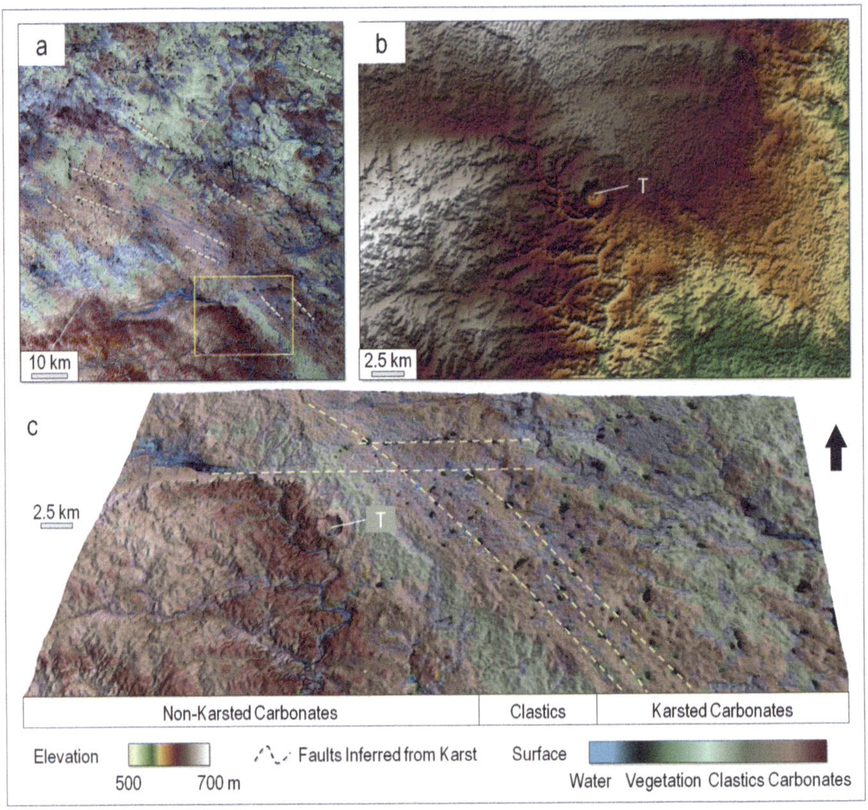

Fig. 12.4 Mapping of structural trends from the detection of surface karst. (**a**) regional overview map from Landsat 8 visible band composite 432 RGB merged with digital elevation model, (**b**) SRTM DEM V3 at 25× vertical exaggeration, (**c**) Landsat 8 432 RGB DRM enhanced image draped over DRM at 10× vertical exaggeration, *T* Talemzane meteor crater. (SRTM DEM V3 and Landsat 8 satellite imagery courtesy of U.S. Geological Survey)

The digital elevation model for the detailed area (b) is rendered at 25 times vertical exaggeration to highlight the subtle structural features. The study area is characterized by a plateau with an erosional valley of a river. In the center of the image, the circular rim of the Talemzane meteor crater (T) stands out as a structural anomaly (Belhai & Sahoui, 2017; Koeberl, 1994).

For the study of the correlation of surface lithology and structural lineaments, we blend the visible band Landsat 8 432 RGB composite with the digital elevation model, drape it onto the digital elevation model and render it at 10× vertical exaggeration (c). Slant illumination from two angles in the northeast perpendicular to the strike of the structural lineaments enhances these trends (dashed lines) and shows how the vegetation-filled karst features are correlated with these lineaments. We also notice that a second trend in west-east direction, which is parallel to the main fold axes of the Atlas Mountains, overprints the northwest-southeast lineaments. Where these two trends intersect, large karst depressions form frequently.

This observation has been recognized in numerous studies of karsted subsurface reservoirs (Sullivan et al., 2006; Kumar et al., 2009; Yang et al., 2010; El-Emam et al., 2013; Laake et al., 2014a).

12.2.3 Salt Tectonics and Salt Glaciers, Zagros Mountains, Iran

Salt glaciers represent a specific type of salt tectonics. They occur in a highly compressional tectonic setting. Under these circumstances the salt is squeezed into faults and extruded at the surface. The most prominent occurrence of onshore salt glaciers has been observed in the Zagros Mountains in central Iran (Talbot, 1981; Talbot & Aftabi, 2004; Zavada et al., 2015; Alizadeh, 2017). With the advent of high-resolution scanning of the seafloor, salt glaciers have also been detected at the seafloor of the Red Sea (Augustin et al. 2016; Hovland et al., 2015), where they pose a challenge for seismic imaging below the seafloor. Most likely, salt glaciers will also manifest as a drilling hazard should exploration yield drillable prospects in this area. Historic salt glaciers have been detected in subsurface formations of the late Triassic in northwest Germany (Mohr et al., 2016).

The onshore salt glaciers of central Iran can, therefore, serve as geologic analogs for recent submarine settings as well as for historic subsurface settings.

Remote sensing can assist mapping the onshore salt glaciers using multi-spectral satellite imagery (Fig. 12.5). We have selected two types of salt glaciers from the Fars province in the Zagros fold belt in south central Iran. The Hormud-e Bagh salt glaciers (a, c) are squeezed out at the flanks of compressional folds and protrude in the basin that formed between the two parallel folds.

The Fadami salt glacier (b, d) has been squeezed out of the center of a faulted fold ridge. The digital elevation models (a, b) allow estimating the thickness of the salt sheets to range between 200 and 400 m. For the lithological delineation of the

Fig. 12.5 Salt glaciers in the Zagros fold belt, south Iran. (**a**) DRM for Hormud-e Bagh salt glaciers, (**b**) DRM for Fadami salt glacier, (**c**) Landsat 8 10-5-1 RGB composite image for Hormud-e Bagh salt glaciers, (**d**) Landsat 8 10-5-1 RGB composite image for Fadami salt glacier. (SRTM DEM V3 and Landsat 8 imagery courtesy of U.S. Geological Survey)

salt glaciers and the adjacent salt-encrusted sediments, we have processed a combination of thermal infrared band 11, VNIR band 5 and visible ultra-blue band 1. The thermal infrared band is sensitive to the different heat capacity of sedimentary rocks and salt, the VNIR band assists with lithology discrimination and the visible blue band is sensitive to moisture. The resulting multiband composite RGB shows non-weathered salt in purple and weathered salt in bright red. Salt-encrusted sediments are highlighted as bright red where the salt crust dominates and turn into cyan where salt and gypsum mix with sand and gravel.

The salt glaciers show a very rough texture towards the edges which is related to dissolution of the salt by rainwater. This phenomenon is called salt karsting, which has been observed not only onshore but also offshore with salt diapirs and salt plugs that reach the seafloor (Talbot & Augustin, 2016; Laake, 2020). Offshore salt karst poses a serious drilling problem the mapping and assessment of which has only recently become available through sonar mapping of the seafloor and high-resolution 3D seismic data.

12.2.4 Salt Tectonics and Diapirs, Louisiana coast, U.S.A.

Isolated salt diapirs that rose close to the surface have been mapped at the Louisiana coast west of the Mississippi delta (Kupfer, 1983). Since the Mississippi delta is very shallow, the subsurface salt diapirs are noticeable as slight surface highs, which are known as "The Five Islands" (Fig. 12.6). Out of the originally five salt mines producing rock salt from the salt domes, three are still in operation.

The high-resolution topobathymetric digital elevation model (USGS, 2014) at 20 times vertical exaggeration (a) shows the central three of the salt domes as island above the otherwise flat coastal plain as is also evident from the elevation profile across the salt domes (b).

Sentinel satellite imagery has been draped over the NGOM digital elevation model to generate a virtual 3D of the salt diapir islands on the coast of Louisiana (c). We have chosen the visible bands 4, 3 and 2 to create a natural color image, that allows to distinguish water, vegetation and settlements. The virtual exaggeration highlight the salt diapirs as elevation anomalies.

The salt diapir at Jefferson Island has largely been dissolved and eroded when an oil drilling rig accidentally drilled into the salt dome which resulted in catastrophic

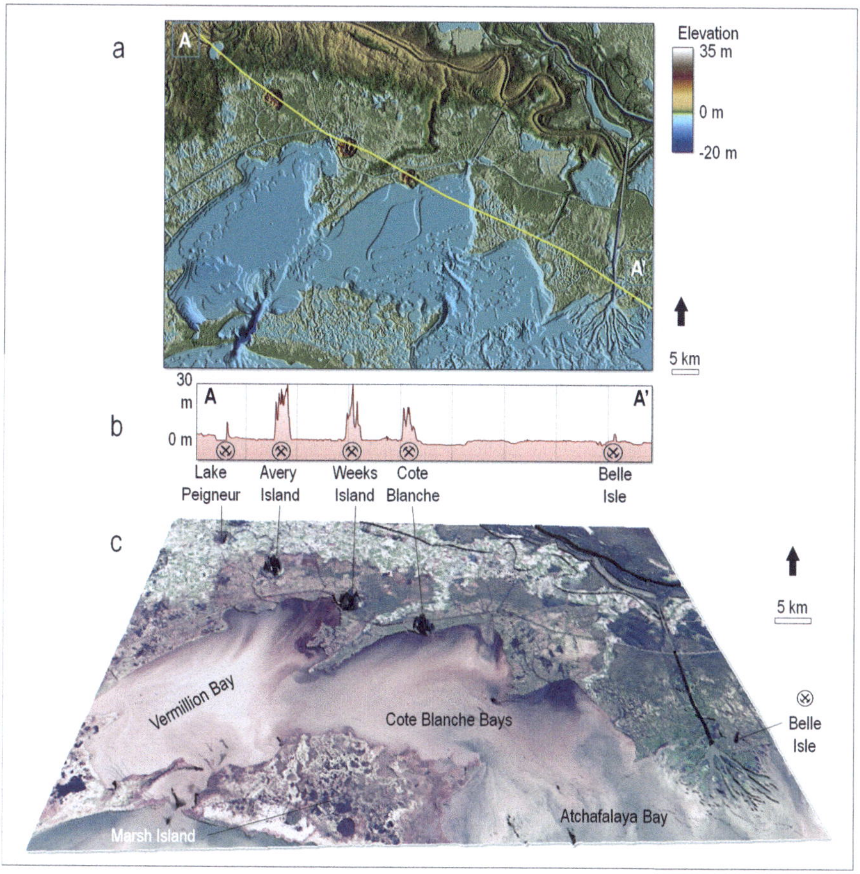

Fig. 12.6 Salt domes on the Louisiana coast. (**a**) high-resolution topobathymetric digital elevation model showing the location of the elevation, (**b**) elevation profile, (**c**) 3D rendering of satellite image at 20 times vertical exaggeration. (DRM from NGOM DEM courtesy of U.S. Geological Survey, Copernicus Sentinel data 2017 courtesy of ESA, elevation profile courtesy of Google Earth)

flooding of the salt mine and the expansion and deepening of Lake Peigneur. The Belle Isle salt diapir never showed a surface expression as obvious as the central three salt diapirs. Mining disasters led to abandoning and flooding of the mine, which may have contributed to the low profile the Belle Isle diapir has today.

The five diapirs on the coast of Louisiana may serve as analogs for isolated salt diapirs.

12.2.5 Rift Tectonics, Afar Triangle, Ethiopia and Djibuti

Exploration along passive continental margins has uncovered substantial offshore reserves. To understand the present setting and the connection between conjugate margins, the first approach is the structural restoration back to the early stage of the rifting.

A prominent example for the early stage of continental rifting is the East Africa Rift. In the Afar Triangle (AT in Fig. 12.7a) three stages of rifting can be studied: the early continental rifting of the East African Rift (EAR), the early marine rifting of the Red Sea (RS) and the mature rift of the Gulf of Aden (GoA). In the Afar

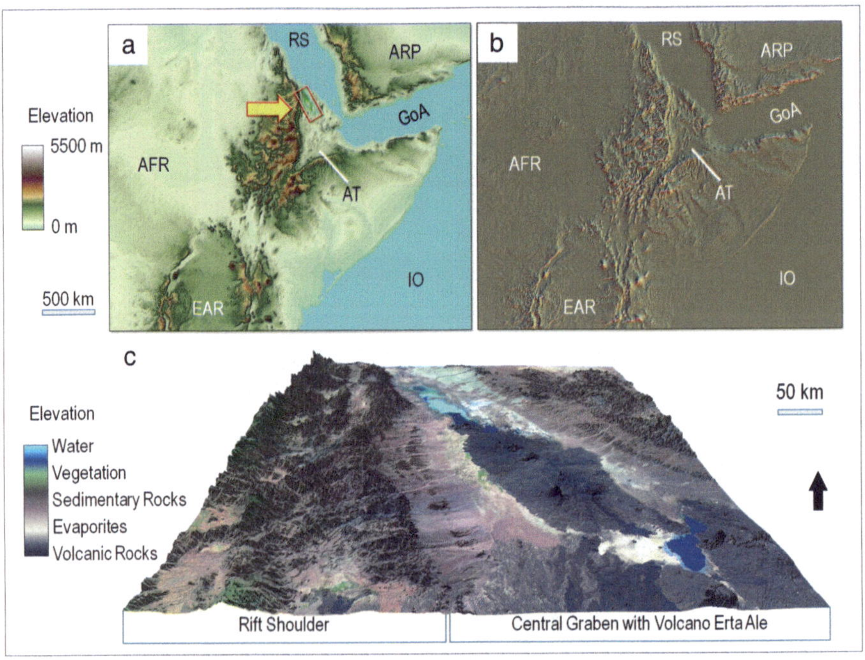

Fig. 12.7 Rift tectonics in the Afar Triangle, east Africa. (**a**) digital elevation model with vertical exaggeration, (**b**) vertically exaggerated surface with directional lighting perpendicular to the main rift orientations, (**c**) Landsat 8 753 RGB image draped over DRM for the onshore portion of the Red Sea rift in Ethiopia, *AFR* African continent, *ARP* Arabian plate, *AT* Afar Triangle, *EA* Erta Ale volcano, *EAR* East African Rift system, *GoA* Gulf of Aden, *IO* Indian Ocean, *RS* Red Sea. (DRM and Landsat satellite imagery courtesy of U.S. Geological Survey)

Triangle transitions from continental to marine rifting phase can be observed, but these are too small for the scale of this study. Remote sensing can the support the understanding of rifting processes through the digital elevation model (a). The topography indicates that the present Red Sea and Gulf of Aden outlines do not fully represent the true extent of the rifted area. Here, the terrain edges should be considered rather than the coastline.

To highlight this, we plot a 100 times vertically exaggerated digital elevation model and illuminate it with different colors perpendicular to main structural orientations found in the area (b). The full extent of the rift and its onshore connection into the East African Rift is much better revealed.

For the modeling of the central rift itself, the area of the Erta Ale volcano can serve as an analog (c). Today, the surrounding of Erta Ale is already a depression with an elevation below sea level, i.e. in geological time the influx of the sea is imminent. Therefore, this part of the Red Sea rift gives good clues for the structure and lithology of the adjacent Red Sea during early stages of active rifting which can also be used when modeling early rifting in the Red Sea and other ancient rift settings.

12.2.6 Pull-Apart Basins, Sinai, Egypt

In strike-slip tectonic setting, pull-apart basins provide space for the deposition of sediments and hence the formation of petroleum systems (Mann et al., 1983). They have been studied in the subsurface (Aribowo, 2017; Gürbüz, 2010; Rojay & Kocyigit, 2012; Chakraborty & Ghosh, 2005), modeled (Hoprich et al., 2009) and studied in outcrop analogs (Hurwitz et al., 2002). We have chosen the pull-apart basins parallel to the Dead Sea strike-slip zone, the setting of which has been described by Moustafa and Khalil (2020).

Remote sensing can assist the structural mapping of pull-apart basins through mapping of terrain edges in the digital elevation model and through lithology discrimination of blocks that have been offset by the pull-apart tectonic processes. For Fig. 12.8a we have overlaid the edges from the terrain gradient on the topography and draped the combined image on the digital elevation model. Vertical exaggeration and slant illumination allow enhancing the fault trends parallel to the strike-slip fault system of the Gulf of Aqaba.

Information about the lithology can be obtained from multi-spectral satellite imagery. We have chosen Landsat 7 and created a multi-band composite 742 RGB focusing on lithology interpretation since the study area is not covered in vegetation (b). Most of the southern Sinai Peninsula shows very dark colors that are corresponding to granites, volcanic rocks and meta-sediments basement from pre-Cambrian and Paleozoic age. The most prominent pull-apart basins are located in the area marked by the yellow rectangle.

For an outline of the structural setting we create the visible band 432 RGB Landsat 7 composite (c), which shows Quaternary sands and gravels in very light

Fig. 12.8 Mapping of pull-apart basins in the eastern Sinai Peninsula, Egypt. (**a**) digital elevation model, (**b**) Landsat 7 742 RGB lithology map, (**c**) Landsat 7 432 RGB natural color composite, (**d**) Landsat 7 (753–642) RGB high contrast lithology composite, all satellite images are draped over the DEM, *WG* Wadi Ghayib, *WR* Wadi Rahab, *KB* Cretaceous block. (SRTM DEM and Landsat 7 imagery courtesy of U.S. Geological Survey)

colors, Cretaceous sandstone in beige to brownish tones and basement rocks in dark brown.

We overlay the fault interpretation for the three pull-apart basins comprising Wadi Ghayib (WG) and Wadi Rahab (WR) as a subset of the strike-slip fault system parallel to the Gulf of Aqaba. The main direction of relative movement is indicated by the arrows. The faults create weak zones in the hard basement rocks, which are subsequently eroded and filled with eolian and fluvial sediments. In the 432 RGB composite image, these valleys show as bright lineaments in an otherwise dark basement environment.

We have a closer look into the lithology to understand the formation of the pull-apart basin and use a Landsat 7 band difference combination of 753 RGB and 642 RGB. With this approach we map the difference between the two RGB composites, which in the case of bare rock proves to significantly enhance the discrimination of the exposed lithology. In comparison to (c), where the basement and sandstone show a very similar signature, the RGB difference image (d) not only allows the clear distinction between sandstone and granite, but also shows areas of different composition within basement rocks. We did even find that the internal texture of the basement rocks is revealed by the band difference image. Applying this method to the interior texture of Wadi Ghayib, we notice in the south an anomalous block (KB) that shows similar bright color as the Cretaceous formation in the northwest of the image. Field work showed that the top of this anomalous block is comprised of Cretaceous sediments that once covered the entire southern Sinai basement rocks. The uplift related to the Red Sea rifting had exposed these soft sediments to weathering, which had removed them entirely from the area adjacent to Wadi Ghayib. However, the formation of the pull-apart basin allowed a block capped with Cretaceous sediments to drop into the opening basin thus being protected from erosion.

Remote sensing can assist in the study of pull-apart basin analogs through both, structural analysis from the gradient of the digital elevation model as well as through lithologic analysis highlighting anomalous blocks in the structural framework.

12.2.7 Fold and Fault Belt, Appalachians, U.S.A.

Tectonic features are an essential component in basin and reservoir model building. Outcrop analogs can assist modeling dipping anticlines as well as the interaction of folding and faulting processes. Faulting can be observed preferably in competent rocks like limestone, hard sandstone and basement rocks whereas softer rocks like shales show less faulting.

The Appalachian Mountains in Pennsylvania, U.S.A. provide a good example to study the interaction of folding and faulting (Fig. 12.9c). Multiple orogenic events starting as early as the Ordovician created a fold belt. Different weathering of the hard Cambrian quartzites and carbonates and hard sandstones of Silurian to Devonian age left ridges, whereas the erosion of the softer shale formations in between the hard rocks created valleys. These are showing in the vertically exaggerated digital elevation model (a).

The surface gradient (b) reveals the tectonic styles even more pronounced. The elongated concentric patterns are associated with a series of plunging anticlines, which are very common trap structures. The surface gradient map gives also clues for faulting. Erosional valleys cutting anticlinal structures perpendicular are an indication for water erosion along faults. The Appalachian Plateaus to the northwest outside the fold belt demonstrate this process even better: faulting has left weak

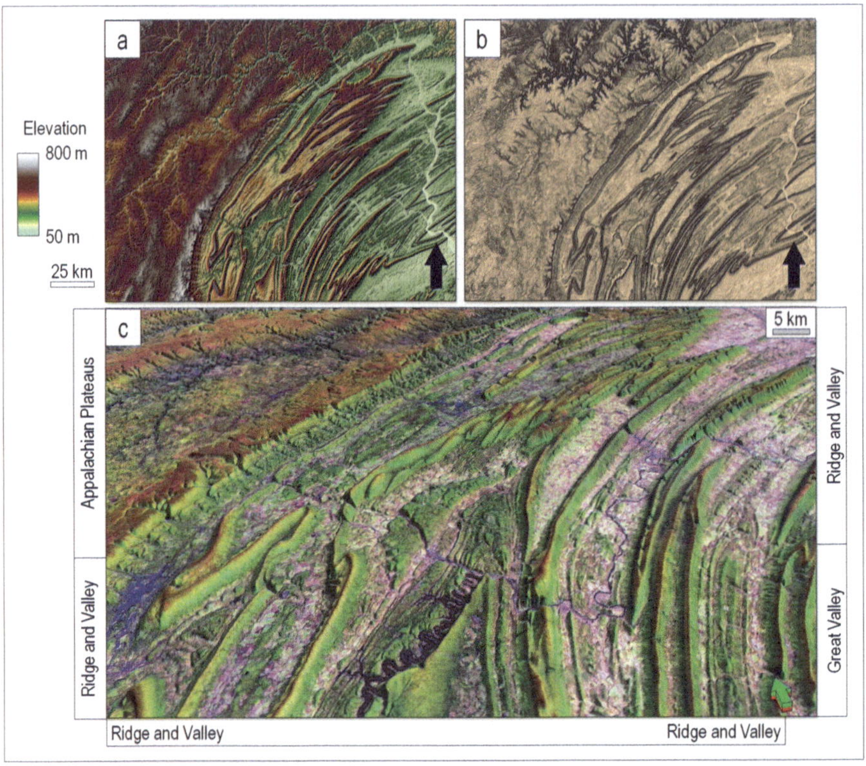

Fig. 12.9 Fold and fault belt of the Ridge and Valley zone of the Appalachian Mountains, PA U.S.A. (**a**) SRTM DEM V3 digital elevation model, 10× vertical exaggeration, (**b**) surface gradient rendered on digital elevation model as surface attribute, (**c**) virtual 3D of Landsat 8 multiband composite 753 RGB merged with digital elevation model and draped on topography. (SRTM DEM and Landsat 8 imagery courtesy of U.S. Geological Survey)

zones in the hard Carboniferous limestone and sandstone. The water subsequently carved deep canyons into the underlying softer shales.

The rendering in 3D highlights the correlation between folds and faults even better (c). We use the Landsat 8 multi-band combination 753 RGB to create a landuse image, which distinguishes farmland (purple), forest and meadows (green) and water (blue). To highlight the ridges, we have blended the satellite image with the color-coded digital elevation shown in (a) and draped the result over the topography. Using this approach, we can nicely distinguish the ridges in light green, yellow and brown tones from the farmland in the valleys which shows as purple. Slant illumination generates shadows at even minor faults that have been eroded into small ravines and canyons. This can be observed particularly at the boundary of the Ridge and Valley region to the Appalachian Plateaus. A variant of this approach is the rendering the bedrock geological map on the topography to better correlate geologic age, dominant rocks and geomorphology.

The interaction of folding and faulting becomes evident in the Shenandoah Valley, where both forks of the Shenandoah River follow the combination of the two processes: the valley trend follows the folding of the Appalachian ridges, whereas the texture of the valley floor shows the impact of fault perpendicular to the ridge and valley axes (Fig. 12.10). The result is a very regular box-style meandering of the Shenandoah River that is visible both in the digital elevation model (a) and in the Landsat 8 753 RGB color composite (b). Both images show flat rendering without vertical exaggeration to place the focus on the structure that is revealed in the images.

We can use this example also to demonstrate the benefit of merging the digital elevation model with its gradient. When we overlay the digital elevation model on its gradient, we obtain a structurally enhanced image (c). Overlaying the satellite image on the gradient of the digital elevation model (d) highlights the faults even further. The strong contrast of Shenandoah River (dark blue) from meadows and forest (light to dark green) enhances the valley course. Shadows in the satellite image contribute to sharper representation of the fault valleys especially in the ridges separating the North from the South Fork of the Shenandoah River.

The options of flat versus vertically exaggerated rendering as well as the blending of the digital elevation model, the surface gradient and the satellite image provide a set of tools to highlight structural styles from the regional scale of the

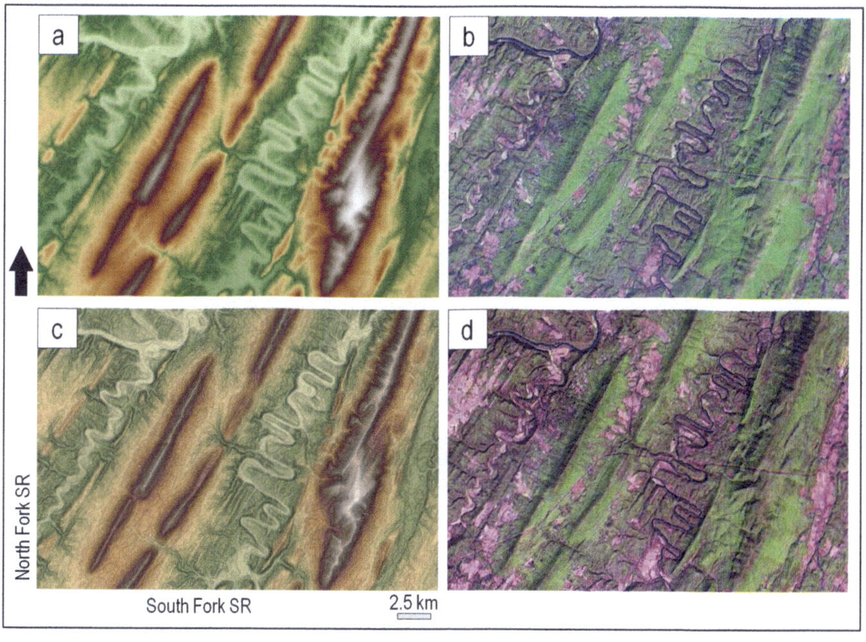

Fig. 12.10 Detailed fold and fault system in the Shenandoah Valley, Ridge and Valley zone of the Appalachian Mountains, PA U.S.A. (**a**) SRTM DEM V3 digital elevation model, no vertical exaggeration, (**b**) Landsat 8 753 RGB composite, (**c**) SRTM DEM overlaid on SRTM DEM gradient, (**d**) Landsat 8 753 RGB composite overlaid on SRTM DEM gradient, *SR* Shenandoah River. (SRTM DEM and Landsat 8 imagery courtesy of U.S. Geological Survey)

Appalachian Mountain belt to local scale showing at the individual meanders of the Shenandoah River.

12.3 Depositional Analogs

12.3.1 Shallow Water and Coastal Arid Environment, U.A.E. Coast

The Abu Dhabi coast of the Arabian Gulf serves as an analog for the Createaceous carbonate reservoirs and evaporite seals in the region. It is a hot and arid desert with a very shallow slope towards the gulf, which allows the seawater to seep into the onshore sediments thus creating coastal sabkha with deposition of salt, gypsum and anhydrite (Fig. 12.11). In the shallow and warm gulf offshore the Abu Dhabi coast coral reefs provide a depositional analog for the deposition of reef and shoal carbonates.

To show both types of depositional analogs we use a blend of two multi-band composite Landsat 8 images. The composite image of 741 RGB uses the SWIR band and VNIR band 4 for infrared discrimination of sand, gypsum and salt. The coastal band 1 has been used instead to the visible blue band to support the deep water penetration used in the second RGB composite. The mapping of reefs that are submerged under water is achieved best with the coastal blue band in combination

Fig. 12.11 Holocene depositional analog for carbonate and evaporite deposition, Abu Dhabi shallow water and coast. Landsat 8 blended composite of 741 RGB and 321 RGB. (Landsat 8 imagery courtesy of U.S. Geological Survey)

with visible bands 2 and 3 using a 321 RGB composite (compare Sect. 2.6). When blending the two images, the 741 RGB image dominates the response onshore, whereas the 321 provides the main contribution to the offshore mapping.

12.3.2 Coastal Arid Environment, Lake MacLeod, West Australia

The second analog for a combined carbonate and evaporite deposition is Lake MacLeod in West Australia (Fig. 12.12). A large shallow depression separated from the coast to the Indian Ocean by the shallow Quobba Ridge does not have a direct connection to the ocean, but saltwater seeps into Lake MacLeod through the narrow ridge near the southern tip of the lake.

For most of the year the lake surface is at about 1–2 m below sea level (a). Occasionally, the lake receives freshwater influx through channels in the Carnarvon

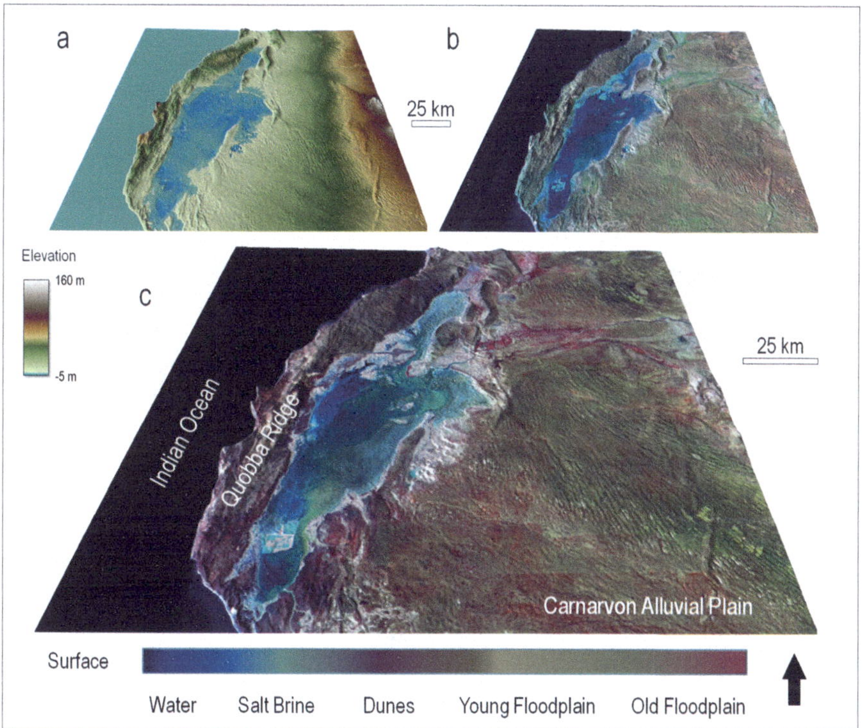

Fig. 12.12 Holocene depositional analog for carbonate and evaporite deposition in a shallow depression near the coast, Lake MacLeod, West Australia. (**a**) digital elevation model, (**b**) Landsat 5 composite 742 RGB, (**c**) Landsat 5 composite 432 RGB, all data draped on the digital elevation model with 25× vertical exaggeration and slant illumination from east. (SRTM DEM V3 and Landsat 5 imagery courtesy of U.S. Geological Survey, geomorphologic interpretation after Whitney & Hengsch, 2015)

Alluvial Plain. The remote sensing mapping in this environment would typically use the Landsat 5 combination 742 RGB, which is the standard for landuse mapping (b). However, even when processing this combination carefully, only limited distinction of freshwater from brine in the lake can be achieved.

Instead we use the combination of 432 RGB which uses the different optical response of water and brine (c). Band 4 assists with vegetation mapping which assists in the distinction of the channel systems in the Carnarvon Alluvial Plain.

12.3.3 Coastal and Deltaic Sabkha, Sabkha Matti, U.A.E.

An example for the transition from coastal to inland sabkha is Sabkha Matti in western Abu Dhabi near the border to Saudi Arabia. Sabkha Matti is the remnant of a wide Miocene fluvial delta. Salt flats remained when the freshwater river dried up. Near the coast, high tide and strong winds push ephemeral brine sheets in the shallow plain of Sabkha Matti leaving salt crusts behind.

To map the sabkha environment we can use the Landsat 8 band combination 753 RGB, that is often employed for landuse mapping (Fig. 12.13). We do not need to use the blue or coastal blue bands for mapping into the water since the sabkha is limited to onshore. This band combination shows wet coastal sabkha in a characteristic cyan tone. Mostly dry inland sabkha appears in dark greenish tones. Both contrast sharply with gravel (red-brown) and active sand sheets and sand dunes (beige to yellow).

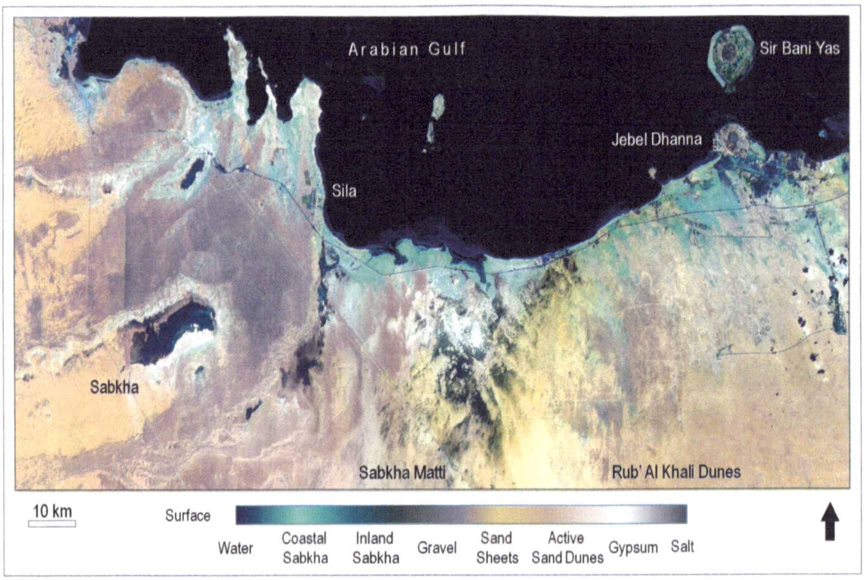

Fig. 12.13 Coastal and deltaic sabkha, Sabkha Matti, U.A.E. Landsat 8 band combination 753 RGB. (Satellite image courtesy U.S. Geological Survey)

Fig. 12.14 Inland sabkha Umm As Samim, northwest Oman. (Landsat 7 composite 742 RGB. Satellite image courtesy of USGS)

12.3.4 Inland Sabkha, Umm As Samim, Oman

We conclude this section with the inland sabkha of Umm As Samim in northwest Oman. Located in a wide and shallow regional depression without an outlet to the sea, this basin collects the runoff from the Oman Mountains. Desertification stopped the influx of freshwater almost entirely, the lake dried up and left a sabkha, that is partially wet from concentrated salt brine.

To distinguish different evaporites, the Landsat 7 band combination 742 RGB provides the lithology discrimination from the SWIR band 7 and VNIR band 4. In this band combination, water and brine appear in a dark greenish tone, which contrasts sharply with the bright white color indicating gypsum (Fig. 12.14). Sand dunes show in intense yellow color which contrast sharply against the gravel plain (purple). Wadi beds, that are filled with sand and clay, appear in beige. During rare flashfloods, the wadi beds lead freshwater from the Oman Mountains into the sabkha basin.

12.3.5 Carbonate Depositional Processes, Caicos Islands

Carbonate rocks play an important role as reservoir rocks especially in the Cretaceous of the Middle East (Alsharhan & Nairn, 1986, 1990; Hollis, 2011). Handford and Loucks (1993) described the relation of the carbonate depositional sequences and their depositional environment. A well-studied modern analog for carbonate depositional environments is the Caicos Islands, which form part of the Bahama archipelago (Logan & Sullivan Sealey, 2013; Dickson & Saller, 2008). Harris et al. (2012, 2014) have used remote sensing to study the flat-topped shallow carbonate platforms of the Bahamas. They used the visible blue band 1 of Landsat 5 to estimate the water depth. In the meantime, Landsat 8 offers the coastal blue band 1 with better penetration into water. This enables a more detailed remote analysis of the water depth and the depositional features. To correlate the findings from remote sensing data we use the detailed study of the sedimentary and diagenetic processes by Bourrouilh-Le Jan (1998).

In Fig. 12.15 we outline the different geologic features we can extract from remote sensing data and relate them as analogs for the petroleum system elements in carbonate reservoirs:

- **Reefs** are the carbonate source. They form high-porosity reservoir rocks.
- **Shoals** are erosional carbonate deposits frequently found in the lagoon. They also form high-porosity reservoir rocks.
- **Tidal channels** provide the hydrodynamic communication between the ocean and the lagoon. Typically deposits from tidal channels show less porosity and, therefore, are not forming preferred reservoir rocks.
- **Tidal flats** and **algal marshes** are very high in organic contents. They can form source rocks.
- **Salinas** originate from evaporation of seawater on exposed parts of the islands in arid climate. They may form seal rocks.

The spatial extent and distribution of depositional features in present carbonate environments are often used as geologic analogs for fossil reservoirs in the subsurface, the spatial distribution of which can only be inferred from seismic and well data. Using remote sensing data from Landsat 8 we can delineate the entire Caicos platform (Fig. 12.16). The SRTM digital elevation model (a) only shows the islands that are exposed. When we use the coastal band 1 of Landsat 8, we obtain sufficient penetration into the shallow water to map the outline of the reefs that delineate the platform (b).

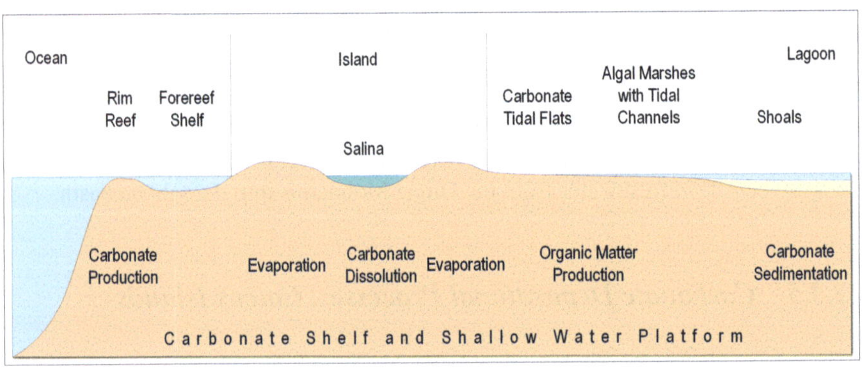

Fig. 12.15 Outline of geologic analogs in carbonate depositional environment

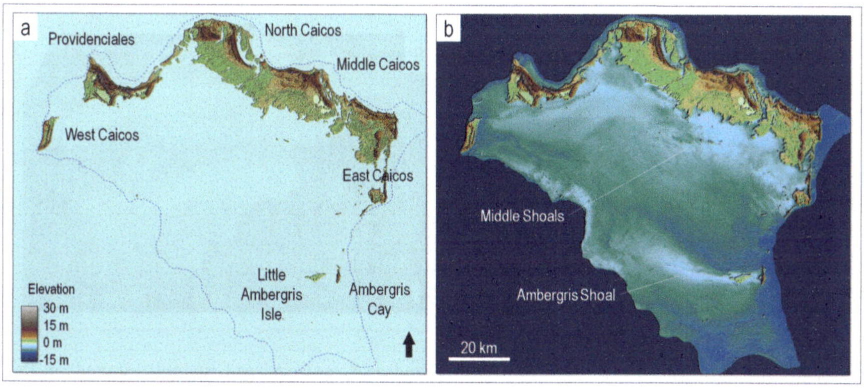

Fig. 12.16 Delineation of Caicos Islands from digital elevation model and satellite imagery. (**a**) SRTM DEM V3, (**b**) bathy-topography elevation model obtained from the merge of the bathymetry from Landsat 8 band 1 and the topography from SRTM DEM V3. (DRM and satellite imagery courtesy of U.S. Geological Survey)

We can also estimate the water depth from the intensity level of band 1, which is related to the absorption of the blue spectral band in seawater (Harris et al., 2014). To achieve a seamless elevation model for onshore and offshore, we need to apply some corrections to the offshore bathymetry model. The direct correlation of absorption and water depth only applies to the areas of the Caicos platform that are covered in carbonate mud. Towards the coast of the islands, the dark color of algal marshes in the tidal zones distorts the depth relation and need to be corrected for. We also need to correct for the errors introduced by the cloud cover in parts of the image. Landsat 8 band 9 can mostly assist with the identification of cloud-covered areas. After these corrections we merge the SRTM DEM with the bathymetry DEM from Landsat 8 band 1 and obtain the bathy-topography elevation model that is shown in (b). The outline of the platform is plotted as a dotted line in (a).

In the next steps, we use the bathy-topographic elevation model to render multispectral RGB composites for the analysis of the carbonate depositional environment (Fig. 12.17). Frequently, the visible band combination 432 RGB is (a) is used to map reef carbonate depositional analogs in shallow water. However, when we use the coastal band in the combination 321 RGB (b) we can resolve more subtle depositional features especially on the central Caicos platform.

The benefits of using the band combination 321 RGB together with vertical exaggeration and slant illumination become even more evident when we drape the image on the bathy-topographic elevation model and render in map view at high vertical exaggeration and slant illumination (Fig. 12.18). Dry land of the islands is represented by dark brown-red colors, algal marshes and tidal channels have lighter brownish-reddish colors. Eroded carbonate sands that are deposited in very shallow water are revealed in very bright colors, whereas the carbonate mud on the Caicos platform shows in beige to greenish tones. The texture of the seafloor on the platform gives clues for the directions of currents.

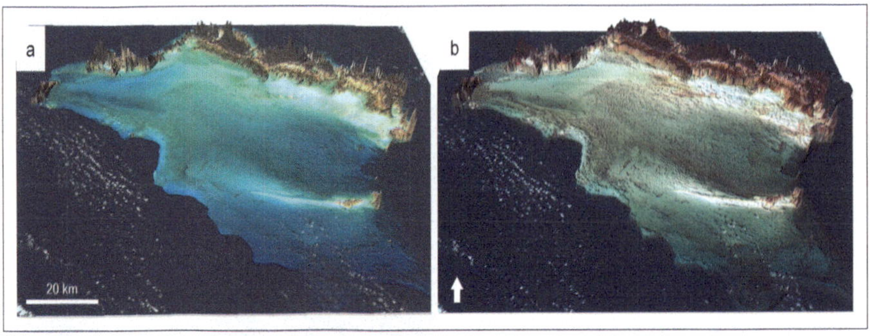

Fig. 12.17 Imaging the Caicos Platform with Landsat 8 multispectral imagery draped on the bathy-topographic elevation model from Landsat 8 band 1 and SRTM DEM V3 at 200× vertical exaggeration and slant illumination. (**a**) 432 RGB composite, (**b**) 321 RGB composite. (DEM and satellite imagery courtesy of U.S. Geological Survey)

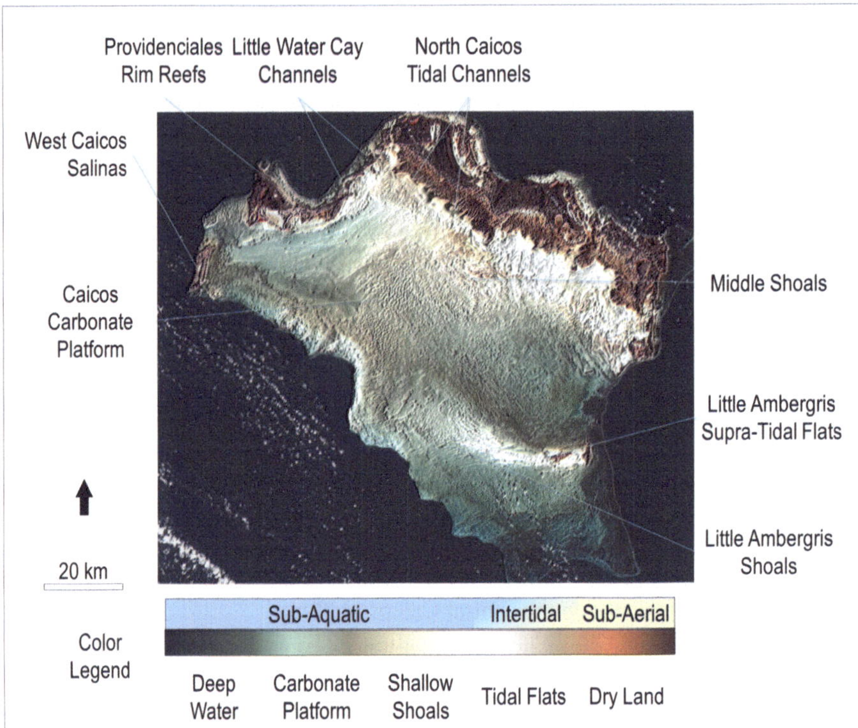

Fig. 12.18 Overview of carbonate reef geologic analogs considered in this study, Landsat 8 multiband 321 RGB composite at 200× vertical exaggeration and slant illumination. (Satellite imagery courtesy of U.S. Geological Survey)

In the following we will study which band combinations provide the best results for the mapping of individual depositional configurations such as rim reefs offshore Providenciales Island, algal marshes and tidal channels on the south shore of North Caicos and shallow shoals of the Middle Shoals west of East Caicos. We will also study sub-aerial features of the supra-tidal algal flats of Little Ambergris and the Salinas of West Caicos.

We start with features on the tidal platform at the southwest coast of North Caicos Island (Fig. 12.19). The SRTM digital elevation model gives elevation information only for those portions of the island that are either sub-aerial or intertidal. For mostly water-covered areas the SRTM DEM shows no values (a). We notice also that tidal flats with channels can hardly be discriminated from algal flats. And finally, the spatial resolution of the SRTM DEM is insufficient to delineate the course of the tidal channels. To improve the mapping of tidal channels, we use the panchromatic band 8 from Landsat 8 (b). Tidal channels are filled with carbonate mud and, therefore, appear brighter than algal flats that appear dark in the visible

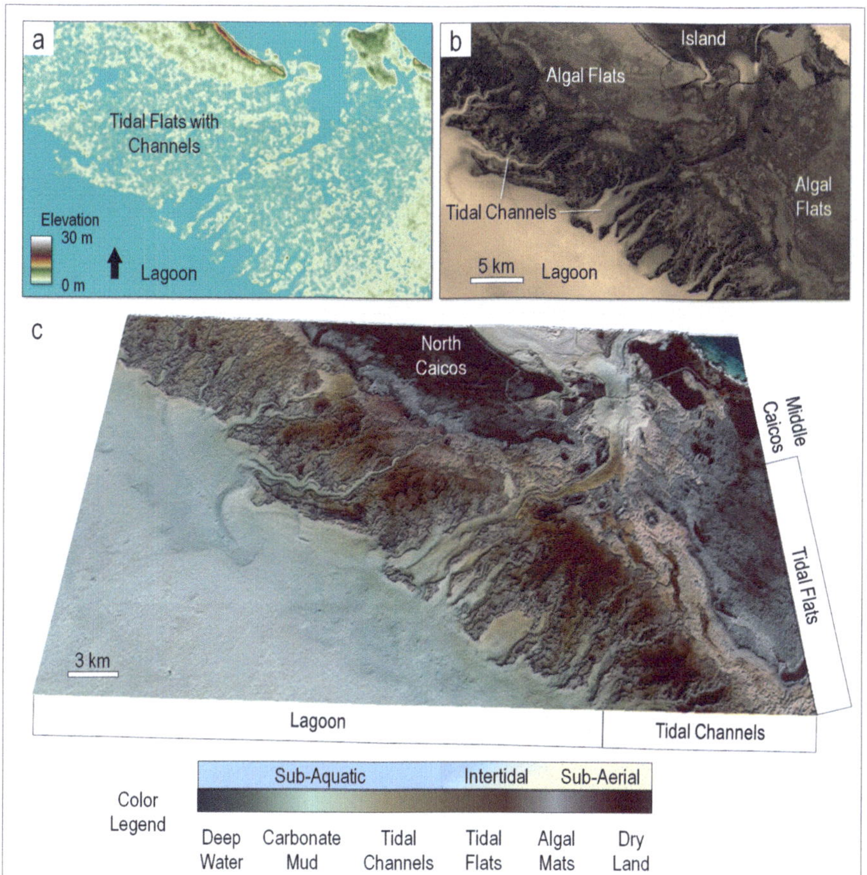

Fig. 12.19 Tidal channels on a tidal platform, southwest coast of North Caicos Island. (**a**) digital elevation model, (**b**) Landsat 8 band 8 as flat grayscale rendering, (**c**) Landsat 8 multiband 321 RGB composite draped over merged surface from SRTM DEM and Landsat 8 band 8 converted to topography, 10× vertical exaggeration. (SRTM DEM V3 and satellite imagery courtesy of U.S. Geological Survey)

bands, that contribute mainly to the panchromatic band. We use this characteristic to convert the intensity of the panchromatic band to elevation and merge it with the SRTM DEM. The result is a sharpened elevation model that enhances the outline of the tidal channels, particularly when we drape the composite image of band combination 321 RGB on to this enhance elevation model (c). The tidal channels and their connection to the texture of the carbonate mud in the lagoon become obvious.

We focus now on carbonate reefs and their erosion products, carbonate shoals (Fig. 12.20). Rocks originating from reefs and shoals are prominent carbonate reservoir rocks because of their high initial porosity. The rim reef of the Caicos Platform is most pronounced at its northern margin offshore Providenciales Island and Little Water Cay, where a forereef shelf can be observed (a). Rim reef and forereef shelf are intersected by tidal channels that connect the lagoon with the open ocean. If the reef carbonate were to become reservoir rocks, these tidal channels would lead to a lithologic reservoir compartmentalization. When wave action erodes the reef, the reef carbonates are crushed into grains the deposition of which follows the local

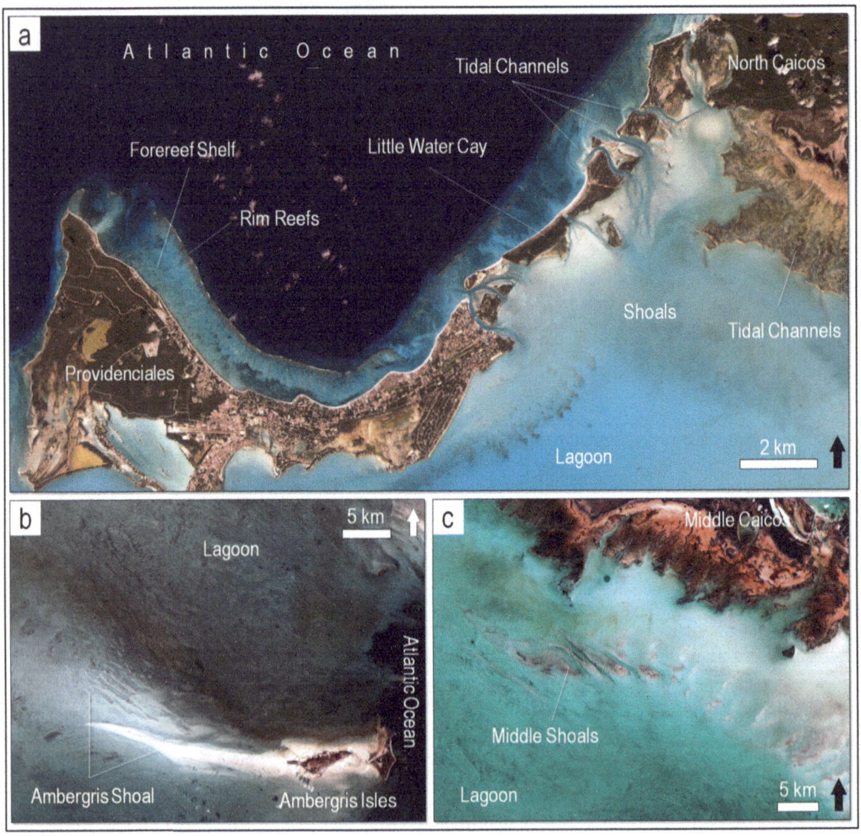

Fig. 12.20 Reefs, tidal channels and shoals of the Caicos reef on Landsat 8 multiband composites. (**a**) rim reefs and tidal channels offshore Providenciales and Little Water Cay islands, 432 RGB, (**b**) Ambergris shoals, 321 RGB, (**c**) Middle Shoals 431 RGB. (Satellite imagery courtesy of U.S. Geological Survey)

currents. We observe this process very clearly at the Ambergris Shoals west of the Ambergris Islands. Wave action, particularly from hurricanes has a preferred direction from the southeast. Hence the carbonate shoals resulting from the wave erosion of the Ambergris Islands are washed into the lagoon in a western direction. Less pronounced than the Ambergris Shoals at the Middle Shoals south of Middle Caicos (c). They are deposited in the lagoon in an area sheltered from direct wave action.

On Little Ambergris Island (Fig. 12.21), supra-tidal flats with algal mats contain significant amount of organic material, that is preserved in its deeper part in an anoxic environment. This depositional environment serves as a geologic analog for a source rock. We attempt the mapping of the algal mats in the context of the reef islands and the carbonate mud of the lagoon. The composite of 321 RGB penetrates the water and provides clues for the context around Little Ambergris Island. To improve the spatial resolution, we pan-sharpen the image with the panchromatic band 8 (a). The degradation for greater water depth is minimal in this case because of the very shallow water in the lagoon. We notice, however, that the sub-aerial island and the algal mats are shown in the pan-sharpened 321 RGB image in very similar brownish tones. The composite 742 RGB, which is often used for lithologic mapping of sub-aerial deposits differentiates well the supra-tidal algal flats (green)

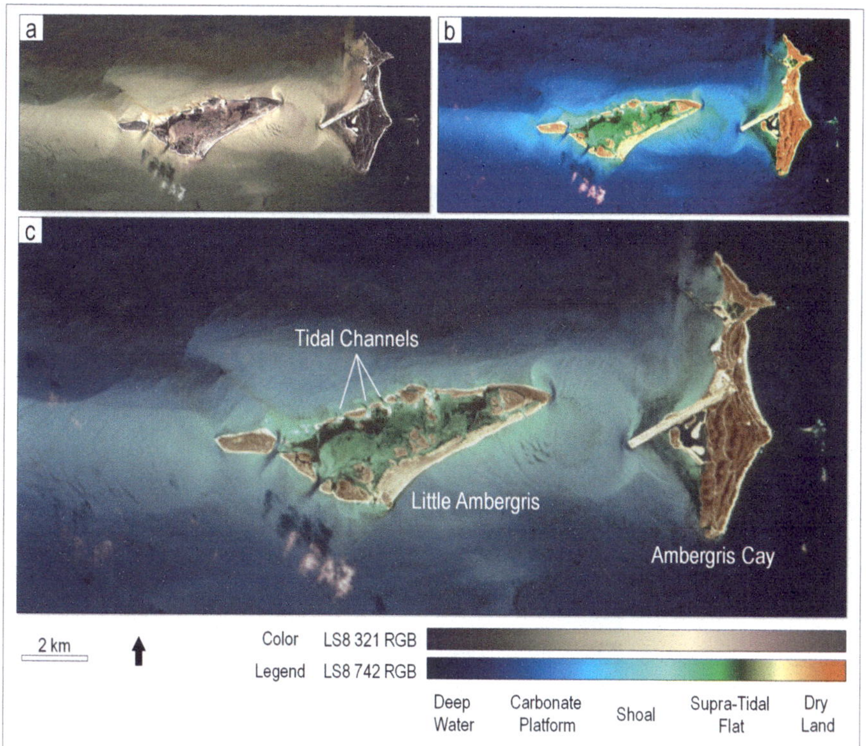

Fig. 12.21 Supra-tidal flats on Little Ambergris Island, Landsat 8 multiband composites without vertical exaggeration. (**a**) 321 RB pansharpened with band 8, (**b**) 753 RGB, (**c**) blend of pansharpened 321 RGB with 753 RGB. (Satellite imagery courtesy of U.S. Geological Survey)

from the dry islands (yellow to orange) but does not map well the sub-aquatic portions of the environment. When we merge both images (c) we benefit from both characteristics. The supra-tidal algal mats show in green, with shallower portions in lighter and deeper portions in darker tones. The currents around Little Ambergris Island, that deposit carbonate shoals are clearly distinguished by light blue-green to mauve tones. Tidal channels that bring oxygen-rich water into the lagoon and disturb the anoxic environment of the algal mats are clearly distinguished by their brighter color.

We have rendered these images as flat images without vertical exaggeration or slant illumination. The depositional environment of the algal mats is characterized by very subtle features that are reflected in subtle changes of the intensity of the green color. If we would introduce the texture of the digital elevation model and might even enhance the texture with slant illumination, we would loose the depositional detail.

So far, we have identified on the Caicos Platform geologic analogs for source and reservoir rocks. On West Caicos Island we can also find an analog for a potential seal rock: rock salt (Fig. 12.22). Salt evaporation pans, called Salinas, can be found where seawater penetrates loose sediments, and forms lakes that evaporate in hot and arid climate. West Caicos, however, shows two types of lakes side by side: saltwater salinas and a freshwater lake resulting from the freshwater phreatic lens (Bourrouilh-Le Jan, 1998).

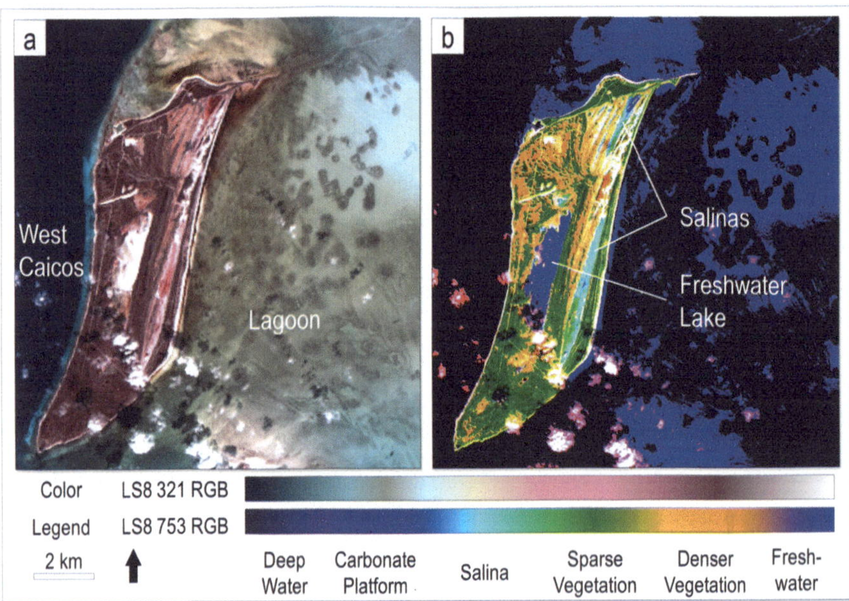

Fig. 12.22 Salinas on West Caicos Island, Landsat 8 multiband composites. (**a**) 321 RB, (**b**) 753 RGB. (Satellite imagery courtesy of U.S. Geological Survey)

The 321 RGB composite shows the freshwater lake in bright white, but only poorly distinguishes the Salinas from the surrounding salt-encrusted sediments (a). This band combination is also insensitive to vegetation. The 753 RGB composite uses shortwave and near infrared bands and hence discriminates the Salinas (cyan) from freshwater (blue) as well as from vegetation with orange indicating denser and green sparse vegetation. Similar to the previous example we also focus on subtle depositional features and hence render the images as flat images without vertical exaggeration.

We conclude the study on mapping geologic analogs in a carbonate depositional environment with the mapping of the floor of the lagoon (Fig. 12.23). We noticed already that the band combination 321 RGB provides the best penetration into the seawater due to the contribution of the coastal band 1 (a). Details of the seafloor texture in the lagoon give clues for the directions of the currents and subtle changes in the lithology.

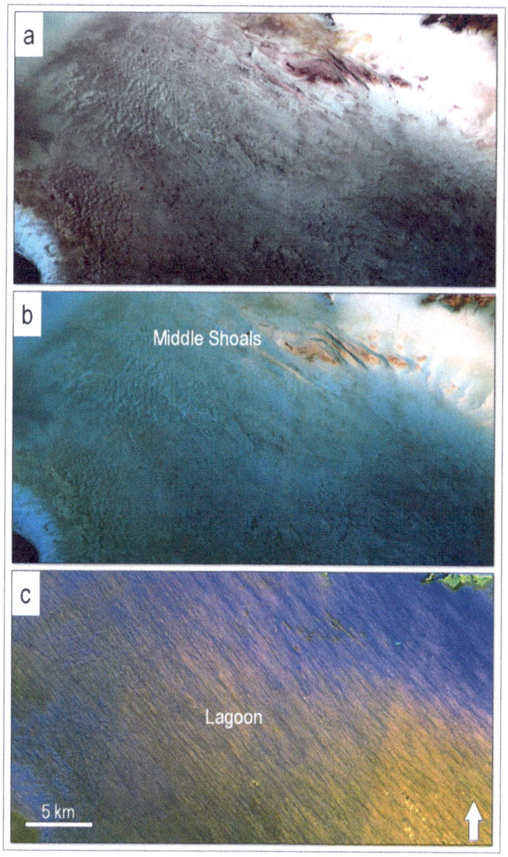

Fig. 12.23 Shoals and texture of carbonate muds on Caicos Platform, Landsat 8 multiband composites. (**a**) 321 RGB, (**b**) 432 RGB, (**c**) 753 RGB. (Satellite imagery courtesy of U.S. Geological Survey)

However, the shallow carbonate shoals of the Middle Shoals are not well distinguished from the carbonates in slightly deeper water. Furthermore, the very shallow shoals are represented by a bright white patch without any texture. This is where the natural color composite 432 RGB helps (b). The shallow shoals show as orange patches which are clearly distinguished from the green-cyan tones of the deeper portions of the lagoon. The very shallow shoals show the imprint of tidal currents perpendicular to the coastline. The lithology composite 753 RGB (c) does not reveal any lithological or depositional features below the water in line with the expectation from the poor water penetration of the bands used in this composite. Instead, it shows the surface texture of the water most likely indicating the wind direction at the time of image acquisition. The understanding of sensitivity of the band combinations to water penetration and lithology is very important for the correct interpretation of satellite images from shallow carbonate platforms for carbonate geologic analogs.

12.3.6 Marine Deltas: Irrawaddi and Sittang Deltas, Myanmar

We focus next on clastic depositional environments and start with the study of river deltas: the onshore Irrawaddy delta and the submerged offshore Sittang delta in Myanmar in Southeast Asia (Fig. 12.24). Deltas play an important role in exploration as high-porosity clastic reservoir rocks (Alsharhan, 1994a; Alsharhan & Nairn, 1997; Ahsan et al., 2011).

For the onshore Irrawaddi delta we use the SRTM DEM V3 elevation model (a). The delta floodplain is extremely flat. Therefore, we zoom into the elevation interval from sea level to 25 m (b) to reveal the detailed texture of the delta plain. We notice that the SRTM DEM can be estimated to have a vertical accuracy of about 1 m judging from the consistent appearance of the Irrawaddy delta channels. In the eastern part at the border to the Sittang delta we observe some noise stripes, which do no affect the analysis of the Irrawaddy onshore delta channels.

The Irrawaddy delta is about 200 km wide and stretches between the Arakan and Pegu Mountains. It is separated from the Sittang delta in the east by the Pegu Mountains which form a high on the western side of the Sagaing Fault (a), that offsets the Sittang delta by 110–220 km north relative to the Irrawaddy delta (Racey & Ridd, 2015a, b).

At Mandalay north of the Pegu Mountains, this movement that occurred in just the last 4 million years could have deviated the Irrawaddy river to the west thus abandoning its former delta between the Pegu Mountains and Shan Planteau, which now forms the Sittang delta.

For the mapping of the onshore delta channels we use two approaches: The extraction from the digital elevation model (b) and the identification of water bodies from multispectral satellite imagery using the 742 RGB composite from Landsat 8 (c). The digital elevation model provides a good estimate for the coastline and the larger inlets (blue features in d). The blue color of the Landsat 8 composite

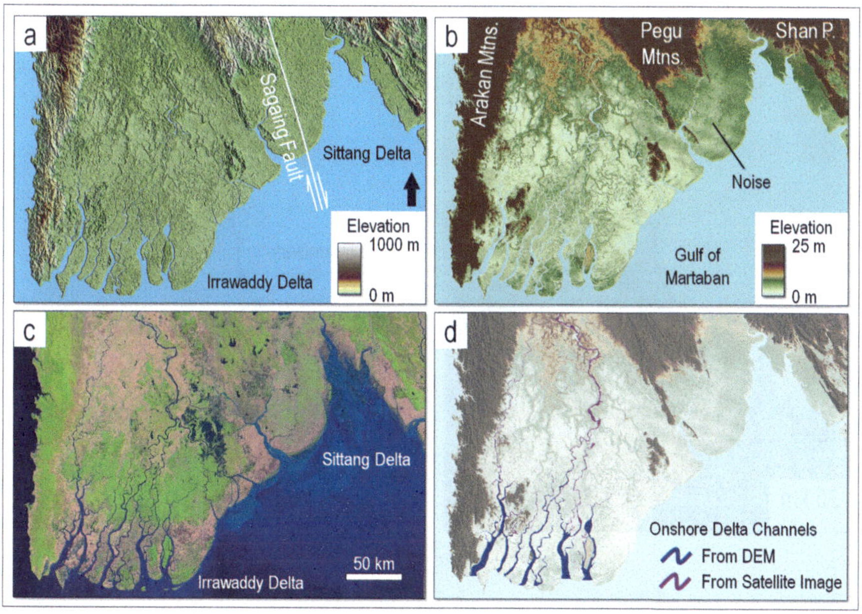

Fig. 12.24 Marine deltas of Irrawaddy and Sittang, Myanmar. (**a**) digital elevation model, (**b**) zoom into elevation interval of 0–25 m, (**c**) Landsat 8 multiband composite 742 RGB, (**d**) delta channels extracted from DEM and satellite imagery. (DEM and satellite imagery courtesy of U.S. Geological Survey)

delineates the delta channels as channel tributaries to the inlets (purple features in d). The two methods complement each other and deliver a map of the onshore delta channel network (d). This channel network represents a geologic analog for an onshore delta.

When we study the Landsat 8 composite 742 RGB offshore the mouth of the Sittang river into the gulf of Martaban, we observe a delicate channel pattern (Fig. 12.25a). We follow the interpretation that some 4 million years ago the Irrawaddy river changed its course north of the Pegu Mountains. The massive sedimentation rate provided by the Irrawaddy has been cut off leaving the Sittang delta sediment-starved and thus submerging. The present seafloor at up to 20 m water depth is dominated by fine sediments such as clays and silts (Ramaswamy & Rao, 2014).

To extract the texture of the submerged Sittang delta, we generate the Landsat 8 742 RGB composite image to capture the offshore depositional texture and convert it to a grayscale image to retain the absorption characteristics. In very shallow water, the grayscale intensity is correlated with the water depth so that we can convert the grayscale image to a pseudo elevation model showing the delicate texture of the offshore delta. For the geomorphological context we insert the offshore pseudo elevation model into the onshore SRTM DEM (b). The sharp straight line in east-west direction is an artefact from the boundary between the two Landsat 8 tiles that have been used to generate the grayscale image.

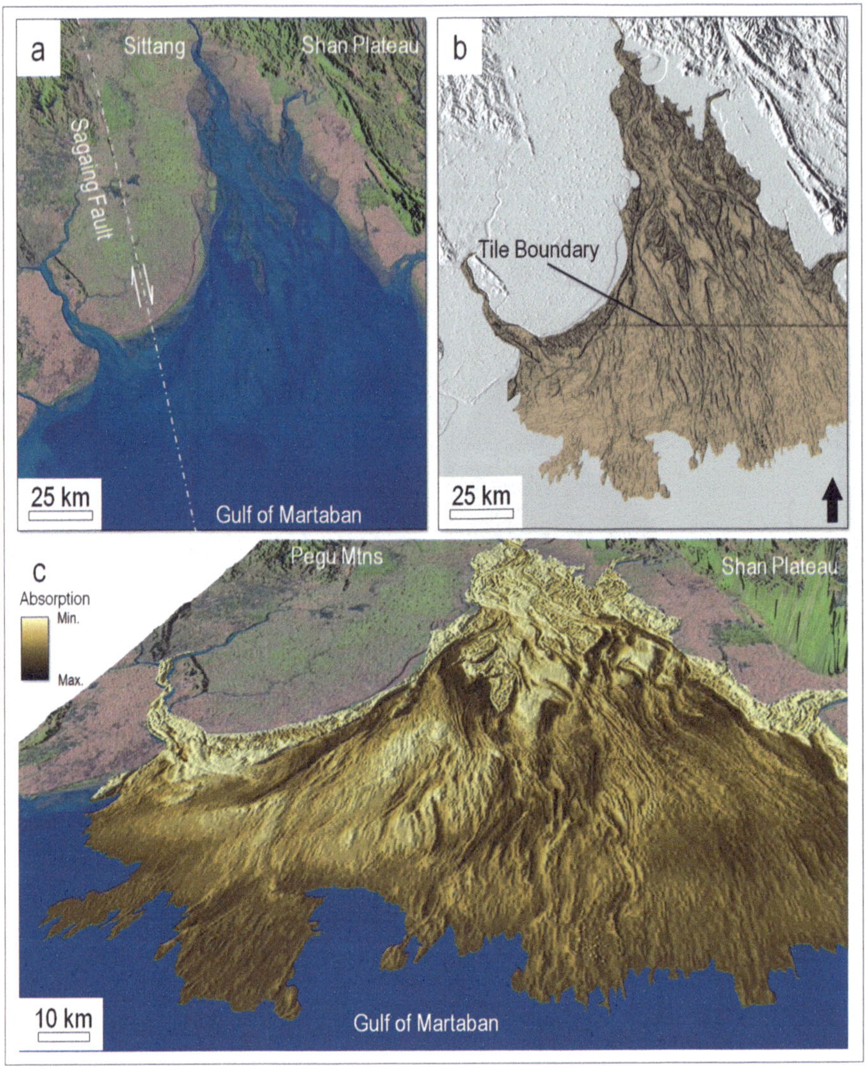

Fig. 12.25 Sittang offshore delta from Landsat 8 imagery. (**a**) 742 RGB composite, (**b**) 742 RGB composite converted to grayscale and rendered at 100× vertical exaggeration, (**c**) image from (**b**) converted to elevation and merged with SRTM DEM. Landsat 8 multiband composite 742 RGB draped over onshore DEM, grayscale of Landsat 742 RGB draped over Landsat 742 RGB grayscale converted to elevation. (DEM and satellite imagery courtesy of U.S. Geological Survey)

The rendering of the pseudo elevation model in 3D gives a good impression of the shape and texture of the delta (c). High values of absorption show as dark brown colors whereas low values of absorption, i.e. very shallow water, are represented as light colors. We use 200 times vertical exaggeration and slant illumination from the west to highlight the channel edges.

To summarize our observations, we compare the 742 RGB composite from Landsat 8 (Fig. 12.26a) with the overlay of the onshore and offshore delta channels on the digital elevation model (b). The size of the submerged Sittang delta appears much larger than the sediment transport the Sittang river can deposit. This

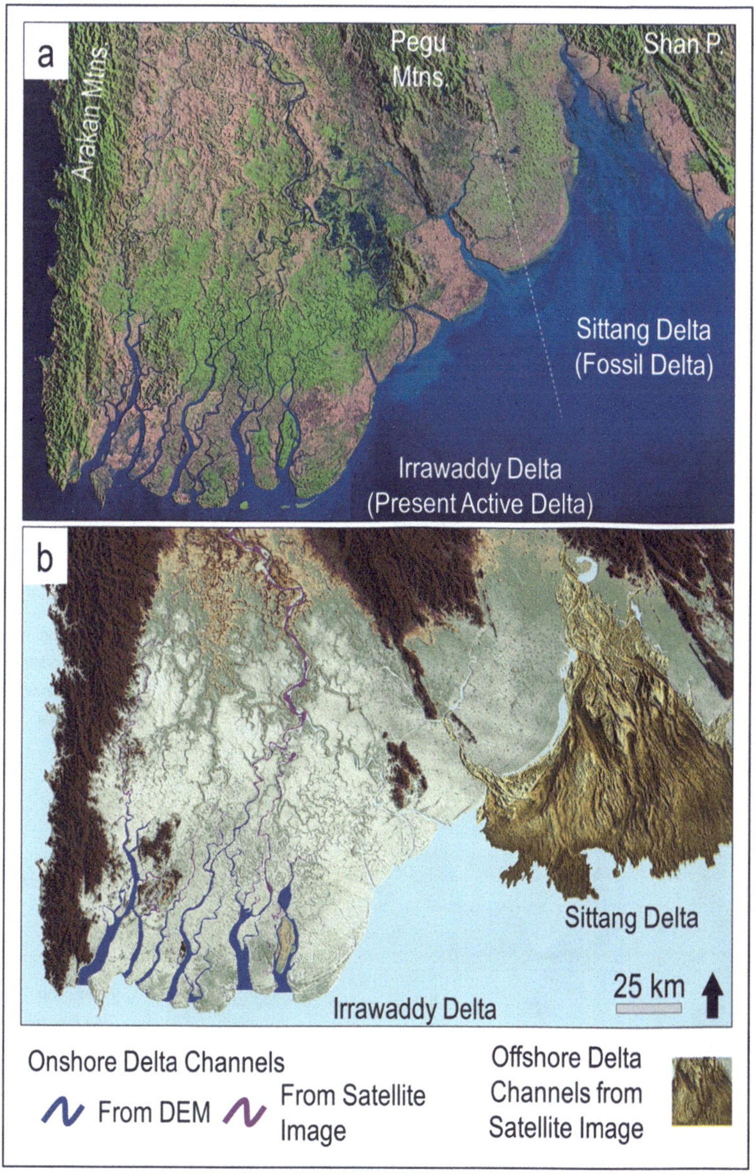

Fig. 12.26 Extraction of geologic analogs for onshore and offshore marine deltas from remote sensing data. (**a**) Landsat 8 multiband composite 742 RGB, (**b**) SRTM DEM zoomed to interval of 0–25 m elevation, delta channels and offshore delta extracted from DEM and satellite imagery overlaid. (DEM and satellite imagery courtesy of U.S. Geological Survey)

observation confirms the concept that the fossil Irrawaddy had used the Sittang valley to deposit the large delta until the Sagain Fault offset the delta and deviated the course of the Irrawaddy to the west. At its present location the modern Irrawaddy created an even larger delta.

12.3.7 Tidal Floodplain, North Frisia Germany

Sand deposits from tidal floodplain form significant reservoirs for example in the Middle East (Al-Eidan et al., 2001; Strohmenger et al., 2006; Filak et al., 2012). The depositional character and channel geomorphologies have been studied extensively on modern analogs (Price, 1963; Reineck, 1972; Middleton, 1991; Johansson, 2019; Fagherazzi et al., 2013; van Straaten, 1961).

We chose the tidal floodplains of the North Sea in North Frisia, Germany as a modern analog to study tidal deposition (Fig. 12.27). The selection of the band

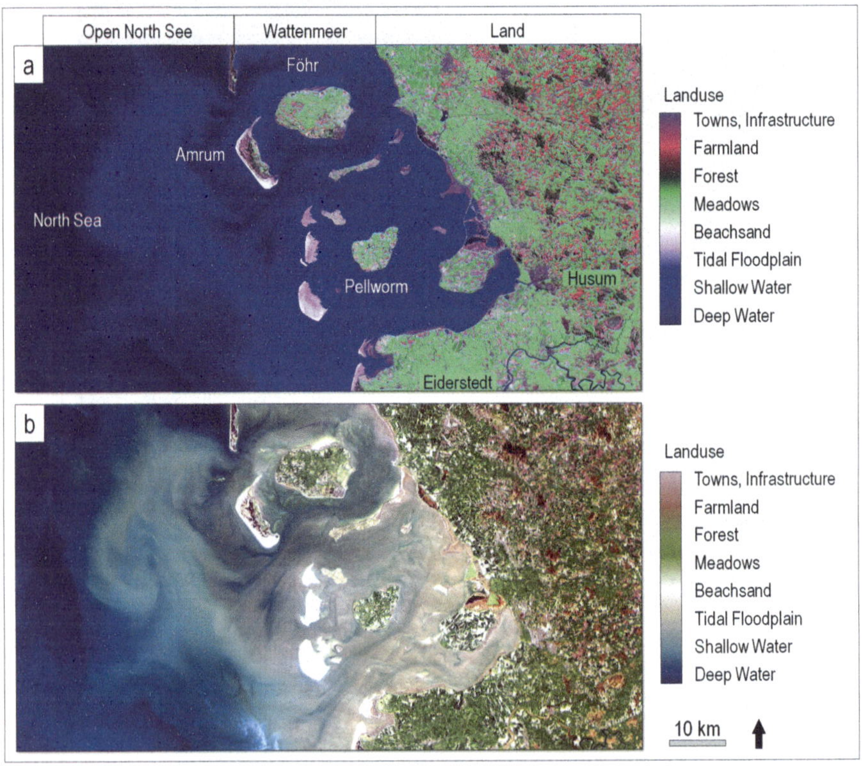

Fig. 12.27 Tidal channels in a wattenmeer environment, North Frisia coast, Germany, Landsat 8 multiband composites. (**a**) 753 RGB, (**b**) 431 RGB. (Satellite imagery courtesy of U.S. Geological Survey)

Fig. 12.28 Tidal channels of the North Frisia coast, Germany, offshore texture inferred from Landsat 431 RGB composite at 20× vertical exaggeration, onshore Landsat 8 432 RGB composite draped on SRTM DEM. (DEM and satellite imagery courtesy of U.S. Geological Survey)

combination 753 RGB of Landsat 8 (a), which is used for the mapping of vegetation, soil and water does not penetrate the water sufficiently and provides only a faint image of the tidal channels.

Focusing on the visible bands and including the ultra-blue coastal band in the multispectral band combination 431 RGB (b) delivers a clear image not only of the tidal channels in the Wattenmeer between the islands of Pellworm, Amrum and Föhr and the mainland, but also shows the currents in the open North Sea.

Due to sufficient penetration of the seawater the hue of the image allows the mapping of tidal floodplains, i.e. watt that is falling dry during ebb tide (beige tones). In deeper water, currents are visualized by the distinguished bright color of the sediment load. The modeling of tidal clastic reservoir sands needs to consider the channels and currents to generate realistic realizations of reservoir models.

To extract the texture of the tidal floodplain and tidal currents we follow the same approach as in the Sittang Delta example. This time we use the multiband composite 431 RGB of Landsat as the input because of the clearer water in the North Sea. We convert the RGB image to grayscale and infer the water depth from the intensity of the grayscale image (Fig. 12.28). For the offshore portion we exaggerate the inferred water depth by a factor 20 and use slant illumination from the northwest. This approach highlights the tidal channel system as well as the offshore currents associated with the currents in these channels.

12.3.8 Flood Plain Channels, Black River, AR U.S.A.

Onshore clastic deposits from flood plain channels play important role as high porosity reservoir rocks in the subsurface of the North Sea Basin (Bullimore & Helland-Hanse, 2009; Vondrak et al., 2018), the Iberian Peninsula (Franzel et al.,

2020), North America (Peacock, 2011; Wang & Plink-Björklund, 2019), and Mexico (San Roman & Yutsis, 2019). They are an essential ingredient in clastic reservoir modelling (Larue & Hovadik, 2006) when considering the complete cycle from source to sink (Sømme et al., 2009).

An important recent geologic analog are the Mississippi Valley and its tributaries. We chose the Black River floodplain in Arkansas which is flooded regularly and hence has retained much of its original character. One typical characteristic of flood plain channels is their perpetual change of the channel course, which leaves its imprint in subtle topographic lineaments. We use LiDAR to map the channel geometries that often amount to less than 0.5 m in elevation (Fig. 12.29). The regional overview (a) shows the floodplain that has an average width of 100 km at only a few me elevation change over a 150 km long stretch of the Black River.

At this large scale, the channels of the different age leave only a faint signature. When we change the colormap to show the subtle elevation changes in the flood plain in rapidly varying colors and leave the flanks in smooth colors, we start seeing more details of the individual channels (b). The best expression of the flood plain channel texture is achieved with vertical exaggeration and slant illumination of the

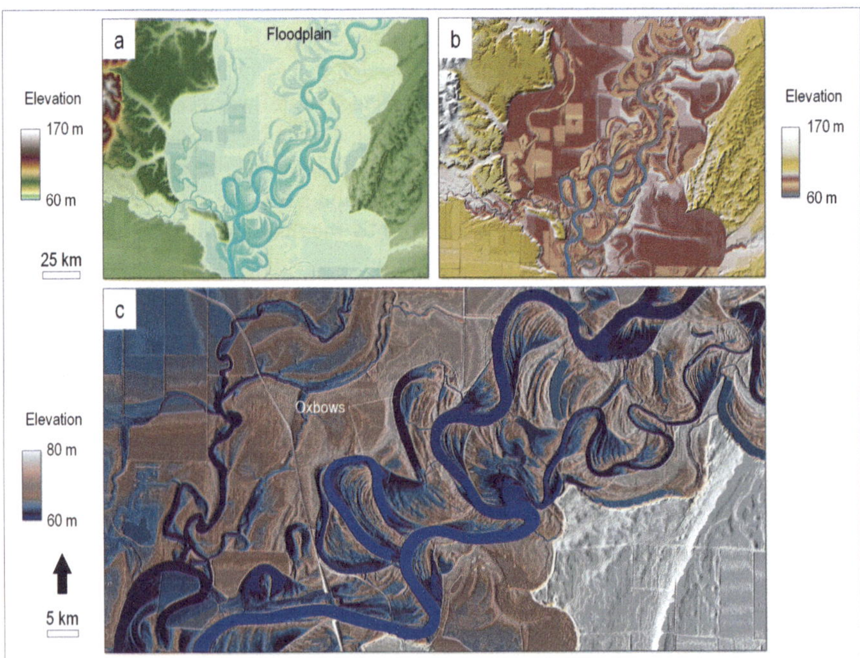

Fig. 12.29 Onshore floodplain channels, Black River floodplain, AR U.S.A., all images show LiDAR data draped on LiDAR DEM. (**a**) LiDAR DEM shown as topographic map without vertical exaggeration, (**b**) LiDAR DEM with focus on elevation range of floodplain at 5× vertical exaggeration, (**c**) zoom into floodplain with colormap highlighting the different levels of channel deposition at 20× vertical exaggeration. (LiDAR DEM courtesy of Arkansas GIS Office)

LiDAR DEM (c). To emphasize the present channels and oxbows, the open water surfaces are colored in blue. This approach reveals clearly the development of meanders into oxbows up to cutting off the abandoned channel. If found in a reservoir sand formation in the subsurface, abandoned channels would represent lithologic compartments in the reservoir.

12.3.9 Intra-Cratonic Deltas of Okavango, Botswana, and Kufra, Libya

Intra-cratonic deltas play an important role in reservoirs originating from clastic deposition in arid climate (Alsharhan, 1994b). For the remote sensing study, we have selected two geologic analogs: the recent Okavango Delta in Botswana and the Neogene Kufra Delta in Libya. For both cases we evaluate different remote sensing data to determine the best option.

The Okavango Delta is located at the edge of a southwest-northeast oriented graben the origin of which is interpreted as rifting associated with the East African Rift system (Gumbricht et al., 2001; Hutchins et al., 1976; Kinabo et al., 2008; McCarthy, 2013; Shemang & Molwalefhe, 2011). The rifting introduced a down throw of several 100 m that was temporarily filled with lakes. The Okavango river drained into these lakes from the northwest. The digital elevation model (Fig. 12.30a) reveals the graben and the graben shoulder with the Okavango incised into the plateau northwest of the graben shoulder. Radar backscatter data (b) highlight the bird-foot delta texture of the Okavango delta but the delta channels are not mapped

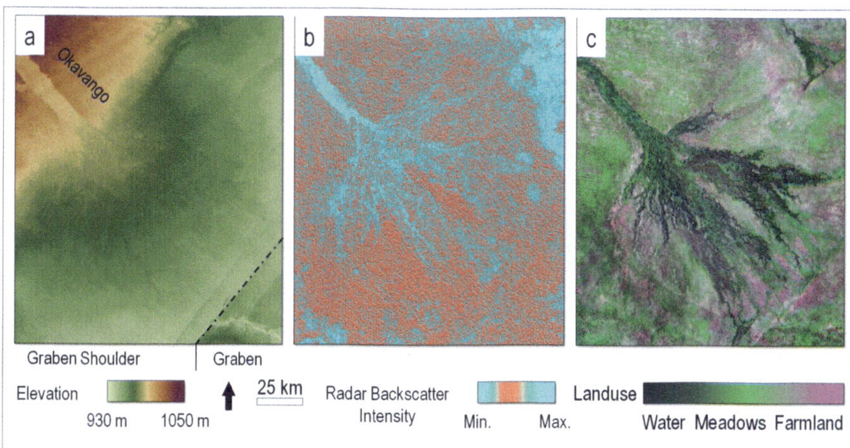

Fig. 12.30 Onshore delta of the Okavango River, Botswana. (**a**) SRTM DEM V3, (**b**) radar back scatter intensity, (**c**) Landsat 7 multiband composite 742 RGB. (DEM and Landsat satellite imagery courtesy of U.S. Geological Survey, Radarsat-1 imagery courtesy of CSA)

continuously because water of different salinity impacts the radar surface backscatter. Areas with lower salinity allow partial penetration of the microwaves into the ground whereas areas with higher salinity attenuate the radar signals almost completely. Optical satellite imagery (c) is significantly better suited to map the Okavango delta. The band combination 742 RGB from Landsat 7 allows a clear discrimination between water (dark), vegetation (different shades of green) and farmland (purple). In the usually arid climate of the surrounding Kalahari Desert, vegetation is associated with the delta channels of the Okavango river and, therefore, enhances the delineation of the channels.

The Neogene inland delta of the Kufra river in the Libyan Sahara desert, however, has been abandoned with the desertification of the Sahara at the end of the last ice age. The Kufra delta has been the focus of several articles using microwave radar to map paleochannels from the different backscattering intensity of sand and clay (see Sect. 2.4, Ghoneim et al., 2007; Paillou et al., 2007, 2009, 2012; Robinson et al., 2005).

The elevation model (Fig. 12.31a) shows two incised channels in south-north direction and a delta feature that is partially masked by Quaternary dune fields. The incised channels cannot be delineated clearly from the topography because of erosion and masking by dunes. Radar imagery (b) provides a very clear image of both, the incised channels (blue) in the Tibesti highlands (green to reddish) as well as the fan delta channel system (light green) in the plains of the Sirt Basin. The contrast of the channels from the surrounding sand is particularly strong. The visible band composite 432 RGB from Landsat 8, in contrast, only gives faint hints for the location of the delta channels and does not reveal the incised channel at all due to eolian deposits that superficially mask these channels.

Fig. 12.31 Pliocene onshore delta of the Kufrah River, central Libya. (**a**) SRTM DEM V3, (**b**) radar back scatter intensity, (**c**) Landsat 8 multiband composite 432 RGB. (DEM and Landsat satellite imagery courtesy of U.S. Geological Survey, Radarsat-1 imagery courtesy of CSA)

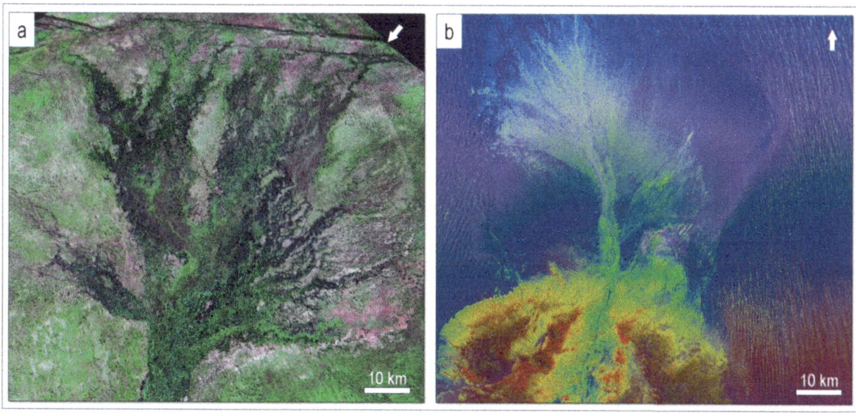

Fig. 12.32 Comparison of recent and fossil onshore deltas mapped using remote sensing imagery. (**a**) Okavango delta, Landsat 7 multiband composite 742 RGB, (**b**) Pliocene Kufrah delta, radar back scatter image merged with digital elevation model. (DEM and Landsat satellite imagery courtesy of U.S. Geological Survey, Radarsat-1 imagery courtesy of CSA)

For the comparison of the two intra-cratonic deltas we render those data that best reveal the delta channels, rotate the images such that both show the depositional direction towards the top and display them at the same scale (Fig. 12.32). For this study we render the respective data as flat maps without vertical exaggeration or slant illumination to concentrate on the surface cover. We notice that the incised Okavango river (a) is two to four times wider than the Kufra river (b) where it is incised in the Tibesti Mountains.

The delta of the Okavango river (a) is almost twice as big as the fossil Kufra delta (b). The birdfoot characteristics, however, is very comparable with dendritic channels developing from the sediment source of the tributary river. When building reservoir models using remote sensing imagery of intra-cratonic deltas at or very near the surface, we need to consider the climatic conditions to choose the satellite imagery that best reveals the channel system.

12.4 Erosional Analogs

12.4.1 Limestone Karsting, Karst Plateau, Slovenia and Italy

Karsting is a chemical erosion process that affects carbonates and evaporites when rainwater that had dissolved carbon dioxide penetrates the rocks in the subsurface. This process is most efficient in pure limestone, specifically along fractures. In humid climate, chemical erosion is complemented with erosion of running water thus creating caves in the subsurface (Onac & van Beynen, 2021; Klimchouk, 2017b; Parise et al., 2018; Phillips, 2017). Limestone karst is an important carbonate reservoir setting because the karsting process creates substantial pore space in the subsurface. Karst-related reservoirs are found in the Middle East (Alsharhan & Whitlle, 1995; Farooq et al., 2020). LiDAR is frequently used to map karst features at the surface (Boersma et al., 2019; Jeffrey et al., 2020; Petrovic et al., 2018; Vinci & Bernardini, 2017).

We have selected the type location for limestone karsting, the Karst Plateau in the northwest Balkan Mountains north and east of Trieste, which gave the erosional process its name. The Karst Plateau is mainly composed of pure limestone. It has been fractured during the Alpine orogenesis and has been exposed to a humid climate during the Pleistocene. The combination of these factors created ideal conditions for karsting on a large scale. Today, almost the entire Karst Plateau and large portions of the Balkan Mountains along the Adriatic coast are heavily karsted.

We begin the study of the surface karsting with the SRTM digital elevation model (Fig. 12.33a), which shows the slightly westward dipping Karst Plateau, that is

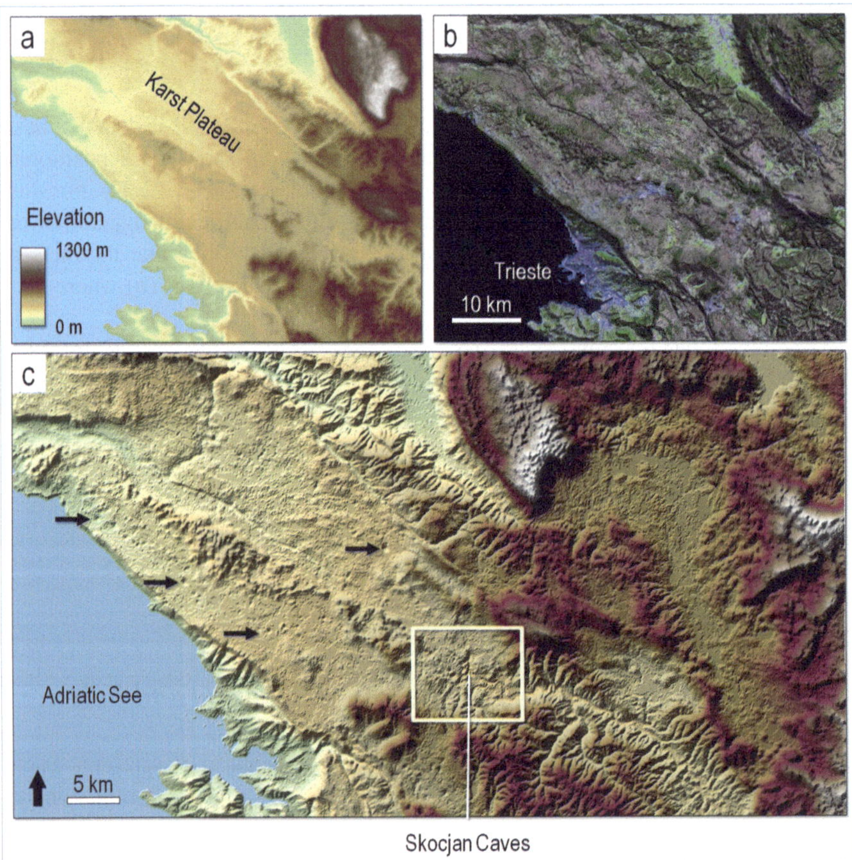

Fig. 12.33 Limestone karsting on the Karst Plateau, Slovenia and Italy. (**a**) SRTM DEM V3 as flat topographic map, (**b**) Landsat 8 multiband composite 742 RGB, (**c**) SRTM DEM V3 at 5× vertical exaggeration. (DEM and satellite imagery courtesy of U.S. Geological Survey)

composed of two northwest-southeast trending sections. The Landsat 8 multiband composite 742 RGB (b) shows the fairly homogeneous character of the plateau, comprising agricultural landuse (gray to pink), forest (dark green), meadows (light green) and infrastructure (bright pink). The water of the rivers and the Adriatic Sea appears black. With both data sets in flat map rendering we cannot detect karst-specific sinkholes. Therefore, we render the digital elevation at 5× vertical exaggeration and light the resulting structures with slant illumination from the north and northeast (c). We notice numerous large sinkholes on the Karst Plateau, two of which are indicated by the arrows. We will now study the area of the Skocjan Caves that is marked by the light yellow box in more detail.

The Skocjan Caves are designated a UNESCO Heritage Site which has been studied extensively both at the surface as well as in the subsurface (Novakovic et al., 2014). Near Skocjan the Karst Plateau is karsted so extensively that large underground caves have been created by underground rivers like the Timavo River. Locally the erosion progressed so far that the roofs of the caves collapsed making room for the subaerial flow of the river.

Figure 12.34 shows the Landsat 8 753 RGB composite draped over the SRTM digital elevation model, which reveals numerous karst holes or dolines west and north of Skocjan whereas in the northwest portion of the image the Karst Plateau has largely sustained the erosion. The satellite image shows the dolines are covered in trees or meadows (dark to light green colors). Most of them are aligned along structural lineaments trending northwest-southeast.

The Timavo River has its source in the caves, flows in a narrow canyon of a collapsed former cave southeast of Skocjan. Some 5 km further east the collapsed area opens up to a meadow plain.

Fig. 12.34 Landsat 8 753 RGB composite draped over the SRTM DEM V3 digital elevation model at 3× vertical exaggeration and slant illumination. (Data courtesy of U.S. Geological Survey)

Fig. 12.35 Detail of river karsting of the Timavo River at Skocjan, Slovenia. SRTM DEM V3 at 5× vertical exaggeration. (DEM courtesy of U.S. Geological Survey)

The heavily karsted surface texture becomes even more evident when rendering the digital elevation model at 5× vertical exaggeration and slant illumination (Fig. 12.35). The karst features are most obvious at the saddle between the Timavo River canyon in the southeast and the valley to the northwest of Skocjan. Most of them are aligned along northwest-southeast trending lineaments.

The reason for the alignment along tectonic lineaments is that faults introduce weak zones into the limestone, which is more susceptible to erosion, particularly to karsting. We observe this very clearly at the karst features on the Hassi R'Mel carbonate plain in the southern foreland of the Atlas Mountains in Algeria (Fig. 12.36). The Landsat 8 composite overview image (a) shows northwest-southeast trending regional structural lineaments from compressional stress perpendicular to the Atlas Mountain ranges.

When we zoom into the yellow area (b) we notice that dolines, the most frequent karst features in this area, are covered in vegetation (green) because their depressions protect the soil from eolian erosion. The carbonate plain itself consists mainly of barren rock, which clearly shows the striations of the tectonic lineaments, that are indicated by yellow lines. Such aligned karst features have been observed also in the Ordovician limestone karst reservoirs in the Tarim Basin in China (Yang et al., 2018; Laake et al., 2014b, c). They form deep fossil case systems that are prolific reservoirs.

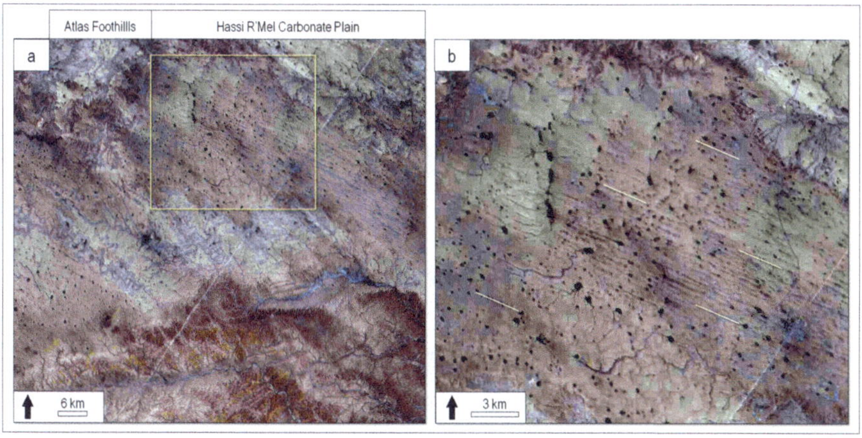

Fig. 12.36 Alignment of karst features along tectonic lineaments, Hassi R'Mel, Algeria. (**a**) overview of the carbonate plain, (**b**) detail of karst features along tectonic lineaments. Landsat 8 753 RGB composite. (Satellite data courtesy of U.S. Geological Survey)

12.4.2 Glacial Valleys, Jutland, Denmark

Reservoirs originating from glacial deposits have been discovered in the Paleozoic strata of North Africa and the Arabian Peninsula (Douillet et al., 2012 for Jordan, LeHeron et al., 2004 for North Africa and Moscariello et al., 2009 for the Arabian Peninsula). Their modern depositional analogs from the Quaternary glaciation, therefore, assist in the modeling of the reservoir facies and depositional environments. Overviews of the reservoir facies and their analogs have been collected by Huuse et al. (2012) and Dowdeswell et al. (2016). The glacial deposition and erosional framework and the related nomenclature can be found in Giles et al. (2017).

In the first section on glacial aspects of reservoir analogues we will study glaciofluvial landforms, both sub-aerial valleys from free-surface glacial runoff as well as subglacial tunnel valleys in Jutland in Denmark. The term tunnel valley was introduced by Madsen (1921) for sub-glacially meltwater-eroded valleys in Jutland, Denmark.

The origin of tunnel valleys has been described in van der Vegt et al. (2012) and Kehew and Kozlowski (2007), studies combining surface and subsurface interpretation of tunnel valleys have been published by Sandersen and Jørgensen (2012), Sandersen et al. (2009), Andersen et al. (2012), and Müther et al. (2012). 3D seismic studies found that the tunnel valleys also extend from onshore into the subsurface of the present-day North Sea (Kristensen & Huuse, 2012).

We analyze the surface elevation and its gradient using the SRTM DEM digital elevation model (Fig. 12.37). With this analysis we need to consider that the present-day distribution of land and sea has changed significantly since the end of the last ice age due to the postglacial sealevel rise of 150–200 m (Lambeck et al., 1998).

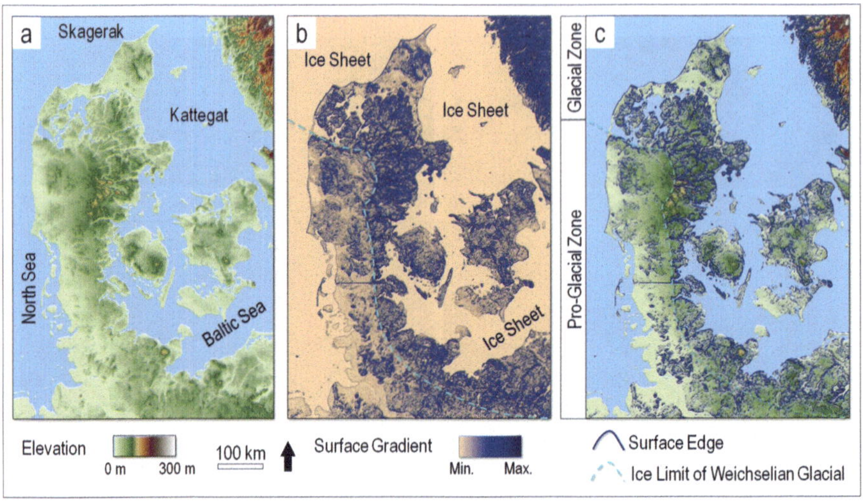

Fig. 12.37 Topographic representation of glaciofluvial features in Jutland, Denmark from SRTM DEM digital elevation data. (**a**) topography, (**b**) surface gradient, (**c**) surface edge on SRTM DEM. (SRTM DEM V3 data courtesy of U.S. Geological Survey)

Therefore, the present-day satellite derived SRTM digital elevation model represents only part of the Quaternary glacial features. Furthermore, the glacial features found in the SRTM DEM are dominated by the overprint of the last stage of ice advance, the Weichselian Glacial, the ice limit of which is evident in the surface gradient data (b). It forms the boundary between a rough area in the east, i.e. the glacial zone, and smooth area in the west, the pro-glacial zone. In (c) we render just the high values of the surface gradient, which are essentially the surface edge showing the distribution of the surface imprint of the sub-glacial tunnel valleys. To the west of the line of the Weichselian ice limit, the valleys are sub-aerial.

We can study this correlation in regional 3D rendering of the SRTM DEM for Denmark and the adjacent areas (Fig. 12.38). Western Denmark and north Germany were ice free at that time with the sub-aerial Elbe meltwater valley overprinting the southern portion of the area. The rough topography to north and east of the Weichselian ice correlates with glacial deposits that are locally eroded into by sub-glacial tunnel valleys. To the far north east the basement rocks of the Scandes are exposed. Most of the subglacial area is presently submerged under the Baltic Sea, Kattegat and North Sea.

With this understanding we return now to Jutland and extract the sub-aerial and tunnel valleys from the surface elevation (Fig. 12.39). Jørgensen and Sandersen

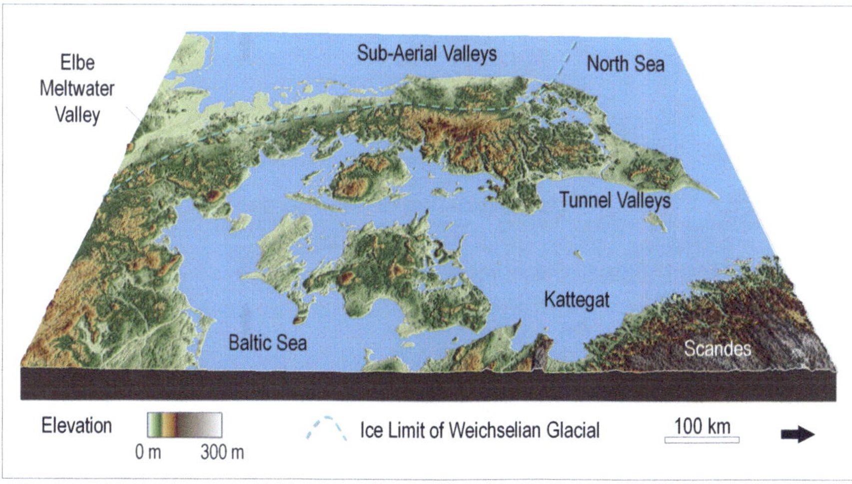

Fig. 12.38 Overview of the surface texture of Denmark and adjacent areas showing the distribution of glacial sub-aerial and tunnel valleys

(2006) have researched selected tunnel valleys in Jutland integrating surface topography with subsurface geoelectric and seismic data.

From the remote sensing point of view we consider only the surface topography obtained from the SRTM DEM V3 (Fig. 12.39). This limits the detection of glacial valleys to those that show a surface expression, such as exposed tunnel valleys. Buried valleys cannot be delineated with this approach and require geophysical methods such as seismic and geoelectric methods. The digital elevation model (a) reveals a highly channelized terrain north and east of the Weichselian Glacial Ice Limit whereas the terrain is largely smooth and flat to the southwestern pro-glacial. Incised valleys with an orientation north to south in the west and east to west in the east (bluish channels in b) are interpreted as tunnel valleys originating from sub-glacial meltwater erosion into the underlying substrate. In addition to the tunnel valleys we observe a second type of channels that are running mainly perpendicular to the tunnel valleys and show a broader meandering shape (reddish channels in b).

We interpret these as sub-aerial glacial channels that originate from meltwater flowing parallel to later stages of the ice advance, when the glaciers had retreated further. These channels overprint the tunnel valley pattern and connect some of the tunnel valleys thus forming a new drainage pattern.

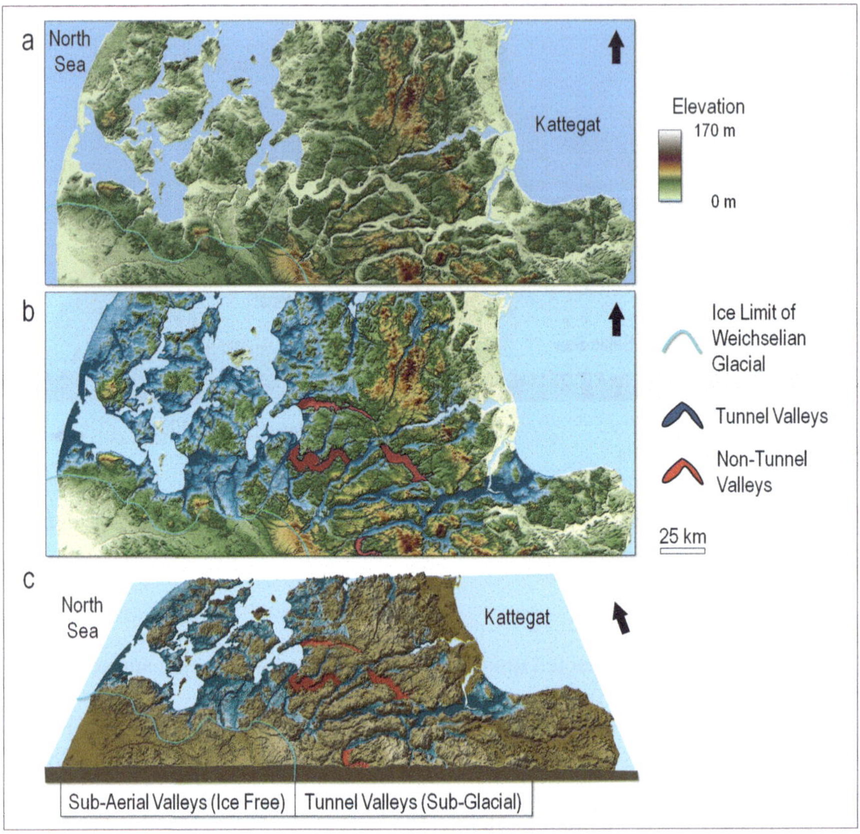

Fig. 12.39 Extraction of sub-aerial and tunnel valleys in Jutland, Denmark. (**a**) digital elevation model, (**b**) overlay of tunnel valleys and non-glacial valleys on digital elevation model, (**c**) idem in 3D at 50× vertical exaggeration. (SRTM DEM V3 courtesy of U.S. Geological Survey)

12.4.3 Regional Glacial Mapping for Moraines, Great Lakes Basin, USA

Glacial moraines play an important role in glaciogenic reservoirs mostly from the Paleozoic in South America, Australia and the Middle East (Huuse et al., 2012). Their Pleistocene analogs are also important as groundwater resources. Early studies of the Quaternary glaciation in North America date back to the turn of the nineteenth to twentieth century with the introduction of the concept of moraines by Taylor (1897).

The regional Pleistocene geology has been compiled as early as the 1900s (Wright, 1900; Leverett & Taylor, 1915). The geology of ice sheets used the mapping of moraines for the delineation of paleo-ice streams (Boulton et al., 1985). In 1987 the Geological Survey of Canada published a series of maps reconstructing the Laurentian ice sheet in northern North America (Dyke & Prest, multiple maps).

In this section we study the moraines left behind by different stages of the Laurentide ice sheet in the Great Lakes area in North America (Fig. 12.40). The digital elevation model (a) is largely based on the SRTM DEM. For the bathymetry of the Great Lakes we amend the bedrock elevation from ETOPO1, which comprises bathymetry measurements in the Great Lakes, thereby providing a continuous surface for the glacial deposition of the Laurentide ice sheet.

The surface gradient of the combined digital elevation model (b) highlights the glacial lineaments particularly of different types of moraines and glacial drainage. The moraines from different stages of the Laurentide ice sheet extent become obvious when rendering the digital elevation model in 3D at 100× vertical exaggeration (c). Arrows indicate the location of large moraine ridges. We have added the water surface of the Great Lakes as semi-transparent blue surfaces using the SRTM DEM elevation model.

Finally, we study details of the moraines from different stages of the Laurentide ice sheet for the southern part of the Great Lakes (Fig. 12.41). In this area the Laurentide ice sheet advanced in three lobes, the Lake Michigan Lobe that carved out Lake Michigan in the west, the Saginaw Lobe in the center and the Erie Lobe at the location of present-day Lake Erie in the east. The digital elevation model gives clues for moraine belts that are indicated by arrows. The retreat of the glaciers starting after the Last Glacial Maximum (LGM) at about 18 kYBP resulted in massive glacial runoff south towards the Mississippi River (MR) that created major glacial river valleys, for example the Illinois River (IR), Wabasha River (WR), and Ohio River (OR).

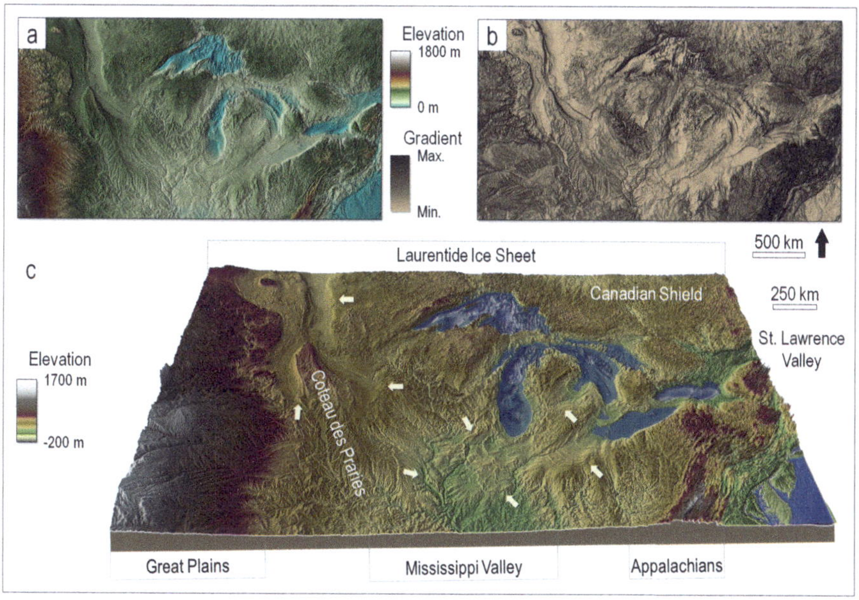

Fig. 12.40 Pleistocene moraines in the Great Lakes Basin, North America. (**a**) combined digital elevation model of SRTM DEM and ETOPO1, (**b**) Surface gradient, (**c**) 3D rendering at 100× vertical exaggeration. (SRTM DEM V3 and ETOPO1 digital elevation data courtesy of U.S. Geological Survey)

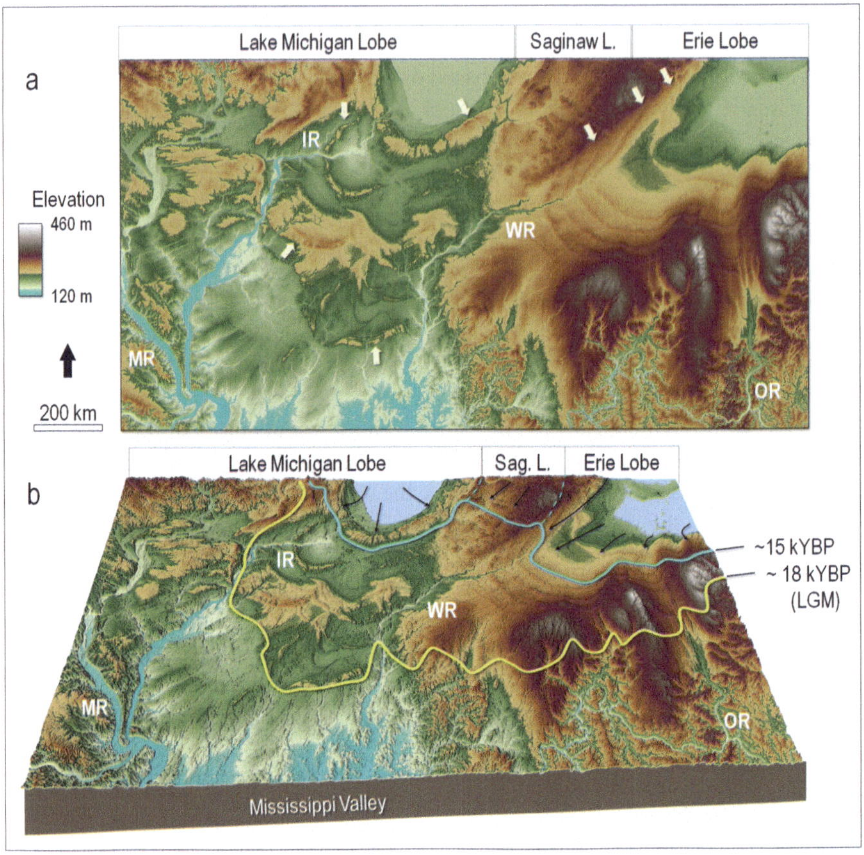

Fig. 12.41 Focus on the glacial moraines of three lobes of the Laurentide ice sheet. (**a**) digital elevation model, arrows point to the moraines, (**b**) 3D view at 100× vertical exaggeration showing ice extent at 18 and 15 kYBP. Rivers: *IR* Illinois, *MR* Mississippi, *WR* Wabasha, *OR* Ohio. *LGM* Last Glacial Maximum. (SRTM DEM V3 courtesy of U.S. Geological Survey)

Rendering the digital elevation model in 3D highlights the moraines that mark at least six different major stages of glacial retreat that have been mapped by Dyke and Prest (1987a, b). The moraine belt furthest south is correlated with the LGM at 18 kYBP when the three lobes formed one ice sheet front. Three thousand years later, the ice had melted such that the Lake Michigan and Erie Lobes had retreated into the present-day lake beds.

Mapping glacial moraines is not only important as a geologic analog. The local basins between the moraine ridges often form important basins for groundwater.

12.4.4 Glacial Valleys, Finger Lakes NY, U.S.A.

To conclude this section, we study elongated valleys that had been carved out by glacier lobes. Records in the subsurface show that these valleys are often filled with porous deposits making them potential reservoirs. A prominent example for such valleys is the Finger Lakes southeast of Lake Ontario (Mullins et al., 1996; Sookhan et al. (2021)). At the last glacial maximum, the Lake Ontario Lobe of the Laurentide ice sheet extended over the Allegheny Plateau. Glaciers carved out long valleys in roughly north-south direction. After deglaciation the depressions were filled with meltwater and the moraines of the Lake Ontario plains provided dams protecting the finger-shaped lakes from drainage (Fig. 12.42).

The digital elevation model (a) shows two distinctly different zones: the finger lakes carved into the Allegheny Plateau and the subglacial plain southeast of Lake Ontario. These are dotted with drumlins which we will study later. The surface gradient (b) highlights the sharp outlines of the finger lakes. At their upper end in the Allegheny Plateau they show dendritic valley patterns that lead into the finger lake basins, most likely the result of post-glacial erosion of rivers flowing into the finger lakes. At their lower end the finger lakes are dammed towards the subglacial plains at Lake Ontario (Chamberlin, 1882). The drainage of the lakes often runs along the edge of the plains before it drains into Lake Ontario. The plains itself are covered in aligned mounds called drumlins, that originate from subglacial sediment displacement. The Landsat 8 753 RGB satellite image (c) allows only the distinction of forested terrain and water surfaces of the finger lakes and Lake Ontario and does not contribute to the terrain analysis.

When vertically exaggerating the digital elevation model we obtain a much better insight into the texture of the drumlin field. The alignment of drumlins along glacial crevasses has been observed in active drumlin fields in Iceland (Johnson et al., 2010). In Fig. 12.43 we render the digital elevation model in map view with 50 vertical exaggeration. To even enhance the textural expression, we add colored slant illumination from two different angles.

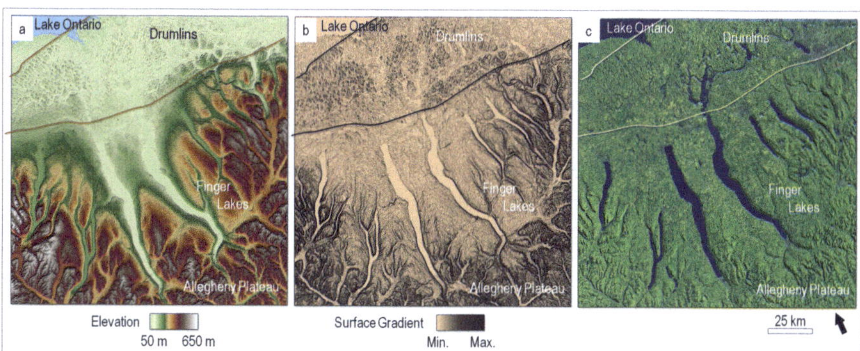

Fig. 12.42 Geomorphologic setting of the Finger Lakes, New York. (**a**) Digital elevation model, (**b**) surface gradient, (**c**) satellite image Landsat 7 742 RGB composite. (SRTM DEM V3 and satellite imagery courtesy of U.S. Geological Survey)

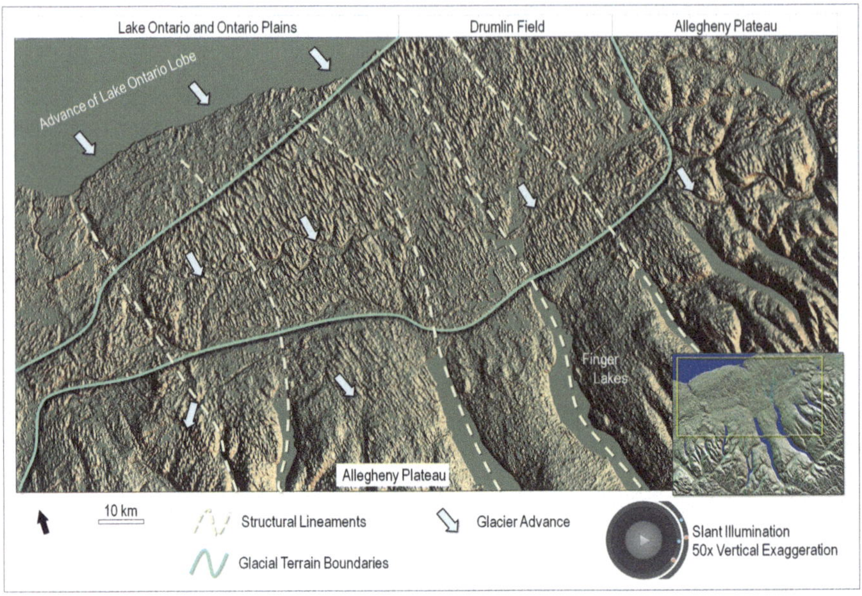

Fig. 12.43 Drumlin field between the Finger Lakes and Lake Ontario. DEM 50× vertical exaggeration and colored slant illumination with glaciologic interpretation. (SRTM DEM V3 courtesy of U.S. Geological Survey)

This method allows distinguishing the three glacial zones present in the area between Lake Ontario and the Allegheny Plateau:

1. Lake Ontario, a glacial basin carved out by the Lake Ontario Lobe of the Laurentide ice sheet,
2. the plains between the lake shore and the foot of the Allegheny Plateau. The sub-glacial texture is represented by drumlins that are arranged along the local advance of the ice sheet. The direction of the ice sheet advance is indicated by arrows.
3. The glacially overprinted northwestern Allegheny Plateau with striations on the surface strata, that are also indicating the direction of the ice sheet advance.

The use of drumlin alignment and surface rock striations to detect the direction of the ice sheet advance can assist in the interpretation of ancient glacial deposits to detect potential reservoir formations, where the Finger Lakes form a continuous body of similar sediment fill which is delimited by basement rock on the sides and moraines at the lake's exit toward the sub-glacial plain.

References

Outcrop Analog: Arches NP

Boulton, G.S., Smith, G.D., Jones, A.S. & Newsome, J. (1985). Glacial geology and glaciology of the las mid-latitude ice sheet. *Journal of the Geological Society London 142*, 447–474.

Chamberlin, T. C. (1882). Preliminary paper on the terminal moraine of the second glacial epoch, in: Powell, J.W., Third Annual report of the United States Geological Survey to the Secretary of the Interior, 1881-1882, 295–402. https://doi.org/10.3133/ar3

Doelling, H. H. (2000). Geologic road, trail, and lake guides to Utah's Parks and Monuments. In P. B. Anderson & D. A. Sprinkel (Eds.), *2000 Utah Geological Association Publication 29*. Utah Geological Survey. 44 p.

Mullins, H.T., Hinchey, E.J., Wellner, R.W., Stephens, D.B., Anderson, W.T.Jr. & Dwyer, T.R. (1996). Seismic stratigraphy of the Finger Lakes: A continental record of Heinrich event H-1 and Laurentide ice sheet instability, in: Mullins, H. T., and Eyles, N., eds., Subsurface Geologic Investigations of New York Finger Lakes: Implications for Late Quaternary Deglaciation and Environmental Change: Boulder, Colorado, Geological Society of America Special, Paper 311.

Rotevatn, A., Fossen, H., Hesthammer, J., Aas, T. E., & Howell, J. A. (2007). Are relay ramps conduits for fluid flow? Structural analysis of a relay ramp in Arches National Park, Utah. *Geological Society of London, Special Publication, 270*, 55–71. https://doi.org/10.1144/GSL.SP.2007.270.01.04

Sookhan, S., Eyles, N., Bukhari, S. & Paulen, R.C. (2021). LiDAR-based quantitative assessment of drumlin to mega-scale glacial lineation continuums and flow of the paleo Seneca-Cayuga paleo-ice stream, Quaternary Science Reviews 263. https://doi.org/10.1016/j.quascirev.2021.107003

Van Gent, H., & Urai, J. L. (2020). Abutting faults: A case study of the evolution of strain at Courthouse branch point, Moab Fault, Utah. *Solid Earth, 11*, 513–526. https://doi.org/10.5194/se-11-513-2020

Subsurface Case Studies of Relay Ramps

Botter, C., Cardozo, N., Lecomte, I., Rotevatn, A., & Paton, G. (2017). The impact of faults and fluid flow on seismic images of a relay ramp over production time. *Petroleum Geoscience, 23*(1), 17–28. https://doi.org/10.1144/petgeo2016-027

Destro, N. (1995). Release fault: A variety of cross fault in linked extensional fault systems, in the Sergipe-Alagoas Basin, NE Brazil. *Journal of Structural Geology, 17*, 615–629. https://doi.org/10.1016/0191-8141(94)00088-H

Dimmen, V., Rotevatn, A., & Nixon, C. W. (2020). The relationship between fluid flow, structures, and depositional architecture in sedimentary rocks: An example-based overview. *Geofluids, 2020*, 3506743. https://doi.org/10.1155/2020/3506743

Khalil, S. M., & McClay, K. R. (2017). 3D geometry and kinematic evolution of extensional fault-related folds, NW Red Sea, Egypt, in: Childs, C., Holdsworth, R. E., Jackson, C. A.-L., Manzocchi, T., Walsh, J. J. & Yielding, G. (eds), The Geometry and growth of normal faults. *Geological Society, London, Special Publications, 439*, 109–130. https://doi.org/10.1144/SP439.11

Laake, A. (2018). Omnidirectionally sampled 3D seismic data give detailed insight into fault-fold relations. In *SEG 88th Annual Meeting* (pp. 1628–1632). https://doi.org/10.1190/segam2018-2995179.1

Torabi, A., Johanneessn, M. U., & Ellingsen, T. S. S. (2019). Fault core thickness: Insights from siliciclastic and carbonate rocks. *Geofluids, 2019*, 2918673. https://doi.org/10.1155/2019/2918673

Outcrop Analog: Hassi R'Mel and Talemzane Meteor Crater

Belhai, D., & Sahoui, R. (2017). Confirmation of the Talemzane structure (Maadna) as a meteoritic impact crater by new criteria. In *80th Annual Meeting of the Meteoritical Society, LPI Contribution No. 1987.*

Klimchouk, A. B. (2017a). Types and settings of hypogene karst. In A. Klimchouk, A. N. Palmer, J. de Waele, A. S. Auler, & P. Audra (Eds.), *Hypogene karst regions and caves of the world* (pp. 1–39). Springer. https://doi.org/10.1007/978-3-319-53348-3_1

Koeberl, C. (1994). African meteorite impact craters: Characteristics and geological importance. *Journal of African Earth Sciences, 18*(4), 263–295.

Surface and Shallow Subsurface Karst

Parise, M., Gabrovsek, F., Kaufmann, G., & Ravbar, N. (2018). Recent advances in karst research: From theory to fieldwork and applications, in: Parise, M., Kaufmann, G. & Ravbar, N., Advances in karst research: Theory, fieldwork and applications. *Geological Society, London, Special Publications, 466*, 1–24. https://doi.org/10.1144/SP466.26

Deep Subsurface Karst and Karsted Reservoirs

El-Emam, A., Ebaid, A., Zahran, W., Al-Ajmi, B., Cunnel, C., Laake, A., Sharaf El Din, R., & Wahab, H. (2013). Delineation of karsts, a new approach using seismic attributes – Case study from Kuwait. In *SEG 2013 Technical Program Expanded Abstracts* (pp. 3846–3850). https://doi.org/10.1190/segam2013-0927.1

Kumar, S., El Gezeery, T., & Al-Anezi, K. (2009). The application of attributes derived from high resolution seismic data in horizontal drilling—A case study from Shuaiba formation West Kuwait. In *International Petroleum Technology Conference, Paper IPTC 14019*. 15 p.

Laake, A., Zhaoming, W., Gengxin, P., Duoming, Z., & Yongfu, C. (2014a). Mapping ultra-deep karst in its geological context. In *SEG 2014 Technical Program Expanded Abstracts* (pp. 2455–2459). https://doi.org/10.1190/segam2014-0746.1

Sullivan, E. C., Marfurt, K. J., Lacazette, A., & Ammermann, M. (2006). Application of new seismic attributes to collapse chimneys in the Fort Worth Basin. *Geophysics, 71*(4), B111–B119. https://doi.org/10.1190/1.2216189

Yang, H., Zue, F. J., Pan, W., Chen, L., Yang, P., Tong, Y., & Zhao, C. (2010). Seismic description of karst topography and caves of Ordovician carbonate reservoirs, Lungu area, Tarim Basin, West China. In *SEG 2010 Technical Program Expanded Abstracts* (pp. 1256–1280). https://doi.org/10.1190/1.3513072

Outcrop Analog: Zagros Mountains, Iran

Alizadeh, S. (2017). Non-diapiric salt domes in the west Zanjan, central Iran. *Open Journal of Geology, 7*, 132–146. https://doi.org/10.4236/ojg.2017.72009

Talbot, C. J. (1981). Sliding and other deformation mechanisms in a glacier of salt, S Iran, in: McClay, K. & Price, N.J. (eds.), Thrust and nappe tectonics. *Geological Society, London, Special Publications, 9*, 173–183. https://doi.org/10.1144/GSL.SP.1981.009.01.16

Talbot, C. J., & Aftabi, P. (2004). Geology and models of salt extrusion at Qum Kuh, central Iran. *Journal of the Geological Society, London, 161*, 321–334.

Zavada, P., Desbois, G., Urai, J. L., Schulman, K., Rahmati, M., Lexa, O., & Wollenberg, U. (2015). Impact of solid second phases on deformation mechanisms of naturally deformed salt rocks (Kuh-e-Namak, Dashti, Iran) and rheological stratification of the Hormuz Salt Formation. *Journal of Structural Geology, 74*, 117–144. https://doi.org/10.1016/j.jsg.2015.02.009

Subsurface Case Studies of Salt Tectonics

Augustin, N., van der Zwan, F., Devey, C. W., Kwasnitschka, T., Feldens, P., Bantan, R. A., & Basaham, A. S. (2016). Geomorphology of the central Red Sea rift: Determining spreading processes. *Geomorphology, 274*, 162–179. https://doi.org/10.1016/j.geomorph.2016.08.028

Hovland, M., Rueslatten, H., & Johnsen, H. K. (2015). Red Sea salt formations – A result of hydrothermal processes. In N. M. A. Rasul & I. C. F. Stewart (Eds.), *The Red Sea, the formation, morphology, oceanography and environment of a young ocean basin* (pp. 187–218). Springer. https://doi.org/10.1007/978-3-662-45201-1

Laake, A. (2020). Interpretation of geohazards related to offshore salt karst. In *82nd EAGE Conference and Exhibition*.

Mohr, M., Warren, J. K., Kukla, P., Urai, J. L., & Irmen, A. (2016). Subsurface seismic record of salt glaciers in an extensional intracontinental setting (Late Triassic of northwestern Germany). *Geology, 35*(11), 963–966. https://doi.org/10.1130/G23378A.1

Talbot, C., & Augustin, N. (2016). Submarine salt karst terrains. *AIMS Geosciences, 2*(2), 182–200. https://doi.org/10.3934/geosci.2016.2.182

Outcrop Analog: Salt Louisiana

Kupfer, D. H. (1983). Methods of rock-salt mining in Louisiana. In *Salt domes of south Louisiana* (3rd ed., pp. 122–143). New Orleans Geological Society.

USGS. (2014). *USGS EROS Archive – Digital elevation – Coastal National Elevation Database (CoNED) Project – Topobathymetric Digital Elevation Model (TBDEM)*. https://www.usgs.gov/centers/eros/science/usgs-eros-archive-digital-elevation-coastal-national-elevation-database-coned?qt-science_center_objects=0#qt-science_center_objects

Subsurface Case Studies of Pull-Apart Basins

Aribowo, S. (2017). The geometry of pull-apart basins in the southern part of Sumatran strike-slip fault zone. *IOP Conference Series: Earth and Environmental Science, 118*, 012002. https://doi.org/10.1088/1755-1315/118/1/012002

Chakraborty, C., & Ghosh, S. K. (2005). Pull-apart origin of the Satpura Gondwana basin, central India. *Journal of Earth System Science, 114*(3), 259–273.

Gürbüz, A. (2010). Geometric characteristics of pull-apart basins. *Lithosphere, 2*(3), 1990206. https://doi.org/10.1130/L36.1

Hoprich, M., Decker, K., Grasemann, B., Sokoutis, D., & Willingshofer, E. (2009). Impact of different detachment topographies on pull-apart basin evolution – Analog modelling and computer visualization. *Geophysical Research Abstracts, 11*, EGU2009–EG12063.

Hurwitz, S., Garfunkel, Z., Ben-Gai, Y., Reznikov, M., Rotstein, Y., & Gvirtzman, H. (2002). The tectonic framework of a complex pull-apart basin: Seismic reflection observations in the Sea of Galilee, Dead Sea transform. *Tectonophysics, 359*, 289–306.

Mann, P., Hempton, M. R., Bradley, D. C., & Burke, K. (1983). Development of pull-apart basins. *The Journal of Geology, 91*(5), 529. https://doi.org/10.1007/978-94-007-6644-0_116-1

Moustafa, A. R., & Khalil, S. M. (2020). Structural setting and tectonic evolution of the Gulf of Suez, NW Red Sea and Gulf of Aqaba rift systems. In Z. Hamimi, A. El-Barkooky, J. M. Frias, H. Fritz, & Y. A. El-Rahman (Eds.), *The geology of Egypt* (pp. 295–342). https://doi.org/10.1007/978-3-030-15265-9_8

Rojay, B., & Kocyigit, A. (2012). An active composite pull-apart basin within the central part of the North Anatolian Fault system: The Merzifon-Suluova Basin, Turkey. *Turkish Journal of Earth Sciences, 21*, 473–496. https://doi.org/10.3906/yer-1001-36

Depositional Analogs

Whitney, B. B., & Hengsch, J. V. (2015). Geomorphologic evidence of neotectonics deformation in the Carnarvon Basin, Western Australia. *Geomorphology, 228*, 579–596. https://doi.org/10.1016/j.geomorph.2014.10.020

de Wickens, H. V., & Bouma, A. H. (1995). The Tanqua basin floor fans, Permian Ecca Group, Western Karoo Basin, South Africa. In K. T. Pickering, R. N. Hiscott, N. H. Kenyon, F. Ricci Lucchi, & R. D. A. Smith (Eds.), *Atlas of deep water environments*. Springer. https://doi.org/10.1007/978-94-011-1234-5_49

Outcrop Analog: Caicos Islands, Turks and Caicos, Caribbean

Bourrouilh-Le Jan, F. G. (1998). The role of high-energy events (hurricanes and/or tsunamis) in the sedimentation, diagenesis and karst initiation of tropical shallow water carbonate platforms and atolls. *Sedimentary Geology, 118*, 3–36.

Dickson, J. A. D., & Saller, A. (2008). Caicos platform shoals: Formation and erosion, Turks and Caicos Islands. In *Developing models and analogs for isolated carbonate platforms—Holocene and Pleistocene carbonates of Caicos Platform, British West Indies, SEPM Core Workshop 22*. ISBN 978-1-56576-138-4.

Harris, P. M., Ellis, J., & Purkis, S. J. (2012). Enabling easy access to analogs for carbonate deposition in early rift settings. In *AAPG Search and Discovery Article #50640 and Poster*. http://www.ellis-geospatial.com/images/2012_AAPG_Rift_Poster.pdf

Harris, P. M., Ellis, J., & Purkis, S. J. (2014). Great Bahama Bank – Part I: Evaluating water-depth variation on a "flat-topped" isolated carbonate platform. In *Poster at AAPG 2014, Houston*.

Hollis, C. (2011). Diagenetic controls on reservoir properties of carbonate successions within the Albian-Turonian of the Arabian Plate. *Petroleum Geoscience, 17*, 223–241. https://doi.org/10.1144/1354-079310-032

Logan, A., & Sullivan Sealey, K. (2013). The reefs of the Turks and Caicos Islands. In C. R. C. Sheppard (Ed.), *Coral reefs of the United Kingdom oversees territories, coral reefs of the World* (4th ed.). Springer. https://doi.org/10.1007/978-94-007-5965-7_9

Subsurface Case Studies of Carbonates

Alsharhan, A. S., & Nairn, A. E. M. (1986). A review of the Cretaceous formations in the Arabian Peninsula and Gulf: Part I. Lower Cretaceous (Thamama Group) stratigraphy and paleogeography. *Journal of Petroleum Geology, 9*(4), 365–392.

Alsharhan, A. S., & Nairn, A. E. M. (1990). A review of the Cretaceous formations in the Arabian Peninsula and Gulf: Part III. Upper Creataceous (Aruma Group) stratigraphy and paleogeography. *Journal of Petroleum Geology, 13*(3), 247–266.

Handford, C. R., & Loucks, R. G. (1993). Carbonate depositional sequences and systems tracts – Responses of carbonate platforms to relative sea-level changes: Chapter 1, in: Loucks, G.L. & Sarg J.F. (eds.), Carbonate sequence stratigraphy: Recent developments and applications. *AAPG Memoir, 57*, 3–41.

Outcrop Analog: Irrawaddy Delta, Myanmar

Racey, A., & Ridd, M. F. (2015a). Myanmar offshore petroleum overview, in: Racey, A. & Ridd, M. F. 2015. Petroleum geology of Myanmar. *Geological Society, London, Memoirs, 45*, 57–62. https://doi.org/10.1144/M45.06

Racey, A., & Ridd, M. F. (2015b). Petroleum geology of the Moattama region, Myanmar, in: Racey, A. & Ridd, M. F. 2015. Petroleum geology of Myanmar. *Geological Society, London, Memoirs, 45*, 63–81. https://doi.org/10.1144/M45.07

Ramaswamy, V., & Rao, P. S. (2014). The Myanmar continental shelf, in: Chiocci, F. L. & Chivas, A. R. (eds) 2014. Continental shelves of the world: Their evolution during the last Glacio-Eustatic cycle. *Geological Society, London, Memoirs, 41*, 231–240. https://doi.org/10.1144/M41.17

Subsurface Case Studies of Marine Deltas

Ahsan, J., Al-Humoud, J., Al-Ajmi, H., Ghosh, D., Datta, K., Chimmalgi, V., Gaxi, N., Haryono, R., Kotecha, R., Gupta, B., & Al Anzi, B. (2011). Burgan multilateral campaign: A success story in development of a complex siliciclastics reservoir in Kuwait. In *International Petroleum Technology Conference, Bangkok 2012, Paper IPTC 14258.*

Alsharhan, A. S. (1994a). Albian clastics in the Western Arabian Gulf Region: A sedimentological and petroleum-geological interpretation. *Journal of Petroleum Geology, 17*(3), 279–300.

Alsharhan, A. S., & Nairn, A. E. M. (1997). *Sedimentary basins and petroleum geology of the Middle East* (843 p). Elsevier.

Outcrop Analog: North Frisia North Sea Coast, German

Fagherazzi, S., FitzGerald, D. M., Fulweiler, R. W., Hughes, Z., Wiberg, P. L., McGlathery, K. J., Morris, J. T., Tolhurst, T. J., Deegan, L. A., & Johnson, D. S. (2013). Ecogeomorphology of salt marshes and tidal flats. *Treatise on Geomorphology, 12*, 201–220. https://doi.org/10.1016/B978-0-12-374739-6.00403-6

Subsurface Case Studies of Tidal Floodplains

Al-Eidan, A., Wethington, W. B., & Davies, R. B. (2001). Upper Burgan reservoir description, Northern Kuwait: Impact on reservoir development. *GeoArabia, 6*(2), 179–208.

Filak, J.-M., Van Lint, J., Guerroue, E. L., Desgoutte, N., Fabre, G., Ali, F., Ma, E., Datta, K., Al-Houti, R., & Madhavan, S. (2012). 3-D modelling of the world's largest siliciclastic reservoirs: Greater Burgan Field, Kuwait. In *AAPG Search and Discovery Article #20169.*

Johansson, M. (2019). Sand body characterization in tidal marine settings: Examples from the Malay Basin. In *SEAPEX Exploration Conference, Singapore*.

Middleton, G. V. (1991). A short historical review of clastic tidal sedimentology, in: Smith, D.G., Reinson, G.E., Zaitlin, B.A. & Rahmani, R.A. (eds.), Clastic tidal sedimentology. *Canadian Society of Petroleum Geologists, Memoir, 16*, ix–xv.

Price, W. A. (1963). Patterns of flow and channeling in tidal inlets. *Journal of Sedimentary Petrology, 33*(2), 279–290.

Reineck, H.-E. (1972). Tidal flats, in: Rigby, J.K. & Hamblin, W.K., Recognition of ancient sedimentary environments. *The Society of Economic Paleontologists and Mineralogists (SEPM), Special Publication, 16*, 146–159. https://doi.org/10.2110/pec.72.02

van Straaten, L. M. J. U. (1961). Sedimentation in tidal flat areas. *Journal of the Alberta Society of Petroleum Geologists, 9*(7), 203–226.

Strohmenger, C. J., Patterson, P. E., Al-Sahlan, G., Mitchell, J. C., Feldman, H. R., Demko, T. M., Wellner, R. W., Lehmann, P. J., McCrimmon, G. G., Broomhall, R. W., & Al-Ajmi, N. (2006). Sequence stratigraphy and reservoir architecture of the Burgan and Mauddud formations (Lower Cretaceous), Kuwait, in Harris, P.M. & Weber, L.J.(eds.), Giant hydrocarbon reservoirs of the world: From rocks to reservoir characterization and modeling. *AAPG Memoir, SEPM Special Publication, 88*, 213–245.

Subsurface Case Studies of Floodplain Channels

Bullimore, S. A., & Helland-Hanse, W. (2009). Trajectory analysis of the lower Brent Group (Jurassic), Northern North Sea: Contrasting depositional patterns during the advance of a major deltaic system. *Basin Research, 21*, 559–572. https://doi.org/10.1111/j.1365-2117.2009.00410.x

Franzel, M., Jones, S. J., Meadows, N., Allen, M. B., McCaffrey, K. M., & Morgan, T. (2020). Basin-scale fluvial correlation and response to the Tethyan marine transgression: An example from the Triassic of central Spain. *Basin Research, 33*, 1–25. https://doi.org/10.1111/bre.12451

Larue, D. K., & Hovadik, J. (2006). Connectivity of channelized reservoirs: A modelling approach. *Petroleum Geoscience, 12*, 291–308.

Peacock, M. J. (2011). Athabasca oil sands: Reservoir characterization and its impact on thermal and mining opportunities. In B. A. Vining & S. C. Pickering (Eds.), *Petroleum Geology: From Mature Basins to New Frontiers, Proceedings of the 7th Petroleum Geology Conference* (pp. 1141–1150). https://doi.org/10.1144/0071141

San Roman, F. A., & Yutsis, V. (2019). Application of spectral decomposition methods to the definition of stratigraphic features associated with channel reservoirs in the Southeast Petroleum Province, Mexico. *Pure and Applied Geophysics, 176*, 873. https://doi.org/10.1007/s00024-018-1999-2

Sømme, T. O., Helland-Hansen, W., Martinsen, O. J., & Thurmond, J. B. (2009). Relationships between morphological and sedimentological parameters in source-to-sink systems: A basis for predicting semi-quantitative characteristics in subsurface systems. *Basin Research, 21*, 361–387. https://doi.org/10.1111/j.1365-2117.2009.00397.x

Vondrak, A. G., Donselaar, M. E., & Munsterman, D. K. (2018). Reservoir architecture model of the Nieuwerkerk Formation (Early Cretaceous, West Netherlands Basin): Diachronous development of sand-prone fluvial deposits, in: Kilhams, B., Kukla, P. A., Mazur, S., McKie, T., Mijnlieff, H. F. & van Ojik, K. (eds) 2018. Mesozoic resource potential in the southern Permian Basin. *Geological Society, London, Special Publications, 469*, 423–434. https://doi.org/10.1144/SP469.18

Wang, J., & Plink-Björklund, P. (2019). Stratigraphic complexity in fluvial fans: Lower Eocene Green River Formation, Uinta Basin, USA. *Basin Research, 31*, 892–919. https://doi.org/10.1111/bre.12350

Outcrop Analog: Okavango Delta, Botswana

Gumbricht, T., McCarthy, T. S., & Merry, C. L. (2001). The topography of the Okavango Delta, Botswana, and its tectonic and sedimentological implications. *South African Journal of Geology, 104*(3), 243–264. https://doi.org/10.2113/1040243

Hutchins, D. G., Hutton, L. G., Hutton, S. M., Jones, C. R., & Loennert, E. P. (1976). A summary of the geology, seismicity, geomorphology and hydrogeology of the Okavango Delta, Republic of Botswana. *Geological Survey Department Bulletin, 7*, 1, 39 p.

Kinabo, B. D., Hogan, J. P., Atekwana, E. A., Abdelsalam, M. G., & Modisi, M. P. (2008). Fault growth and propagation during incipient continental rifting: Insights from a combined aeromagnetic and Shuttle Radar Topography Mission digital elevation model investigation of the Okavango Rift Zone, northwest Botswana. *Tectonics, 27*(3), TC3013. https://doi.org/10.1029/2007TC002154

McCarthy, T. S. (2013). The Okavango Delta and its place in the geomorphological evolution of Southern Africa. *South African Journal of Geology, 116*(1), 1–54. https://doi.org/10.2113/gssajg.116.1.1

Shemang, E. M., & Molwalefhe, L. N. (2011). Geomorphic landforms and tectonicsm along the eastern margin of the Okavango Rift Zone, north western Botswana as deduced from geophysical data in the area. In E. Sharkov (Ed.), *New frontiers in tectonic research – General problems, sedimentary basins and island arcs*. InTech Open. ISBN 978-95305952, 169-182.

Outcrop Analog: Kufra Delta, Libya

Ghoneim, E., Robinson, C., & El-Baz, F. (2007). Radar topography data reveal drainage relics in the eastern Sahara. *International Journal of Remote Sensing, 28*(8), 1759–1772. https://doi.org/10.1080/01431160600639727

Paillou, P., Lopez, S., Farr, T., & Rosenqvist, A. (2007). Mapping subsurface geology in Sahara using L-band SAR: First results from the ALOS/PALSAR imaging radar. *IEEE Journal of Selected Topics in Applied Earth Observations and Remote Sensing, 3*(4), 632–636. https://doi.org/10.1109/JSTARS.2010.2056915

Paillou, P., Schuster, M., Tooth, S., Farr, T., Rosenqvist, A., Lopez, S., & Malezieux, J.-M. (2009). Mapping of a major paleodrainage system in eastern Libya using orbital imaging radar: The Kufrah River. *Earth and Planetary Science Letters, 277*, 327–333. https://doi.org/10.1016/j.epsl.2008.10.029

Paillou, P., Tooth, S., & Lopez, S. (2012). The Kufrah paleodrainage system in Libya: A past connection to the Mediterranean Sea? *Comptes Rendus Geoscience, 344*(8), 406–414. https://doi.org/10.1016/j.crte.2012.07.002

Robinson, C. A., El-Baz, F., Al-Saud, T. S. M., & Jeon, S. B. (2005). Use of radar data to delineate paleodrainage leading to the Kufra Oasis in the eastern Sahara. *Journal of African Earth Sciences, 44*(2), 229–240. https://doi.org/10.1016/j.jafrearsci.2005.10.012

Subsurface Case Studies of Intra-Cratonic Deltas

Alsharhan, A. S. (1994b). Albian clastics in the Western Arabian Gulf Region: A sedimentological and petroleum-geological interpretation. *Journal of Petroleum Geology, 17*(3), 279–300.

Outcrop Analog: Karst Plateau, Slovenia and Italy

Klimchouk, A. B. (2017b). Types and settings of hypogene karst. In A. Klimchouk, A. N. Palmer, J. de Waele, A. S. Auler, & P. Audra (Eds.), *Hypogene karst regions and caves of the world* (pp. 1–39). Springer. https://doi.org/10.1007/978-3-319-53348-3_1

Onac, B. P., & van Beynen, P. (2021). Caves and karst. In R. C. Selley, L. R. M. Cocks, & I. R. Plimer (Eds.), *Encyclopedia of geology* (2nd ed., pp. 495–509). https://doi.org/10.1016/B978-0-12-409548-9.12437-6

Phillips, J. D. (2017). Landform transitions in a fluviokarst landscape. *Zeitschrift fuer Geomorphologie, 61*(2), 109–122. https://doi.org/10.1127/zfg/2017/0452

Subsurface Case Studies of Karsting

Alsharhan, A. S., & Whittle, G. L. (1995). Carbonate-evaporite sequences of the Late Jurassic, Southern and Southwestern Arabian Gulf. *AAPG Bulletin, 79*(11), 1608–1630.

Boersma, Q., Prabhakaran, R., Bezerra, F. I., & Bertotti, G. (2019). Linking natural fractures to karst cave development: A case study combining drone imagery, a natural cave network and numerical modelling. *Petroleum Geoscience, 25*, 454–469. https://doi.org/10.1144/petgeo2018-151

Farooq, U., Meyer, A., Al Obeidli, A. K., Ben Maroof, M., Baloch, S. A., Kumar Dey, S., & Almarzooqi, M. J. (2020). Characterization of karst development using an integrated workflow in an Upper Cretaceous carbonate reservoir from onshore field, United Arab Emirates. In *Society of Petroleum Engineers, SPE-202995-MS*.

Jeffrey, Z. E., Penn, S., Giles, D. P., & Hastewell, L. (2020). Identification, investigation and classification of surface depressions and chalk dissolution features using integrated LiDAR and geophysical methods. *Quarterly Journal of Engineering Geology and Hydrogeology, 53*, 620–644. https://doi.org/10.1144/qjegh2019-098

Laake, A., Zhaoming, W., Gengxin, P., Duoming, Z., Yongfu, C., & Hui, Z. (2014b). Mapping ultra-deep karst in its geological context. In *Society of Exploration Geophysicists 2014 Annual Meeting, Proceedings* (pp. 2455–2459). https://doi.org/10.1190/segam2014-0746.1

Novakovic, G., Mlekuz, D., Rozman, L., Lazar, A., Peric, B., Cerkvenik, R., Peternelj, K., & Eric, M. (2014). New approaches to understanding the world natural and cultural heritage by using 3D technology: UNESCO's Skocjan Caves, Slovenia. *International Journal of Heritage in the Digital Era, 3*(4), 629–641.

Petrovic, A. S., Calic, J., Spalevic, A., & Pantic, M. (2018). Relations between surface and underground karst forms inferred from terrestrial laser scanning, in: Parise, M., Grabovsek, F., Kaufmann, G. & Ravbar, N. (eds) Advances in karst research: Theory, fieldwork and applications. *Geological Society, London, Special Publications, 466*, 107–120. https://doi.org/10.1144/SP466.23

Vinci, G., & Bernardini, F. (2017). Reconstructing the protohistoric landscape of Trieste Karst (north-eastern Italy) through airborne LiDAR remote sensing. *Journal of Archaeological Science: Reports, 12*, 591–600. https://doi.org/10.1016/j.jasrep.2017.03.005

Yang, W., Gong, X., & Li, W. (2018). Geological origins of seismic bright spot reflections in the Ordovician strata in the Halahatang area of the Tarim Basin, western China. *Canadian Journal of Earth Sciences, 55*, 1297–1311. https://doi.org/10.1139/cjes-2018-0055

Glacial Geology

Dowdeswell, J. A., Canals, M., Jakobsson, M., Todd, B. J., Dowdeswell, E. K., & Hogan, K. A. (Eds.). (2016). Atlas of submarine glacial landforms: Modern, quaternary and ancient. *Geological Society, London, Memoirs, 46*. https://doi.org/10.1144/M46

Giles, D. P., Griffiths, J. S., & Evans, D. J. A. (2017). Chapter 3 Geomorphological framework: Glacial and periglacial sediments, structures and landforms, in: Griffiths, J. S. & Martin, C. J. (eds) 2017. Engineering geology and geomorphology of glaciated and periglaciated terrains – Engineering group working party report. *Geological Society, London, Engineering Geology Special Publications, 28*, 59–368. https://doi.org/10.1144/EGSP28.3

Regional Glaciology Europe

Laake, A., Olsen, H., Frizzell, B., & Schmidt, I. (2014c). Geologically validated stratigraphic earth models from seismic data. In *Conference Proceedings, 76th EAGE Conference and Exhibition 2014, paper Tu G107 07* (pp. 1–5). https://doi.org/10.3997/2214-4609.20140841

Lambeck, K., Smither, C., & Johnston, P. (1998). Sea-level change, glacial rebound and mantle viscosity for norther Europe. *Geophysical Journal International, 134*, 102–144. https://doi.org/10.1046/j.1365-246x.1998.00541.x

Madsen, V. (1921). *Terrainformerne pa° Skovbjerg Bakkeø. Danmarks Geologiske Undersøgelse Meddelelser fra Dansk geologisk Forening*. Bd. 6. Nr. 5. Trykkes tillige som Danmarks geologiske Undersøgelse. IV.R. Bd.l. Nr.1 1–24.

Müther, D., Back, S., Reuning, L., Kukla, P., & Lehmkuhl, F. (2012). Middle Pleistocene landforms in the Danish Sector of the southern North Sea imaged on 3D seismic data, in: Huuse, M., Redfern, J., LeHeron, D.P., Dixon, R.J., Moscariello, A. & Craig, J. (eds) 2012, Glaciogenic reservoirs and hydrocarbon systems. *Geological Society, London, Special Publications, 368*, 111–127. https://doi.org/10.1144/SP368.7

Regional Glaciology North America

Dyke, A. S., & Prest, V. K. (1987a). *Late Wisconsinan and Holocone retreat of the Laurentide Ice Sheet. Geological Survey of Canada Map 1702A*. https://doi.org/10.4095/122842

Dyke, A. S., & Prest, V. K. (1987b). *Paleogeography of Northern North America, 18,000 - 5,000 years ago, Geological Survey of Canada Map 1703A 1:12,500,000*. https://doi.org/10.4095/133927

Leverett, F., & Taylor, F. B. (1915). The Pleistocene of Indiana and Michigan and the history of the Great Lakes. In *Monographs of the United States Geological Survey* (Vol. LIII).

Wright, G. F. (1900). *The ice age in North America and its bearings upon the antiquity of man*. D. Appleton.

Glacial Tunnel Valleys

Andersen, T. R., Huuse, M., Joergensen, F., & Christensen, S. (2012). Seismic investigation of buried tunnel valleys on- and offshore Denmark, in: Huuse, M., Redfern, J., LeHeron, D.P., Dixon, R.J., Moscariello, A. & Craig, J. (eds) 2012, Glaciogenic reservoirs and hydrocarbon systems. *Geological Society, London, Special Publications, 368*, 129–144. https://doi.org/10.1144/SP368.9

Douillet, G., Ghienne, J.-F., Geraud, Y., Abeuldas, A., Diraison, M., & Al-Zoubi, A. (2012). Late Ordovician tunnel valleys in southern Jordan, in: Huuse, M., Redfern, J., LeHeron, D.P., Dixon, R.J., Moscariello, A. & Craig, J. (eds) 2012, Glaciogenic reservoirs and hydrocarbon systems. *Geological Society, London, Special Publications, 368*, 275–292. https://doi.org/10.1144/SP368.4

Jørgensen, F., & Sandersen, B. E. (2006). Buried and open tunnel valleys in Denmark – Erosion beneath multiple ice sheets. *Quaternary Science Reviews, 25*(11–12), 1339–1363. https://doi.org/10.1016/j.quascirev.2005.11.006

Kehew, A. E., & Kozlowski, A. L. (2007). Tunnel channels of the Saginaw Lobe, Michigan, USA. In P. Johansson & P. Sarala (Eds.), *Applied Quaternary research in the central part of glaciated terrain, Geological Survey of Finland, Special Paper* (Vol. 46, pp. 69–78).

Kristensen, T. B., & Huuse, M. (2012). Multistage erosion and infill of buried Pleistocene tunnel valleys and associated seismic velocity effects, in: Huuse, M., Redfern, J., LeHeron, D.P., Dixon, R.J., Moscariello, A. & Craig, J. (eds) 2012, Glaciogenic reservoirs and hydrocarbon systems. *Geological Society, London, Special Publications, 368*, 159–172. https://doi.org/10.1144/SP368.15

LeHeron, D., Sutcliffe, O., Bourgig, K., Craig, J., Visentin, C., & Whittington, R. (2004). Sedimentary architecture of Upper Ordovician tunnel valleys, Gargaf Arch, Libya: Implications for the genesis of a hydrocarbon reservoir. *GeoArabia, 9*(2), 137–160.

Moscariello, A., Spaak, P., Jourdan, A., & Azzouni, A.-H. (2009). The Ordovician Glaciation in Saudi Arabia — Exploration challenges Part 1: Geology (outcrop, subsurface, analogues). In *AAPG Search and Discovery Article #50175.*

Sandersen, P. B. E., & Jørgensen, F. (2012). Substratum control on tunnel-valley formation in Denmark, in: Huuse, M., Redfern, J., LeHeron, D.P., Dixon, R.J., Moscariello, A. & Craig, J. (eds), Glaciogenic reservoirs and hydrocarbon systems. *Geological Society, London, Special Publications, 368*, 145–157. https://doi.org/10.1144/SP368.12

Sandersen, P. B. E., Jørgensen, F., Larsen, N. K., Westergaard, J. H., & Auken, E. (2009). Rapid tunnel-valley formation beneath the receding Late Weichselian ice sheet in Vendsyssel, Denmark. *Boreas, 38*, 834. https://doi.org/10.1111/j.1502-3885.2009

Van der Vegt, P., Janszen, A., & Moscariello, A. (2012). Tunnel valleys: Current knowledge and future perspectives, in: Huuse, M., Redfern, J., LeHeron, D.P., Dixon, R.J., Moscariello, A. & Craig, J. (eds), Glaciogenic reservoirs and hydrocarbon systems. *Geological Society, London, Special Publications, 368*, 75–97. https://doi.org/10.1144/SP368.13

Moraines

Taylor, F. B. (1897). Moraines of recession and their significance in glacial theory. *The Journal of Geology, 5*(5), 421–465.

Drumlins

Johnson, M. D., Schomacker, A., Benediktsson, I. Ö., Geiger, A. J., Ferguson, A., & Ingolfsson, O. (2010). Active drumlin field revealed at the margin of Mulajökull, Iceland: A surge-type glacier. *Geology, 38*(10), 943–946. https://doi.org/10.1130/G31371.1

Glaciogenic Subsurface Reservoirs

Huuse, M., Redfern, J., LeHeron, D. P., Dixon, R. J., Moscariello, A., & Craig, J. (2012). Glaciogenic reservoirs and hydrocarbon systems. *Geological Society, London, Special Publications, 368*, 75. https://doi.org/10.1144/SP368.0

Glossary

Absorption of electromagnetic waves Amplitude attenuation as a result of interaction with matter.

Algal mat Mat created from blue-green algae, typically forming source rock for hydrocarbons.

Altimetry Measurement of height or elevation, often based on runtime measurements of electromagnetic signals.

Altimetry, laser Altimetry using airborne or ground-based laser technology.

Altimetry, radar Altimetry using satellite or airborne radar technology.

Analog (geology) Observable geologic setting that is used as input to analogue modeling of unknown subsurface settings.

Analog, depositional Analog for depositional setting.

Analog, erosional Analog for erosional setting.

Analog, structural Analog for tectonic setting.

Anoxic environment Environment in which oxygen is absent, typically resulting in preservation of organic matter.

Anticline Type of arch-like folding that has its oldest beds at its core.

Anticline, plunging Anticline that is tilted downwards along its axis.

Canyon Deep erosional valley with steep flanks.

Carbonate platform Calcareous sedimentary body which possesses topographic relief that is composed of deposits of carbonate erosion.

Celerity Rapidity of motion, for sand dunes: speed of sand dune migration.

Channel Bed of a river or surface or submarine stream.

Channel, buried Paleo-channel that is buried by sediments.

Channel system, braided Channelized drainage pattern consisting of braided channels.

Cirrus clouds Genus of high altitude atmospheric clouds that distort part of the atmospheric electromagnetic transmission spectrum.

A. Laake, *Remote Sensing for Hydrocarbon Exploration*, Springer Remote Sensing/Photogrammetry, https://doi.org/10.1007/978-3-030-73319-3

Climate

Arctic climate Characterized by long, cold winters and short, cool summers.

Arid climate Dry climate, synonymous with desert climate.

Humid climate Generally moist climate with hot and wet summers and cold to mild winters, typically between 25° and 40° latitude.

Tropical climate Climate with temperatures of at least 18 °C and typically abundant rainfall.

Coastal plain Flat, low-lying land adjacent to the sea coast, typically the result of long-term deposition.

Competent rock Rock that, because of its physical and geological characteristics, is capable of sustaining openings without structural support.

Convergent plate boundary Area on earth where two or more lithospheric plates collide which typically results in the formation of mountains.

Co-rendering In visualization: simultaneous rendering of two or more data sets.

Deep water Water depth in oceans of more than 300 m.

Delta, birdfoot Type of river or stream delta composed of isolated channels that look like a bird's foot.

Desert varnish Colored or black coating of rocks that are exposed to desert climate. Desert varnish needs to be considered when interpreting satellite imagery of desert areas as it alters the electromagnetic reflection and absorption spectra.

Diffuse reflection (em waves) Reflection of electromagnetic waves in more than one direction typically resulting from surface that rough compared to the wavelength.

Divergent plate boundary Area on earth where two lithospheric plates move away from each other typically occurring at mid ocean ridges.

Doline Closed surface depression resulting from subsurface chemical evacuation (suffosion) of carbonate rock, typical surface karst feature.

Drainage Removal of surface water from an area with excess of water.

Drainage pattern Pattern of channels through with water is drained.

Dendritic drainage Specific shape of drainage pattern in the shape of a tree.

Paleo drainage Fossil drainage that is buried in the subsurface and no longer used to drain surface water.

Draping (visualization) Render a first data set on a virtual 3D representation of second data sets. Example: display a satellite image on top of a digital elevation model.

Dune Sand accumulation in a desert.

Barchan dune Sand dune in crescent shape typically produced by wind coming from one direction.

Dune encroachment Migration of dunes towards an area of human use.

Dune field Accumulation of several dunes in a desert.

Giant sand dune, megadune Sand dune with height in excess of 60 m.

Mobile dune Sand dune that is moving as a result of directional wind.

Seif dune Long, narrow sand dune or chain of dunes generally oriented in the direction parallel to the prevailing wind.

Star dune Sand dune resembling star shape with multiple arms resulting from a complex wind pattern from different directions.

Duricrust Hard layer at the surface typically occurring in deserts.

Elevation model, bathy-topographic Elevation model combined of bathymetric (offshore) and topographic (onshore) data.

Eolian Wind related.

Ephemeral Lasting a very short time.

Erosion Geologic process of wearing away rock and transporting away the resulting debris.

Escarpment Sharp terrain edge.

Evaporites Sediment resulting from evaporation of water-soluble minerals out of an aqueous solution.

False color RGB color composite using a color assignment other than visible red-green-blue to RGB colors.

Fan delta Deposition of sediments from a river or channel into a body of standing water in the shape of a fan.

Fan, alluvial Onshore accumulation of sediments shaped like a section of a shallow cone.

Fault Fracture or fracture zone between two blocks of rock.

Fault throw Vertical offset between two blocks of rock at a fault.

Normal fault Fault resulting from dip-slip motion in an extensional setting where the hanging wall has moved downward relative to the footwall.

Reverse fault Fault resulting from dip-slip motion in a compressional setting where the hanging wall has moved upward relative to the footwall.

Strike-slip fault Fault resulting from strike-slip or lateral movement in a shear setting.

Thrust fault Type of reverse fault with a dip of less than 45°.

Transform fault In a rift setting, fault along a plate boundary where the dominant motion is horizon, also: strike-slip fault.

First break (seismic) First arrival of seismic wave energy in a seismic shot gather.

Flood basalt Massive basalt deposition resulting from giant volcanic eruption, occurs both onshore and offshore.

Floodplain Area of land adjacent to a river between the river banks and the enclosing valley walls which is flooded during periods of high water discharge of the river.

Fold Bending of a stack of rock layers as a result of tectonic stress.

Fold belt Series of folded structures aligned in a belt following regional tectonic stress.

Geomorphology Science of landforms with an emphasis on origin and evolution and the dependency on subsurface layering and the resistance of the layers to erosional forces.

Glacial terminus End of a glacier.

Glacial till Sediment resulting from erosion by the glacier.

Gradient, lateral (terrain) Terrain dip mapping the steepness of the terrain.

Gravel Loose aggregation of rock fragments, typical cover in flat deserts.

Gravity Force of attraction between any two masses. The measurement of gravity gives clues for the composition of the subsurface.

Bouguer gravity anomaly Gravity anomaly corrected for the height at which it is measured and the attraction of terrain.

Free air gravity anomaly Gravity anomaly corrected for the height at which it is measured.

Grayscale (visualization) Rendering of data in a color scale from black via gray to white.

Ground roll (seismic) Surface waves in seismic data, mostly composed of Rayleigh waves.

Heliportable operation Seismic acquisition operation using helicopters for logistics.

Ice margin Edge of an ice shield.

Illumination, slant (visualization) Artificial illumination at a shallow angle of incidence that is typically used to highlight subtle terrain elevation changes.

Imagery, stereoscopic Imagery obtained from stereoscopic images acquired under two different look directions.

Intertidal Zone at the seashore above the water level at low tide and below the water level at high tide.

Intra-craton Feature occurring in the interior of an old and stable piece of the continental lithosphere.

Isochron On a map, the outline of an area of similar age of rock.

Isopach On a map, the outline of an area of similar thickness for one formation.

Karst Topography formed from the dissolution of soluble rocks, named after the Karst area in the northern Balkans.

Epigenic karstification karstification is a result of carbonate dissolution by carbonic acid H_2CO_3.

Hypogenic karstification Karstification is a result of carbonate dissolution by sulfuric acid H_2SO_4.

Thermokarst Topography formed in permafrost setting resulting from the interaction of warming climate and precipitation with melting permafrost in the ground. The features resemble carbonate dissolution karst.

Lagoon Shallow body of water near or communicating with a larger body of water.

Land use (satellite imagery) Surface cover of the earth as detected on satellite imagery, comprises water, rock, soil, vegetation and human impact.

Lineament (structural geology) Linear feature resulting typically from tectonic processes.

Magnetic anomaly Local variation of the earth's magnetic field resulting from variations in the rock composition and its magnetization.

Reduced to pole magnetic anomaly Correction for the magnetic anomaly of the direction of magnetization is known, useful to interpret local variations.

Total magnetic anomaly Uncorrected magnetic anomaly.

Margin, conjugate (plate tectonics) Edges of plates that result from rifting and seafloor spreading, reconstruction by reverting the seafloor spreading leads to close fit of the plate boundaries.

Mesa Isolated flat-topped hill that is bounded by steep escarpments, typical erosion feature in deserts.

Moraine Accumulation of unconsolidated debris created by the motion of glaciers.

Ground moraine Moraine at the base of a glacier characterized by numerous shallow lakes after glacier melting.

Lateral moraine Moraine at the side of a glacier.

Terminal moraine Moraine left by the glacial terminus.

Move out (seismic wave) Time offset between seismic events in a seismic shot record, move out allows the computation of the seismic wave velocity.

Multiband composite (visualization) Composition of three or more images from multiple spectral bands in one RGB image, also: multispectral composite.

Multispectral imagery Imagery obtained from reflections in multiple spectral bands.

Multi-temporal sensing Acquisition and processing of images at different times used for the study of temporally changing processing.

Mylonite (tectonics) Fine-grained rock resulting from tectonic movements thus reducing the grain size of the rock, mylonite typically occurs in fault zones.

Near-surface (seismic) Shallow rock layers below the surface, in seismic records "near-surface" typically means the depth interval in which surface waves propagate.

Outcrop Visible exposure of rock layers at the surface.

Overburden Rock layers overlying a rock layer of interest.

Oxbow Abandoned section of a meandering river that has been separated from the flow of water.

Panchromatic band (imagery) Imagery recorded over a spectral band that covers more than one analysis bands, typically the panchromatic band is recorded at higher lateral resolution.

Panchromatic sharpening (imagery) Process of increasing the lateral resolution of an image using a panchromatic band.

Permafrost Permanently frozen ground the temperature of which remains continuously below 0 °C.

Photogeologic map Co-rendering of a remote sensing image and a geologic map, shows the correlation between colors of the image and the geologic units on the map.

Radar backscatter imagery In remote sensing, the technique of using the reflection of microwave waves from the surface.

Reflectance (em waves) Effectiveness of a material to reflect electromagnetic radiant energy.

Reflected wave (seismic) Seismic wave that is reflected at a subsurface lithologic boundary.

Refracted wave (seismic) Seismic wave that is refracted at a subsurface lithologic boundary, a subsequent reflection returns the refracted wave signal back to the surface.

Resolution (data) Spatial characteristics of data determining the minimal size of features that can be detected.

RGB (visualization) Technique combining three grayscale images through assignment to colors red-green-blue in a color image.

Rift Linear zone where the lithosphere is pulled apart by extensional tectonic forces.

Fracture zone Fracture zone that is typically perpendicular to the orientation of the rift.

Rift-parallel fault Fault zone that is parallel to the orientation of the rift.

Rim reef Portion of a reef that protrudes the water.

Risk, vehicle manoeuvre Logistic risk related to challenges of manoeuvring vehicles.

Risk, vehicle roll-over risk Logistic risk of vehicles rolling over.

Sabkha Salt flat.

Coastal sabkha Supra-tidal coastal salt flat.

Dry sabkha Sabkha with a dry and hard crust at the surface (duricrust).

Ephemeral sabkha Sabkha lasting only for very short time.

Inland sabkha Sabkha occurring within deserts, also: continental sabkha.

Wet sabkha Sabkha covered with highly concentrated brine.

Scattering, surface backscattering Scattering of electromagnetic waves at the surface resulting in a reflection signal.

Subsurface backscattering Scattering of electromagnetic waves at a subsurface interface resulting in a reflection signal.

Volume backscattering Scattering of electromagnetic waves in the subsurface resulting in a reflection signal.

Sediment fairway Potential pathway for hydrocarbons along a sediment deposition feature.

Sediments, eolian Sediments deposited as a result of transportation by wind.

Sediments, fluvial Sediments deposited as a result of transportation by water in a river.

Seismic wave, body Seismic wave travelling as a compressional wave through the body of subsurface layers.

Seismic wave, direct Seismic wave travelling directly between seismic source and receiver.

Seismic wave, surface Seismic wave travelling at and near the surface, mostly composed of Rayleigh waves.

Shoal Submerged ridge of unconsolidated material typically resulting from erosional processes.

Specular reflection Mirror-like reflection of waves at an interface.

Static correction (seismic) Correction applied to seismic data to compensate for irregular topography and near-surface lithology.

Texture (geology) Surface quality of a geological surface, for example the earth's surface.

Texturing (visualization) Process of creating a virtual surface texture using a representation of data.

Thalweg On a map, the line of the lowest elevation within a valley.

Thermokarst Topography formed in permafrost setting resulting from the interaction of warming climate and precipitation with melting permafrost in the ground. The features resemble carbonate dissolution karst.

Tidal Related to or caused by tides.

Subtidal Zone below the tidal zone.

Supratidal Zone above the tidal zone.

Tidal channel Channel resulting from tidal flow.

Tidal flat Flat created by tidal processes.

Tidal zone Zone at the seashore above the water level at low tide and below the water level at high tide.

Top of basement Top of deepest zone considered in a geologic model, basement may be the deepest layer of economic interest (economic basement) or the base of the sedimentary fill of a basin (geologic basement).

Transmission (em waves) Propagation of electromagnetic waves through a material.

Travel time (seismic) Time a seismic wave front requires from entry into a rock body to exiting from it.

True color Visualization.

Vertical exaggeration (visualization) Virtual exaggeration of vertical properties in 3D visualization.

Vibroseis Technology of generating onshore seismic waves using a heavy controlled shaker mounted on a truck.

Volcanic vent Opening in the earth's crust through which magma reaches the surface.

Wadi Valley or dry riverbed, may be flooded temporarily by flash floods.

Wadi, paleo Fossil buried riverbed.

Water table Upper surface of water saturated zone in the near-surface, boundary affecting the propagation of seismic waves.

Wattenmeer Intertidal zone in the southern North Sea forming a shallow water body with tidal flats and wetlands, also: wadden sea.

Index

A. Laake, *Remote Sensing for Hydrocarbon Exploration*, Springer Remote Sensing/Photogrammetry, https://doi.org/10.1007/978-3-030-73319-3

The manufacturer's authorised representative in the EU is Springer Nature Customer Service Centre GmbH, Europaplatz 3, 69115 Heidelberg, Germany. If you have any concerns regarding our products, please contact ProductSafety@springernature.com

Printed and bound by CPI Group (UK) Ltd, Croydon, CR0 4YY
30/09/2026
02225504-0015